MICHIGAN FERNS AND LYCOPHYTES

MICHIGAN
Ferns & Lycophytes

A GUIDE TO SPECIES OF THE GREAT LAKES REGION

Daniel D. Palmer

University of Michigan Press
Ann Arbor

Published in the United States of America by the
University of Michigan Press
Printed and bound by CPI Group (UK) Ltd, Croydon, CR0 4YY

2021 2020 2019 2018 4 3 2 1

A CIP catalog record for this book is available from the British Library.

Library of Congress Cataloging-in-Publication Data

Names: Palmer, Daniel D. (Daniel Dooley), 1930– author.
Title: Michigan ferns and lycophytes : a guide to species of the Great Lakes region / Daniel D. Palmer.
Description: Ann Arbor : University of Michigan Press, [2018] | Includes bibliographical references and index. |
Identifiers: LCCN 2017059950 (print) | LCCN 2017061258 (ebook) | ISBN 9780472123650 (e-book) | ISBN 9780472037117 (pbk. : alk. paper)
Subjects: LCSH: Ferns—Michigan—Identification.
Classification: LCC QK525.5.M5 (ebook) | LCC QK525.5.M5 P35 2018 (print) | DDC 587/.4—dc23
LC record available at https://lccn.loc.gov/2017059950

The University of Michigan Press gratefully acknowledges financial support for this volume from the Clarence R. and Florence N. Hanes Fund of Kalamazoo, MI.

Cover illustrations courtesy of the W. H. Wagner Slide Collection, University of Michigan Herbarium.

PREFACE

In offering this first comprehensive review of Michigan's fern flora in more than half a century, I understand the disquiet expressed by the early-eighteenth-century writer, John Ray, in the preface to his own book, *The Wisdom of God Manifested in the Works of the Creation.*

> In all Ages wherein Learning hath flourished, Complaint hath been made of the Itch of Writing, and the Multitude of worthless Books, wherein importunate Scriblers have pester'd the World. . . . I am sensible that this Tractate may likely incur the Censure of a superfluous Piece. . . . There having been so much so well written of this Subject, by the most learned Men of our Time. . . . *First* therefore, in Excuse of it, I plead, That there are in it some Considerations new and untouch'd by others: Wherein if I be mistaken, I alledge; *Secondly* that Manner of Delivery and Expression may be more suitable to some Mens Apprehension, and facile to their Understandings. If that will not hold, I pretend *Thirdly*, That all the Particulars contained in this Book, cannot be found in any one Piece known to me, but lie scattered and dispersed in many; and so this may serve to relieve those Fastidious Readers, that are not willing to take the Pains to search them out.

May it be so for this book also. This work is indebted to the scholarship of many earlier authors, who greatly expanded our knowledge of Michigan ferns. Among these are Joseph Beitel, Cecil Billington, Don Farrar, Dale Hagenah, Bob Preston, Florence Wagner, and *most notably* Warren "Herb" Wagner, late of Ann Arbor. This book is written for a wide range of readers that includes both interested amateurs and professional botanists. As such it will not be entirely satisfactory to all readers.

ACKNOWLEDGMENTS

Warren H. (known as Herb to all acquainted with him) and Florence Wagner, who were my tutors in decades past, assisted me in gaining the skills needed to produce a fern flora. Herb was a remarkable and enthusiastic teacher, and the passionate sharing of his deep knowledge of ferns stimulated many to study this neglected group of plants. Florence also contributed much to the knowledge of ferns, especially in the area of chromosome studies. Without her, members of the Wagner family could not have pursued their teaching and research.

Joseph M. Beitel, with his enthusiastic interest in the ferns and particularly in the lycopods, introduced me to basic fern taxonomy.

Christiane Anderson, Don Farrar, Chad Husby, James Montgomery, Robert Preston, and Alan Smith were very helpful in my attempts to make this book taxonomically correct.

Dick Fidler and Julie Medlin reviewed the work in progress and gave much-needed editorial advice and guidance at various stages. Their input, advice, and persistent nagging to get the work done were invaluable.

Tony Reznicek at the University of Michigan Herbarium was very helpful, and his extensive knowledge of the Michigan flora was very useful. His extensive editing very much improved the quality of this book. Susan Fawcett (University of Vermont, Burlington) was very helpful with providing a modern evolutionary context for our Michigan ferns. Beverly Walters (University of Michigan Herbarium) provided many of the specimen scans on which the illustrations in this book were based. Michael R. Penskar, research investigator at the University of Michigan Herbarium and botanist for the Michigan Natural Features Inventory (retired) helped with access to information regarding endangered plants and provided material for use in the illustrations. Natalia Vasquez produced the illustrations from computer scans, as well as other artworks and figures. Her careful work, professional advice, and assistance are much appreciated.

My frequent phone calls, faxes, and e-mails to all these people must have been a burden. Their help is greatly appreciated.

My wife Helen patiently tolerated the long hours I spent writing this book.

CONTENTS

Part 4 Lycophytes

INTRODUCTION

It has been over half a century since *Ferns of Michigan* by Cecil Billington was published. This was the only comprehensive review of Michigan ferns and lycophytes. It was a fine book for students of the fern flora of the state at the time; however, since its publication in 1952 there have been great advances in the understanding of the ferns and lycophytes. Changes in taxonomy have often made the classifications recognized by Billington as to family, genus, and species outdated and inaccurate. New genera and species have been recognized, and more information regarding distribution of these plants has been gathered. Many species have been recognized as a different taxon or placed in different genera. Genera have been placed in different families. Several families are now recognized from within the single large family Polypodiaceae as Billington treated it. Similar changes in the taxonomy of horsetails, quillworts, and clubmosses have occurred. A new book was needed.

This book treats 104 species of ferns and lycophytes found in Michigan (121 taxa including hybrids and varieties). These include 103 indigenous species, and one naturalized species (*Marsilea quadrifolia*). Eighty-two species of ferns are found in the state when the Ophioglossaceae and *Equisetum* are included (95 taxa including varieties and hybrids). There are 55 species of ferns if the Ophioglossaceae and *Equisetum* are excluded (65 taxa including hybrids). There are 22 species of lycophytes found in Michigan (27 taxa including hybrids). Four additional entities have been recorded historically, making a total of 108 species.

The grouping of families, genera, and species in this book follows the Pteridophyte Phylogeny Group (PPG 1, 2016), with a few minor exceptions. This classification system is the result of the efforts of an international community of nearly one hundred pteridologists working together to incorporate recent advances in molecular phylogenetics to make decisions on the recognition, rank, and circumscription of taxa. Although fern classification systems have historically been highly problematic and controversial, the advent of the PPG represents a move toward long-term stability and accord among pteridologists. As new data are published, and more species sampled, minor changes should be anticipated, pertaining primarily to poorly sampled tropical groups, though the major groups will likely remain stable.

The Life Cycle of Ferns and Lycophytes

Although ferns and lycophytes, loosely termed "pteridophytes," do not form a natural group—that is, they do not share a common ancestor (ferns are more closely related to seed plants than to lycophytes)—they are recognized together because they share a similar life history [see "Evolutionary History of Ferns and Lycophytes"]. Pteridophytes are plants that reproduce by spores rather than seeds as in all other vascular plants. Ferns and lycophytes have two distinct free-living plant forms that alternate. The large plants that we commonly see (in ferns, typically recognized by morphological features such as fronds) are the sporophyte generation.

Through meiosis, sporophytes produce millions of tiny spores with only one set of chromosomes. This process occurs in the plants' sporangia, usually found in small, round, or linear clusters called sori that are often sheltered by a protective flap, the indusium, covering the sorus, or protective hairs called paraphyses. Marginal sori are often covered by the recurved leaf margin (a false indusium). The sporangium of most ferns is usually a thin-walled case on a stalk. This case has a row of thick-walled cells called the annulus that serves to open the case and actively expel the spores (usually 64 spores, between 30 to 50 microns long (a micron is 1/1000 of a millimeter) (see the fern life cycle illustration).

When these spores find a suitable habitat, they germinate to form a small, inconspicuous, and seldom-observed gametophyte. The typical fern gametophyte grows above the ground and is heart-shaped, flat, green, and photosynthetic. (However, some ferns produce filamentous or straplike gametophytes; some of these reproduce asexually by producing gemmae, small propagules on the microphylls that fall off to create new plants identical to the parent.) On their undersurfaces, the gametophytes form gametangia; antheridia, which produce many sperm; and/or archegonia, each of which produce a single ovum. Fertilization occurs after the archegonia are mature and when a thin film of water is present on the undersurface of the gametophyte. The archegonia open and the flagellate sperm (with one set of chromosomes) actively swim to them. A single sperm is allowed to enter and fertilize the ovum (with one set of chromosomes). The fertilized ovum or zygote, now with two sets of chromosomes, grows into a sporophyte. Thus, the gametophyte represents the sexual generation, and the sporophyte, and the spores it produces, are asexual.

This pattern involving two free-living life forms is referred to as the alternation of generations. No other vascular plants reproduce by spores and have alternation of free-living generations. Nature has developed mechanisms that usually prevent self-fertilization (a sperm from a gametophyte fertilizing an ovum from the same gametophyte).

Variations in Fern and Lycophyte Reproduction

Over the course of their evolution, the typical fern life cycle has been modified in various ways, resulting in both adaptive advantages and diversification. The following paragraphs describe the life cycle of the more typical ferns, but many different patterns are found in nature, including asexual reproduction and self-fertilization.

The classic fern life cycle diagram depicts what amounts to intragametophytic selfing, but this is only one of four ways in which ferns can reproduce, and one that is infrequently employed. Intragametophytic selfing occurs when a single (hap-

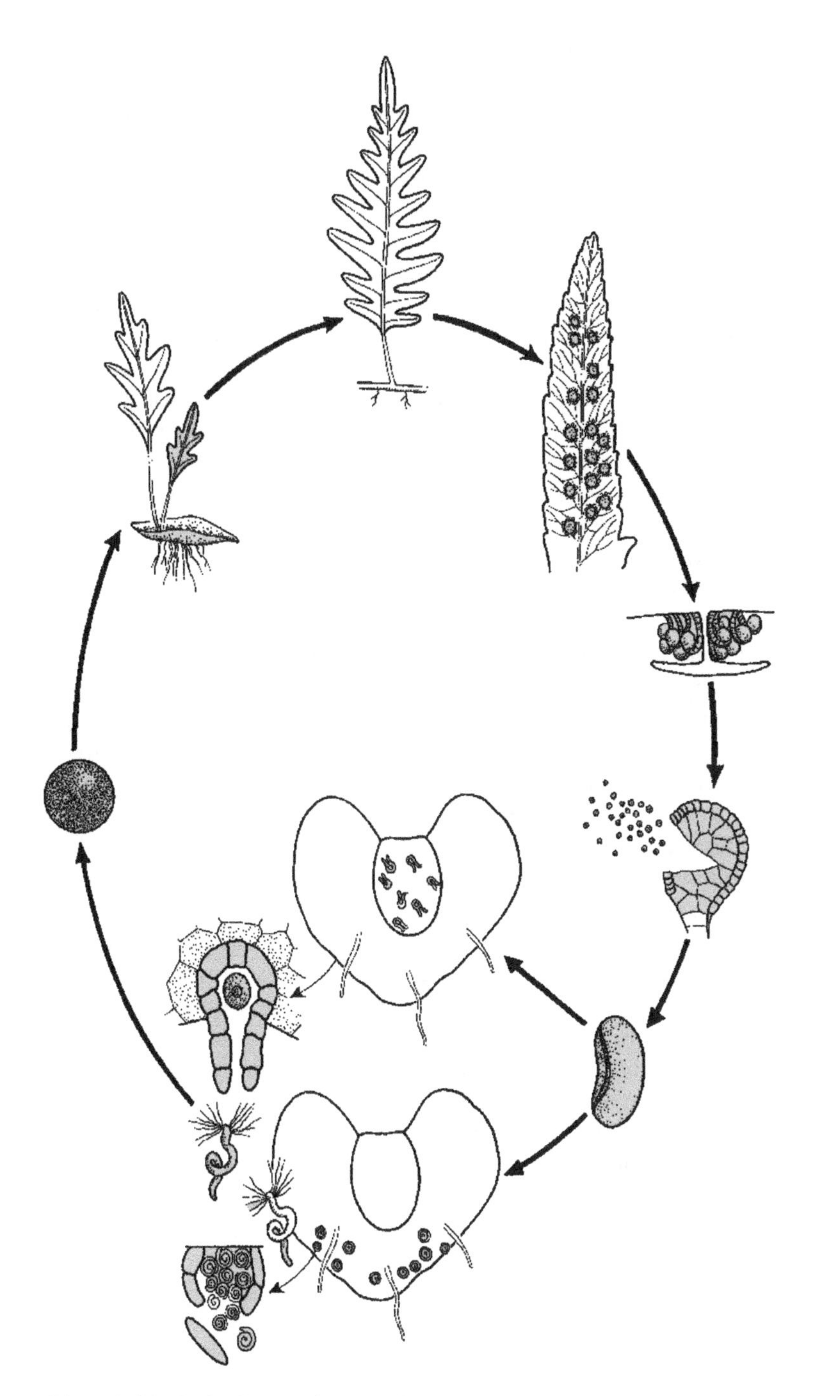

Fern life cycle (*clockwise from top*): sporophyte; pinnule abaxial surface showing sori; detail of sorus and indusium; sporangium expelling spores; spore; gamentophyte undersurface with antheridium and archegonia; antheridia releasing sperm; sperm; archegonia; fertilized egg; young sporophytes arising from gametophyte.

loid) gametophyte provides both sperm and egg—this is the most extreme form of inbreeding in nature. All loci will be homozygous in the resulting (diploid) sporophyte, and there is no reproductive equivalent in flowering plants. The benefit of this strategy is that a fern colony may be established by the dispersal and establishment of a single spore. Genomes of higher ploidy—common in ferns—may buffer such homozygosity by carrying greater standing variation.

Intergametophytic selfing occurs when a gametophyte is fertilized by another gametophyte produced by the same sporophyte parent. Although less extreme than intragametophytic selfing, it is the reproductive equivalent of autogamy or geitonogamy in flowering plants—fertilization from within a single flower or different flowers on the same plant. It allows for some recombination since different gametophytes are not genetically identical.

Intergametophytic crossing occurs when a gametophyte is fertilized from sperm from a gametophyte produced by a different sporophyte parent—the equivalent of outcrossing in angiosperms.

A fourth possibility is apogamy (or apomixis), a form of asexual reproduction by spores. This typically results from a mutation during gametogenesis in which the diploid cells are not reduced, resulting in diploid gametophytes that can produce sporophytes without fertilization. Like intragametophytic selfing, this is a successful strategy for the colonization of new territories with a single spore, a strategy frequently employed by ferns in xeric habitats, where the scarcity of water limits sperm dispersal. This is essentially a form of clonal reproduction and is equivalent to vegetative reproduction.

Some ferns and lycophytes rely heavily on vegetative reproduction. Bracken ferns (*Pteridium aquilinum*) rarely produce spores, relying primarily on vegetative spread by rhizomes. The walking fern (*Asplenium rhizophyllum*) reproduces by proliferous buds, generated at the tips of its long leaves, which send down roots. Among lycophytes, *Huperzia* spp. produce gemmae. Because of these mechanisms, large colonies of ferns or lycophytes may have originated from a single spore and then spread asexually.

The sex life of the ferns is often complicated and outrageous by human standards. Hybridization of closely related species may be common in some genera, such as *Dryopteris*. Hybridization between more distantly related species is less common, but even intergeneric hybrids have been found. Some species are formed when chromosome numbers are doubled or even multiplied six times. Herb Wagner used to state that some ferns are bastards and some ferns are real bastards.

For more detailed information regarding the often complicated life cycles of ferns and fern allies, the reader is referred to David B. Lellinger's *A Manual of the Ferns and Fern-Allies of the United States and Canada* (1985).

Evolutionary History of Ferns and Lycophytes

Ferns have a reputation as "ancient plants," and this designation is well earned. Fossil evidence suggests that these plants first appeared 400 million years ago in

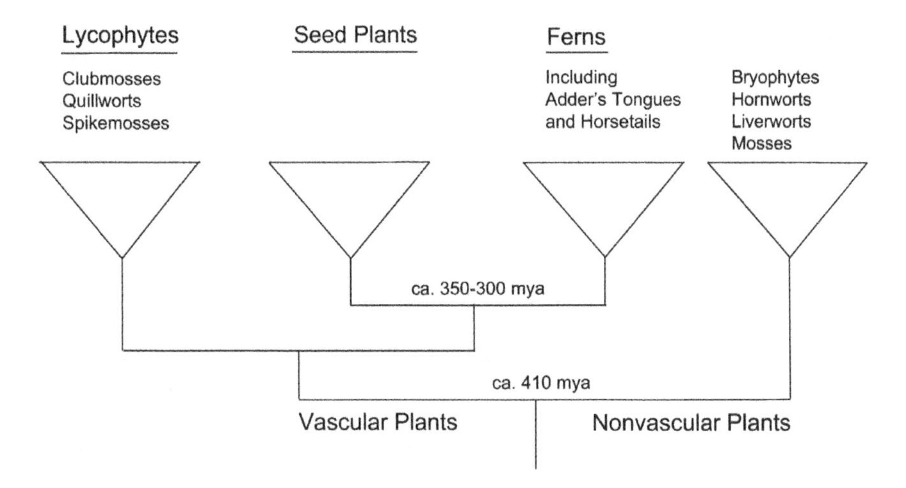

Time scale of the evolution of lycophytes and ferns.

the Devonian era. Some authors estimate an origin more than 430 million years ago in the mid-Silurian. Despite this ancient origin, many species of ferns are relatively young. Most extant families originated only about 200 million years ago and diversified since the late Cretaceous, about 65 million years ago, after flowering plants (angiosperms) had already come to dominate the planet. Our modern fern flora evolved in the shadow of angiosperms—diversifying in the shady forest understory and in the treetops, where they occupied epiphytic niches. Even though ferns have been present on Earth for hundreds of millions of years, they continue to diversify, and new species are constantly evolving.

After emerging on Earth, the ferns diverged into lineages with distinct morphologies. The horsetails, Equisetaceae, are the earliest diverging lineage and the most distinct morphologically. First appearing during the late Devonian, their ancient relatives are well known from fossil evidence and include the tree-sized *Calamites*, and woody, shrublike *Sphenophyllum*. The adder's tongue family, Ophioglossaceae, diverged later, but almost nothing is known about their ancient history on Earth because their soft tissue fossilized poorly. In spite of being one of the oldest lineages of land plants, Ophioglossaceae also includes some of the youngest ferns, with some species in the genus Botrychium likely originating since the late Pleistocene. The leptosporangiate ferns—the largest group of extant ferns—are believed to have begun diversifying about 360 million years ago. Within the leptosporangiate ferns, the family Osmundaceae was the earliest to diverge. Fossil evidence from the upper Triassic has revealed a plant nearly indistinguishable from modern-day *Osmunda claytoniana*, indicating that this species has existed relatively unchanged for over 200 million years!

Although they are treated in this book, lycophytes represent an evolutionary lineage distinct from that of the ferns, with unique morphology and no close relatives. They are as distantly related to the ferns as they are to seed plants. The lycophytes belong to the oldest existing vascular plant group (tracheophytes) and first appeared in the mid-Devonian era about 410 million years ago.

The lycophytes exhibit some of the least altered characters of vascular plants, with extant species sharing traits with early fossils. As unassuming as the clubmosses and quillworts are today, the ancestors of these plants played an incredibly important role in the history of Earth, and they have the oldest fossil record of any surviving plant lineage. The extinct Devonian *Baragwanathia longifolia* was probably a creeping herbaceous plant similar to modern-day *Huperzia*. *Asteroxylon mackiei*, an extinct plant fossilized in the Rhynie Chert, has been compared to contemporary *Huperzia* and *Lycopodium*, although the resemblance may be superficial. The first plants resembling *Selaginella* appear in the fossil record in the early Carboniferous, around 350 million years ago. The diminutive *Isoetes* is all that remains of a lineage that included the arborescent lycophytes. Well known from the fossil record, these giant trees with branching crowns originated in the middle to late Devonian and dominated the coal swamps that covered much of Earth during the Carboniferous period, ultimately forming much of the basis for today's fossil fuels. Much of this diversity disappeared during the Permian/Triassic extinction event, leaving smaller forms like modern-day *Isoetes* to carry on.

This book's title, *Michigan Ferns and Lycophytes*, reflects the evolutionary relationships discussed in the previous paragraphs, which were not widely recognized in the last century. In the past, this book would have likely been titled *Michigan Ferns and Fern Allies*.

Fern Ecology in Michigan

As with plant species generally, fern and lycophyte diversity increases as latitude decreases (Greenland has 30 species, Michigan 108, Florida 113, and Guatemala 652). The number of species of vascular plants per county in Michigan ranges from fewer than 1,000 in the Upper Peninsula to 1,000 to 1,200 in the northern half of the Lower Peninsula and 1,300 to 1,700 in the lower half of the Lower Peninsula. Michigan also has many fern and lycophyte species that reach their southernmost distribution in the Upper Peninsula and others that reach their northern limits in the southern tiers of counties.

This book finds that there are 96 species of pteridophytes in the Upper Peninsula (24 species essentially limited to the Upper Peninsula, 8 of these with minimal collections in the far northern Lower Peninsula counties or the Manitou Islands). Of these, 74 are ferns and 22 are lycophytes. There are 84 species found in the northern half of the Lower Peninsula and 77 in the lower half. Seven are found only in the lower three tiers of counties. Of all species, 84 are ferns and 24 are lycophytes. Of the lycophytes, 22 are found in the Upper Peninsula (4 of these essentially limited to the Upper Peninsula and 1 with a few collections in the tip of the Lower Peninsula), 18 in the upper half of the Lower Peninsula and 17 in the lower half of the Lower Peninsula.

Much of Michigan is covered with forest, and, as is the case throughout the world, ferns are most diverse in forest understories. Michigan's geography is divided into sections dominated by particular forest communities. The lower half of the Lower Peninsula is occupied by southern deciduous forests, with important upland forest types, including oak-hickory and beech-maple, depending on site and soil. Wetter settings in this area contain various types of hardwood and, less commonly, conifer swamps. The northern half of the Lower Peninsula features uplands with mixed "northern hardwood–conifer forest," primarily pine-oak forests in drier sites. Here again, various types of hardwood or conifer swamps dominate wetland settings. In the northernmost portions of Michigan, there is increased conifer importance as the forest communities transition to the great boreal forests north of Lake Superior.

Ferns occur in the understories of all these forest types. Characteristic ferns of upland southern deciduous forests include *Athyrium filix-femina*, *Dryopteris* spp., *Botrypus virginianus*, and *Phegopteris hexagonaptera*; in mesophytic sites, *Polystichum acrostichoides*; and in drier settings, *Pteridium aquilinum*, *Sceptridium* spp., and *Diphasiastrum* spp. Many species occur in wetter forests, including sometimes dense stands of *Matteucia struthiopteris* in floodplains, as well as *Athyrium filix-femina*, *Onoclea sensibilis*, and *Thelypteris palustris* in various types of hardwood swamps. Wet forests with lower pH soil tend to support species of Osmundaceae. Open wetlands may have *Thelypteris palustris*, *Equisetum* spp. if there is mineral soil, and Osmundaceae in acid soils. *Woodwardia virginica*, the Virginia chain fern, is unique to bogs and boggy, open forests, while fens and calcareous shores may have the tiny, mosslike *Selaginella eclipes*.

Similar components can be found in hardwood forests of the northern Lower Peninsula. Traveling northward, *P. hexagonaptera* and *P. acrostichoides* disappear, while *Phegopteris connectilis*, *Gymnocarpium dryopteris*, species of Lycopodiaceae, and *Thelypteris noveboracensis* become more numerous. Here again, dry forests will have *P. aquilinum*, *Sceptridium* spp., and various Lycopodiaceae. Conifer swamps are more abundant and have a diverse fern and Lycophyte understory, including *Dryopteris* spp., *Gymnocarpium dryopteris*, *T. palustris*, Osmundaceae, *Huperzia lucidula*, and *Equisetum scirpoides*. Very few ferns and Lycophytes occur in sunny, dry habitats, but particularly in northern Michigan such areas with sandy soils (e.g., dune openings in jack pine plains) support *Equisetum* spp., *Diphasiastrum tristachyum*, and *Selaginella rupestris*.

Some ferns and lycophytes occupy specialized habitats in Michigan. *Asplenium platyneuron*, species of *Botrychium*, *Sceptridium*, *Ophioglossum pusillum*, *Diphasiastrum* spp., *Lycopodiella* spp., and *Equisetum* spp. can be found in early successional habitats such as roadside banks and ditches, abandoned sand and gravel pits, and old fields.

Rock outcrops and cliffs (both sunny and, more typically, shady) are another habitat that supports largely ferns in their vascular floras. Examples range from boulders in forests to bedrock "pavement" to cliffs. Different species prefer different types of outcrops; some are confined to calcareous rocks (limestones and dolomites) or more acidic rocks (granites, other rocks of the Precambrian shield, or some sandstones). A large proportion of Michigan's rare ferns occur in this habitat, including all but one of the state's *Asplenium* species, both *Pellaea* spp., both *Cryptogramma* spp., several *Cystopteris* spp., and all *Woodsia* spp.

The sole submergent aquatic among Michigan ferns and lycophytes is the genus *Isoetes*, which is most abundant in clear, low-nutrient lakes in Michigan's Upper Peninsula.

Unique to ferns and lycophytes, the occurrence of these plants in various habitats is dependent on the availability of microsites for safe establishment and growth of the gametophyte generation. In forests, ferns are typically associated with microsites that were free of leaf litter at the time the tiny photosynthetic gametophytes flourished. These necessary microsites are often the result of soil disturbance, and because of this ferns are often found on banks (either actively eroding or steep enough to shed debris), at the base of stumps and trees, on tip-up mounds, on old logs (sometimes long rotted away), and associated with boulders and outcrops. Interestingly, this appears to be at least partly true even in those species with subterranean, nongreen gametophytes.

The History of Fern Study in Michigan

The description of Michigan fern species began in the eighteenth century. Carl Linnaeus (1707–78) published 28 (51 percent of 55 species, hybrids not counted) of the Michigan fern species and 12 of the lycophytes (55 percent of the 22 species, hybrids not counted) found in Michigan. He published 47 percent of all the species treated in this book (most under different generic names)—all based on materials from Europe, Canada, and the eastern and southeastern United States—none of them from Michigan material. Other Europeans published descriptions of ferns or lycophytes found in this book. These include André Michaux (1746–1802/1803), a French botanist who spent twelve years in North America and described 9 species of ferns found in Michigan; S. G. Gmelin (1744–74), a German who became a professor of botany in Saint Petersburg, Russia, and described 4 species; and Carl Ludwig von Willdenow (1765–1812), a German botanist, pharmacist, and plant taxonomist who published 1 species, 3 varieties, and 1 hybrid.

Other botanists of this early era who contributed one or two taxa treated in this book include Kurt Sprengel (1766–1833), a German botanist and physician; Ambroise Marie François Joseph Palisot, Baron de Beauvois (1752–1820), a French naturalist; Nicaise Auguste Desvaux (1784–1856), a French botanist; and Edward Hitchcock (1793–1864), a noted American geologist and the third president of Amherst College. (The first American fern described by an American was *Botrychium simplex* by him in 1823).

The early botanical and geological explorations of Michigan by Thomas Nuttall (1810), Lewis Cass with Henry Schoolcraft and David Bates Douglass (1820), Douglass Houghton (1831, 1832), and others resulted in many new plant discoveries but added no new fern species. From the nineteenth century to the last half of the twentieth, there were few new species added. Oliver Farwell, who collected botanical specimens from the Keweenaw Peninsula and elsewhere in the state from the 1890s through the 1930s, published numerous additions of previously described taxa to the Michigan fern flora (some of doubtful venue).

In last part of the twentieth century, Warren H. Wagner (1920–2000) authored or coauthored descriptions of eight new species of *Botrychium, Sceptridium,* and *Lycopodiella* based on Michigan material. His wife, Florence Wagner, and his colleagues and students have described or coauthored descriptions of several new species or contributed to the advancement of knowledge of these Michigan plants. A few new species—*Dendrolycopodium hickeyi* W. H. Wagner, Beitel & R. C. Moran; *Lycopodiella margueritae* J. G. Bruce, W. H. Wagner & Beitel; *Lycopodiella subappressa* J. G. Bruce, W. H. Wagner & Beitel; and several species of *Botrychium*—were described by them. Some new species split from recognized species after careful study using newer methods. Some were described from populations found in other states, Europe, Canada, and Japan and later recognized to be present in Michigan. Other major contributors to the knowledge of Michigan ferns include Joseph Beitel, J. G. Bruce, Donald Farrar, Dale Hagenah, John Mickel, James Montgomery, Robbin Moran, and Robert Preston.

Organization of This Book

This manual includes technical details for the botanist (synonyms, basionyms, somewhat technical keys, adequate generic descriptions, etc.), as well as illustrations and nontechnical descriptions aimed at a general audience. Throughout the text, descriptions deemed particularly important for distinguishing between taxa or especially helpful for nonspecialist readers are printed in boldface.

The species within each genus are arranged alphabetically. Thus, closely related genera will not necessarily be near each other in the text. It was felt that a phylogenetic arrangement—would not be the most efficient organization for most readers and would certainly guarantee a trip to the index to find the location of the desired genus, and, unlike the alphabet, would be subject to future changes, resulting from continuing research. The family name is given following the generic name. For genera in families with two or more Michigan representatives, the reader is referred to the appendix in which the family is briefly described and a key to the genera is given.

The descriptive portion of this book divides the pteridophytes into two parts: The ferns, including three groups (Polypodiopsida (the later diverging ferns) plus Equisetaceae (horsetails and scouring rushes) and Ophioglossaceae (adder's tongues, moonworts, grapeferns, and rattlesnake ferns); and the lycophytes (Lycopodiaceae, the lycopods, Isoetaceae, quillworts, and Selaginellaceae, spikemosses). Each group and subgroup will have its own keys to the genera or species, and in the ferns each subgroup will have its own illustrated glossary. Synonyms (older names) can be found in the index.

Taxonomic problems still present among Michigan Ferns are discussed and the common hybrids found in Michigan are noted. Since the taxonomy of ferns and lycophytes has historically been unstable, many different names have been used to refer to a given taxon. These names are listed under each taxon's synonymy and are also found in the index.

How to Use Keys

Keys are lists of paired, contrasting partial descriptions, or couplets, that help to separate plants for identification purposes. To use a key, read the first couplet to see which statement agrees best with the unidentified fern. Most ferns will agree with one of the choices in a couplet and disagree with the other. The statement selected will provide either the name of the fern or the number of the next couplet from which to make a choice. The numbers in parentheses next to the statement numbers are those of the referring couplets, making it possible to backtrack through the key all the way back to the beginning.

If it is unclear which statement of the couplet applies to your fern, choose the more likely one. If this leads to an incorrect choice, go back and try the other choice. Statements that give distribution, habitat, and rarity can help to include or exclude species.

If you are still unable to identify your plant, you may be in the incorrect genus. Go back to the keys to families of ferns and fern-allies. This is also the starting point if you have no idea which family or genus your fern is in. Once you key the plant to the proper family, the next step is to determine to which genus it belongs. Check the conspectus to see how many genera are treated in the family. If the family includes two or more genera, a key to genera in that family is included in an appendix in the back of the book. Once you key out the correct genus, you can go to the genus treatment in the main body of the book to track down the species.

While keys are useful in identifying a fern, they provide only a limited description. Refer to the illustrations and descriptions of that species for further confirmation of the identity. Immature, abnormal, or hybrid plants may not be identifiable from the keys.

Rare Taxa

This book recognizes 116 taxa of native ferns and lycophytes in Michigan. Species collected only once many decades to over a century ago, and not collected since, are not described in this book. Some species with questionable venue have not been included. A few species recognized by some, but not considered as distinct by me, are not included.

These include:

Botrychium acuminatum found to be a variant of *B. matricariifolium*.

Diphasiastrum alpinum, a widely disjunct occurrence of an arctic-alpine species, known from a single 1895 collection from the Keweenaw peninsula

D. ×*sabinifolium*. A hybrid between *D. tristachym* and *D. sitchense* (the latter not known in Michigan), this was discovered in a sand pit in Chippewa County in 1957, but is apparently no longer extant.

Equisetum telmateia (giant horsetail) collected August 25, 1890 and June 1, 1895 on the Keweenaw Peninsula.

Woodwardia areolata (Netted chain–fern) collected twice in Van Buren County in 1880; looked for but not collected since.

Other species, not reported from Michigan, that may be expected to occur here include:

Thelypteris simulata, known from the driftless region of Wisconsin, and areas of Ontario at similar latitudes, in low pH forested wetlands.

Woodsia glabella, occurring north of Lake Superior

Woodsia scopulina, occuring north of Lake Superior

Nomenclature and Synonymy

The main entry for each genus, species, and infraspecific taxon is printed in boldface. Within each genus, species are arranged alphabetically and numbered sequentially. Synonymous and misapplied names are listed under the specific name, and are arranged with the basionym first, followed alphabetically by other appropriate binomials.

For most species a minimal number of synonymous names are listed, even though in some cases there are a great many. For some species several synonymous names are listed. In these cases, the numerous names serve two functions. Many were used in older fern books, such as Billington's *Ferns of Michigan* or Eaton's *Ferns of North America*, and are included here so that the reader will be able to relate those older treatments to the currently used names. In some cases several scientific names are listed to show the difficulties that botanists have had wrestling with the various concepts of generic and family placement.

Bibliographic information for published names can usually be located in the series *Index Filicum* (Christensen 1905, 1906–33; Pichi Sermolli 1934–60; Jarrett 1961–75; Johns 1976–90, 1991–95). There are occasional errors and omissions, and names after 1995 are not included in these publications. The online International Plant Names Index (IPNI), http://www.ipni.org, provides up-to-date access to names.

Etymology (Scientific and Vernacular Names)

Derivations of genus, species, and infraspecific scientific names have been included because they often have descriptive value and may help the reader to associate names with the plants and thus remember them more easily. They are also just plain interesting. These are found under the main entries for genus, species, or infraspecific taxa following any synonyms and common names. The names involving the surnames of people are simply interesting historically.

Primary references used include *Botanical Latin* (Stearn 1995), *Composition of Scientific Names* (Brown 1985), *Webster's Third International Dictionary, Unabridged*, *Cassell's Latin Dictionary*, *Standard Greek Dictionary*, and diverse other sources.

Most ferns have multiple vernacular names. In this book the most commonly used or preferred name is placed first, followed by other names that have also been applied to the plant.

The derivations of genus, species, and infraspecific vernacular names have been included because they often have descriptive value in describing shape, color, indument, habitat, or distribution and may help the reader to associate names with the plants.

Description of Taxa

This book is designed to help the reader identify *Michigan* ferns, and thus the generic descriptions cover the members of the species in the genera as they are found in Michigan. These descriptions will, in many cases, be inadequate to describe the genus worldwide, and statements covering all species in genera with widespread distributions would lose much of their usefulness in the state. For example, in Michigan, *Dryopteris* fronds are up to 3-pinnate-pinnatifid and up to 1.2 m long, while worldwide some species have 5-pinnate fronds that are up to 3 m long. To use a broad definition of this genus would make it more difficult to recognize the genera as they occur in Michigan.

The species descriptions in this book cover the normal range of variation, but occasionally one finds abnormally large, deformed, or discolored plants, as well as ferns growing outside the expected environment.

All descriptions, with few exceptions, are limited to characters that can be seen with the naked eye or a 10× hand lens. Chromosome numbers, if known, are given following the descriptions.

Observations correlating frond fertility with season, temperature, or rainfall (phenology) have been made in only a few instances. This aspect of fern biology in Michigan has not been studied in detail.

Measurements use the metric system and encompass the normal range of variation for the majority of the plants in the taxon being described. Some species occasionally produce atypically large or small fronds; these atypical findings are largely ignored, although occasionally they are included parenthetically in the measurements. Technical information regarding spore morphology (except in the *Isoëtes* treatments), isozyme, and DNA data is not included.

Generic and specific descriptions begin with general statements such as "Plants large, evergreen" or "Plants small, deciduous." These statements should immediately help readers to recognize whether the fern they are attempting to identify might be found in the subsequent description. In this context "small" would be up to about 15 cm, "medium-sized" would be roughly 15 to 90 cm, and "large" would be more than 90 cm long. Of course these are general guides, not precise statements, and there is much variation of frond size within species.

Commonly recognized interspecific hybrids, if any, are included at the end of each generic treatment. No intergeneric fern hybrids have been found in Michigan.

Although most hybrids have binomial names, all recognized hybrids are treated here under their parental combinations, with the two parent taxa cited alphabetically. Any published binomial names are treated as synonyms and are listed with the synonymies. These names will also be found in the index. Examples include the hybrid *Dryopteris cristata* × *D. intermedia*, which designates the two parental species in the hybrid named *Dryopteris* ×*boottii* (Tuckerman) Underwood; similarly, *Dryopteris carthusiana* × *D. intermedia* was named *D.* ×*triploidea* Wherry; and *Equisetum laevigatum* × *E. variegatum* was named *E.* ×*nelsonii* (A. A. Eaton) J. H. Schaffner.

Distribution and Habitat

Distribution and habitat data were collected from many sources, including the University of Michigan Herbarium, Michigan State University Herbarium, and Cranbrook Institute of Science Herbarium, as well as knowledgeable persons, literature searches, fieldworkers, and my own experience. Elevation ranges are not emphasized because Michigan has low topographical relief; however, longitudinal information is described in general in the species distribution.

A general description of the plant's habitat is given, and additional special habitats (e.g., swamps, deciduous forests, damp, streamside banks, or sand dunes) are sometimes noted. Often an indication of the taxon's frequency in the wild is given (e.g., common, occasional, or rare). Plants are occasionally found in habitats unusual for that taxon or at extreme disjuncts from other known locations. These atypical findings are not mentioned except sometimes parenthetically.

Beyond the dot maps, concise distributions are given in general terms such as Upper Peninsula, upper Lower Peninsula, lower tier of counties, and so on, summarized in table 4. In some cases, usually with rare taxa, specific counties are named, but without reference to location in the county. Distributions, of course, are only an approximation of actual distributions. In many cases, species are present in counties where they are not noted. In other cases, they are present only in isolated locations within a county, making them appear to be far more common and widespread than they are. We still have much to learn about the distribution of Michigan ferns and Lycophytes.

There is a marked change in climate as latitude decreases from north to south. Mean annual temperatures vary from 4.4° C (40° F) in Keweenaw County to 10° C (50° F) in the lower tier of Lower Peninsula counties. Annual length of growing season (frost-free period) varies from 60 days in the southwestern Upper Peninsula to 160 to 170 days in the southern Lower Peninsula. Similar variation is found in rainfall and average minimum and maximum daily temperatures. A thorough review of the climate, geology, vegetation, biogeography, and plant communities of Michigan may be found in *Michigan Trees* (Barnes and Wagner 2004).

Chromosome Numbers

Each somatic cell of the sporophyte (the plant we recognize as the fern) contains two sets of chromosomes—one from each parent. This is labeled $2n$ or the *diploid* chromosome number. Fern spores contain one set of chromosomes labeled n or the *haploid* chromosome number. The gametophytes produced by the spores are haploid and produce an egg-producing archegonium (n) and sperm-producing antheridium (n). When the egg is fertilized it becomes diploid and produces the sporophyte—the plant we recognize as the fern ($2n$).

The base number for a genus, the genome, is labeled x. This number is the lowest haploid number found in any species in that genus. Species within a genus, while often diploid, may sometimes be polyploid with tetraploid (with 4 four sets of chromosomes), hexaploid (6 six sets of chromosomes), octoploid (8 eight sets of chromosomes), or sometimes even much higher ploidy levels are found. Sterile hybrids are often triploid (3 chromosomes, or $3n$). The x for the genus *Dryopteris* is 41. In this genus are found sterile triploids (e.g., *D. ×triploidea* $2n = 123$), fertile tetraploids (e.g., *D. carthusiana* $2n = 164$), and a fertile hexaploid (*D. clintoniana* $2n = 264$). Polyploidy in ferns is very common.

Illustrations

Most species and infraspecific taxa are illustrated to some extent. An attempt has been made to place the illustrations as close as possible to the written descriptions, and to arrange them in related groups of species so as to illuminate their differences.

Many species are quite variable in size, frond cutting, scale distribution, and presence or absence of hairs and glands. It was not possible to illustrate the wide spectrum of variability found in some species. In most cases, a typical or average frond is illustrated; the reader should bear in mind that the illustration represents the author's experience, and may vary from the experiences of others. In other cases, a series of pinnae showing the variation within that species is included in the illustration. Hybrids were illustrated only if common and widespread.

The author produced the basic illustrations by scanning fresh material, when available. When fresh specimens were not available, dried material was used. In general, smaller fronds produced better quality illustrations because they didn't have to be as severely reduced to fit the book format, and because of this the average size of living plants is often larger than what is represented by the illustration.

Glossaries

A general glossary defining terms is found in the appendix at the back of the book, while an illustrated glossary showing major morphological features particular to each group can be found near the beginning of the four sections.

Words have real value, and scientific words are often very helpful. A single word often conveys an idea that might otherwise require several words, or even a sentence, to explain. (Look up *clathrate*, a type of scale that helps to identify the genus *Asplenium*, or *acicular*, which describes a type of hair found in the family Thelypteridaceae.)

It is important to be able to understand the basic division of the fern blade. Simple descriptions, such as simple, 1-pinnate, and 2-pinnate-pinnatifid, acroscopic, basiscopic, proximal, and distal are essential. A short study of the illustrated glossaries and the main glossary should clarify these important descriptive words. It is difficult to identify many ferns without understanding these descriptive terms.

For more information on the structure of ferns and fern-allies, see Lellinger (1985).

Fertile Hybrids

Hybrids between different species may be sterile or fertile, but they always exhibit characteristics intermediate between the species. (See also the discussion under the *Dryopteris* hybrids.) A fertile diploid hybrid may be produced when a haploid spore from one species fertilizes a haploid ovum from the other. The chromosomes must be quite similar to each other for fertilizaton to occur. Generally, they are not similar enough to pair during meiosis and the hybrid sporophyte will not produce viable spores. However, a hybrid such as *Dryopteris ×triploidea* (*D. carthusiana × D. intermedia*), with one set of chromosomes from one parent and two from another (triploid), will be sterile because the production of spores requires that the chromosomes pair during the meiotic cell division that is necessary to produce spores.

A sterile triploid hybrid (with three sets of chromosomes) may produce billions of spores, almost all distorted and sterile. However, on very rare occasions instead of putting the three sets of chromosomes into two separate abortive spores, they are by chance all placed in one spore, producing a hexaploid (6 with six sets of chromosomes). The chromosomes can now pair in meiosis and produce fertile plants, and potentially give rise to a new and different fertile species.

The species produced in the normal course of Darwinian evolution are called *orthospecies*. The species produced by chromosome doubling are called *nothospecies*, and while they are good species, defined by reproductive isolation, and often distinct morphologies, their origin is entirely different from that of orthospecies.

The most common type of polyploid northospecies in Michigan are allopolyploids, which arise when sets of chromosomes from different species, derived from interspecific hybrids, are doubled to produce the fertile plant. Such polyploid nothospecies are common in ferns, especially in the genera *Asplenium* and *Dryopteris*.

Abbreviations

cm	centimeter (1 cm = 0.39 inch, 2.56 cm = 1 inch)
diam.	diameter
f.	forma
hybr.	hybrid
m	meter (1 m = 39.37 inches)
mm	millimeter (1 mm = .039 inch)
sp.	species (singular)
spp.	species (plural)
subg.	subgenus
subsp.	subspecies
var.	variety
vars.	varieties

PART 1

Equisetum (Equisetaceae)

EQUISETUM Linnaeus Equisetaceae

Horsetail and Scouring Rush Family

Equisetaceae is a nearly worldwide family with one genus, *Equisetum*. The treatments of this family and genus are identical.

Etymology. Latin *equus*, horse, + *seta*, bristle. Dioscorides thought the roots of *Equisetum fluviatile* resembled a horsetail.

Plants. Monomorphic or dimorphic; evergreen or dying back in winter; aerial stems arising from extensive persistent subterranean rhizomes.

Rhizomes. Similar to aerial stems; branching at internodes and sheaths, with or without canals; to 2 (–5) m below surface in some species; sometimes bearing tubers that store nutrients and are capable of producing new plants; aerial stems in the spring arise from overwintering buds at the nodes of the rhizomes.

Stems. Aerial stems hollow with 3 types of canals or tubes: centrum (center canal), vallecular canals (under the valleys or grooves), and carinal canals under the ridges (Latin *carina* = keel); presence, absence, or size of the canals varies with each species; ridges extending length of the internodes and continuing into sheaths, terminating in sheath teeth.

Stomata. Superficial or recessed (usually visible with 20× magnification); scattered or in lines.

Branches. When present arising at nodes through base of the sheaths.

Sheaths. Composed of fused, reduced, and modified leaves in whorls originating at nodes, sheathing the base of the internode above, the teeth being the tips of the leaves.

Cones. Borne on tips of fertile green stems, or in some species on tips of light brown, chlorophyll free, fertile stems; composed of whorls of small peltate fertile sporangiophores; tips rounded or pointed.

Sporangia. Usually 6 (5–10) per sporangiophores, borne pendant on undersurface of sporophores, elongate, opening along a longitudinal slit to release spores.

Spores. Smooth, green, with elaters; (same size in all species); sterile hybrid spores are white.

Silica tubercles. Present on ridge tops or absent.

Habitat. Wet woodlands, riverbanks, lakeshores, roadsides, seepage areas, ditches, meadows, and marshes.

Distribution. Mostly in the northern hemisphere; some are found in tropical America; 1 is found in southern Africa, tropical Asia, and Indonesia; they are not found on Pacific islands, in Australia, or in New Zealand.

Chromosome #: $2n = 216$

Illustrated Glossary to Equisetum

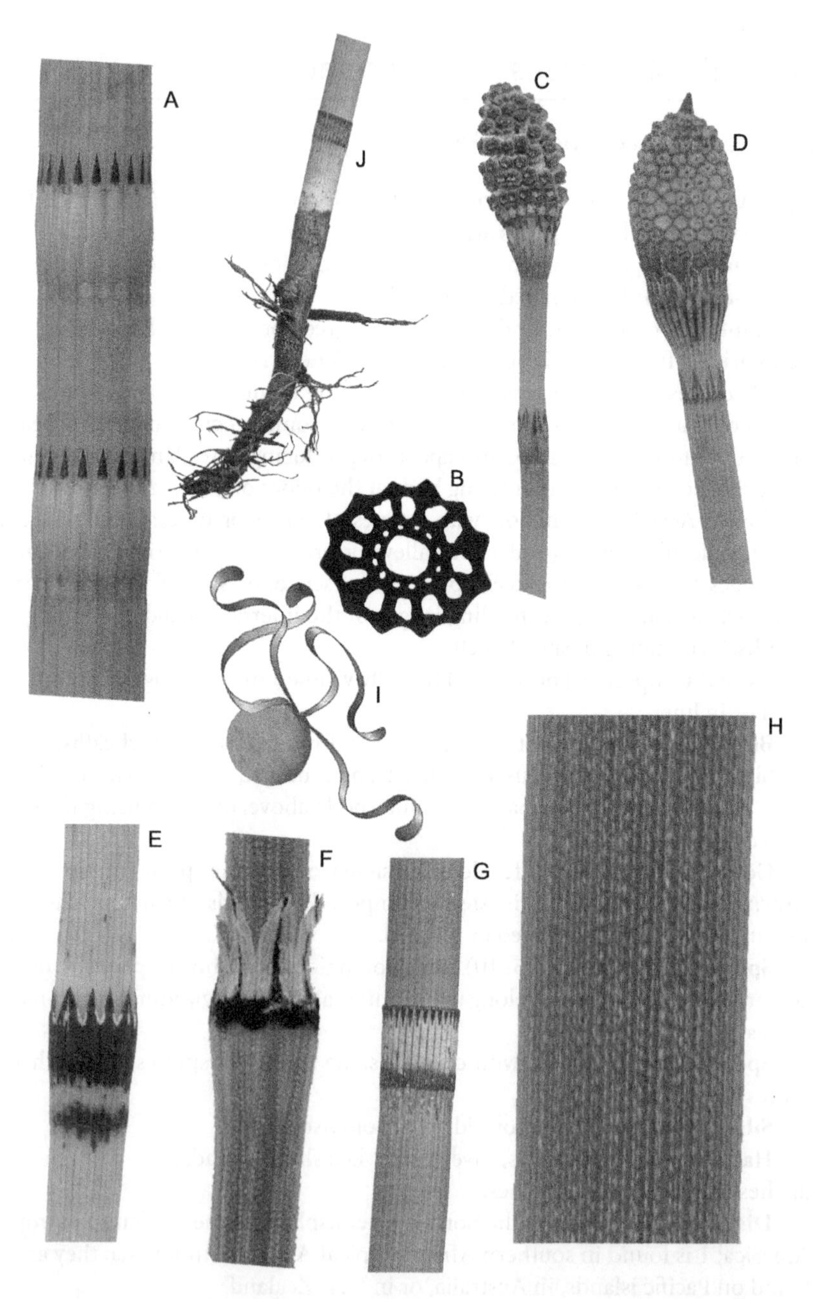

Equisetum glossary: A. Stem showing sheaths, nodes, and internode. B. Cross section of stem. C. Cone with rounded tip. D. Cone with apiculate tip. E. Sheath—teeth stiff, black, some coherent. F. Sheath—teeth black with wide white margins. G. Sheath—gray band with black margins, teeth shed. H. Enlargement of stem showing silica spicules and stomata. I. Spore with elaters. J. Roots.

The fossil record of the Equisetaceae goes back 350 million years to the Upper Devonian Age. In the Carboniferous, 325 million years ago, Equisetaceae reached their highest diversity and were represented by dense thickets of the large and abundant treelike plants *Calamites* and *Pseudobornia*. Some Carboniferous fossils (around 300 million years old) are so similar to today's *Equisetum* species that paleobotanists have placed them in the same genus. The 15 extant species of horsetails and scouring rushes are members of the only surviving genus, *Equisetum*, in the family Equisetaceae. Eleven species are treated in *Flora of North America*, vol. 2. Nine species and 4 hybrids are found in Michigan. (*Equisetum telmateia* was apparently collected twice in Keweenaw County in 1890 and 1895; it has not been collected since.)

While for many this family sits uncomfortably with the ferns, recent phylogenic studies involving DNA and morphologic characteristics have shown that *Equisetum* is in a lineage closely related to the ferns. It is clearly far removed from the lycophytes, a lineage that includes the Lycopodiaceae (clubmosses), *Isoëtes* (quillworts), and *Selaginella* (spikemosses) and experienced an evolutionary separation much earlier than the ferns in geologic time.

In *Equisetum*, the aerial stems are upright branches of extensive branched subterranean rhizomes that share the same anatomy. In some species, the rhizomes produce tubers. The sheaths at the internode are actually whorls of reduced and fused leaves having only a midrib for venation, the teeth being the tips of leaves. Longitudinal ridges on the stems extend up the internode and continue along the sheath and end at teeth tips. Buds appear between the leaves at the base of the sheaths. In some species these regularly break through to form branches; in others branches develop only as a result of stress or injury.

Cones form either on green stems or, in some species, on specialized, nongreen, unbranched stems. In one species the nongreen stem is evanescent and vanishes in early summer. In some the nongreen stem becomes green, sheds its cone, and resembles the sterile stems. The cones consist of an axis bearing whorls of 4 to 6 angled peltate sporoangiophores with vascular stalks and flattened tops bearing usually 6, but between 5 and 10, elongate sporangia on their undersurfaces. When the spores ripen the stalks of the cones lengthen to separate the sporangiophores and the sporangia split longitudinally to release the spores into the wind.

Developmental evidence and comparison with fossil members of the group show that the sporangiophores are reduced stems rather than leaves. The term *cone* is used instead of *strobilus* because the central stem has side branches of stem tissue to modified peltate leaves with sporangia. The strobili found in the lycophytes are very different, being composed of microphylls with a single axillary sorus that opens with a transverse slit. As commonly used the terms, *cone* and *strobilus* are interchangeable. The term *cone* is used in this book for *Equisetum* and *strobilus* for the Lycopodiaceae to emphasize the difference between the 2 structures.

The spores of all *Equisetum* species are spherical, green, thin-walled without ornamentation, bear elaters, and appear identical. However, in spite of appearing identical, they produce unisexual, male or female gametophytes. They must germinate in favorable habitats in a few days to a few weeks because they lack energy stored in fat globules and rely only on photosynthesis.

The elaters form when an outer layer of the spore wall dries in a spiral manner and splits into 2 ribbonlike straps. These long bands are attached at their middle at one point to the spores. Their tips have spoonlike expanded ends. The elaters are very sensitive to humidity and coil around the spore when the air is moist and uncoil when the air is dry. Their function is not precisely known. They may dig the spore into the soil or tangle spores together during dispersal, thereby creating a group of gametophytes after germination and increasing the probability that gametophytes will be close enough to ensure fertilization. They may aid in wind dispersal of the spores. Whatever their function is, elaters must be valuable, having been present for at least 250 million years.

Equisetum species have the same sexual life cycle as the other ferns. Sexual reproduction occurs when the short-lived green spores find ideal conditions for germination; conditions for spore germination are limited and relatively rare and depend much on atmospheric humidity. They must be dispersed and germinate within 5 to 15 days after release.

Once formed, the sporophyte expands and reproduces by asexual means—a primary means of *Equisetum* reproduction. It produces an extensive subterranean network of rhizomes. Tubers sometimes form on the underground stems and may become new plants when detached. The extensive subterranean rhizome system helps the plants to tolerate acute and severe environmental stress. Roots develop at the bases of lateral branch buds, both on rhizomes and, when conditions are right, on erect shoots. Detached aerial stems may produce new plants from their internodes if laid in a moist environment with proper soil.

Equisetum species are among the strongest accumulators of silicon among the higher terrestrial plants, and only the horsetails among all terrestrial plants have been definitively shown to require silicon as an essential, not simply beneficial, mineral nutrient. The silica found in the aerial stems of *E. arvense* makes up to 20% of its dry weight and up to 25.3% of *E. palustre*. Little silica is found in the rhizomes and none in the roots. It has been proposed that the silica serves a protective as well as an important biomechanical role. *E. arvense* when grown in silica-free hydroponic nutrient solutions produces very weak stems that appear collapsed, while those with even low concentrations of silica grow normally. The silica spicules may also serve as protection against herbivores, insects, and fungal attacks. In some species the silica is deposited in tubercles on the tops of the stem ridges, while in others the ridges remain bare of tubercles and are smooth. *Equisetum hyemale* is called the scouring rush because the extensive deposits of silica on its ridges allowed the stems to be used for scouring pots and pans.

Various *Equisetum* species have been assigned therapeutic benefits, including treatment of urinary problems such as kidney stones, edema, wounds, tuberculosis, and hepatitis (in Asia). It is said to strengthen connective tissue and produce diuretic, anti-inflammatory, antioxidant, sedative, and anticonvulsant effects. There is little clinical research supporting these claims.

Equisetum species are commonly associated with wet places such as lakeshores, wet meadows, riverbanks, marshy areas, ditches, seepage slopes, and moist woods. This connection is reflected in the Latin names of some species: *fluviatile* (of river), *palustre* (of marsh), and *pratense* (of meadows).

Many species of *Equisetum* may actively secrete water through specialized stomata inside the leaf sheaths of branches and branchlets. When plants growing on wet soils are examined on still mornings, a careful observer can see thousands of water droplets.

Equisetum species show plasticity of their characters, sometimes showing extreme variability. Environmentally induced, as well as genetic, factors are probably involved in the diverse morphologies within each species. This variation is reflected in the confusing number of specific, varietal, and subspecific botanical names and synonomies that have been assigned to them. For example, at least 36 published scientific names have been applied to *E. hyemale* var. *affine*, 32 to *E. laevigatum*, and at least 30 to *E. variegatum*.

The *Equisetum* hybrids have also had a profusion of published names attached to them—*E.* ×*ferrissii* with at least 17 and *E.* ×*mackayi* with about 28. It was only in the 1950s and 1960s that Richard Hauke made sense of the diversity and assigned all the synonymous species names to 15 species worldwide. Nine of these species and 4 sterile hybrids are discussed in this book.

EQUISETUM SUBGENERA

There are two distinct groups in this genus: subgenus *Equisetum* and subgenus *Hippochaete*. In the past they have been treated by some as separate genera, but the current consensus is to recognize the groups at a subgeneric level. They have distinct growth habits and morphology. Their spores appear identical, and the chromosome number of both subgenera is 108. (The chromosomes of subgenus *Equisetum* appear smaller than those of subgenus *Hippochaete*.) Hybrids between species in each subgenus are common, but no hybrids between species of different subgenera have been found.

SUBGENUS *EQUISETUM* L. IN MICHIGAN HORSETAILS

Sterile aerial stems annual, dying back during winter; usually branched in regular whorls from the internodes (unbranched forms are found in *E. fluviatile* and *E. palustre*); fertile stems of dimorphic species unbranched, brown and evanescent in *E. arvense*, or unbranched, brown becoming green and branched after cones are shed (*E. pratense* and *E. sylvaticum*); stomata superficial without outer chamber, scattered in the furrows between ridges or in bands on surface. Cone tips round, borne on green vegetative stems or on nonchlorophyllous fertile stems.

Subgenus *Equisetum* has 8 species worldwide. Represented in Michigan by 5 branching species (*Equisetum arvense, E. fluviatile, E. palustre, E. pratense,* and *E. sylvaticum* [*E. fluviatile* and *E. palustre* have unbranched forms]) and the uncommon sterile hybrid *E.* ×*litorale* (*E. arvense* × *E. fluviatile*).

SUBGENUS *HIPPOCHAETE* MILDE IN MICHIGAN SCOURING RUSHES

Aerial stems perennial, lasting more than a year (except *Equisetum laevigatum*); usually unbranched (branched only when stressed or injured); sterile and fertile stems similar; stomata in single lines, sunken with an outer chamber; cone tips pointed (except *E. laevigatum*). Always borne on green vegetative shoots.

Subgenus *Hippochaete* has 7 species worldwide. It is represented in Michigan by 4 species: *E. hyemale*, *E. laevigatum*, *E. variegatum*, and *E. scirpoides*.

Three species in this subgenus produce three interspecific hybrids found in Michigan: *E. ×ferrissii* (*E. hyemale* × *E. laevigatum*), *E. ×nelsonii* (*E. laevigatum* × *E. variegatum*) and *E. ×mackayi* (*E. hyemale* × *E. variegatum*). The first two are very common. No *E. scirpoides* hybrids have been found in Michigan.

KEY TO THE SPECIES OF *EQUISETUM* IN MICHIGAN

Equisetum species have green, spherical, smooth-surfaced spores. Their common sterile hybrids have white and distorted spores and are not included in this key. Most species are quite variable—genetic variation, habitat, stress, and trauma may lead to alterations in size and branching.

A key including both the species and hybrids could be made. The hybrids are among the most common in Michigan, but their variable morphology and detailed structure would make it very difficult to use and impractical for most readers.

The 4 known interspecific hybrids are quite variable in appearance and difficult to include in an accurate key. Check the hybrids if a plant suggests a species but does not quite seem to fit the treatment.

1. Unbranched fertile stems bearing cones, appearing only in early spring in colonies of sterile stems, pinkish to light brown, soon dying and vanishing, or persisting and changing into branched sterile stems (2).

1. Stems branched or unbranched, bearing cones or not, appearing in spring in colonies of similar stems, green, persistent unchanged through the summer (4).

2(1). Stems pink to pale to dark brown, usually soon vanishing after spores are discharged; fertile stems mostly gone before branched green sterile stems appear . . . . *E. arvense* (fertile stem)

2. Initially tan to brown, becoming green and branching; persistent after spores are discharged and cones shed; remaining somewhat smaller and with larger sheaths than sterile stems (3).

3(2). Always found in colony of sterile stems of *E. pratense*; teeth dark brown with white margins . . . . *E. pratense* (fertile stem)

3. Always found in colony of sterile stems of *E. sylvaticum*; pale brown; teeth reddish-brown, often fused in groups of 3 to 4 resembling large teeth . . . . *E. sylvaticum* (fertile stem)

4(1). Stems unbranched (5).

4. Stems branched in regular whorls from the internodes (10).
5(4). Stems, inclined, tortuous, twisted into clusters, small, mostly less than 30 cm tall, 6 stem ridges, sheaths with 3 teeth . . . . *E. scirpoides*
5. Stems erect, straight, 6–50 stem ridges, to 150 cm long, 6–30 teeth persistent or not (6).
(6). Stems dark green; teeth persistent, white with dark center, widest stems (measured between the sheaths) 0.8–1.8 (–2.2) mm . . . . *E. variegatum*
6. Stems light green to green, teeth persistent or deciduous, of various sizes and shapes, stems usually wider than 1.8 mm. (7).
7(6). Stems rough, dark girdle near base and dark wavy upper rim, with gray to brown band between; teeth usually soon deciduous . . . . *E. hyemale* subsp. *affine*
7. Stems smooth, green, teeth persistent or deciduous (8).
8(7). Sheaths green with an upper black rim, flared—slightly widened conically upward; teeth wither early, leaving dark rim on upper sheath; central canal ca. 3/4 stem diam. . . . . *E. laevigatum*
8. Sheaths with black teeth or black teeth with white margins, not flared or widened conically upward; teeth persistent; central canal ca. 9/10 or 1/6–1/3 of stem diam. (branched stems are also found) (9).
9(8). Central canal occupying 9/10 of stem diam.; stems easily compressed between fingers; sheaths about as wide as long; 12–24 black teeth; almost always in standing or slow-moving water. *E. fluviatile*
9. Central canal occupying 1/6–1/3 stem diam.; stems not easily compressed; sheaths about twice as long as wide; 5–10 black teeth often with white margins . . . . *E. palustre*
10(4). Branches always branched again; teeth large, reddish-brown, often fused in groups of 3–4 . . . . *E. sylvaticum*
10. Branches not branched again (*E. arvense* when robust occasionally branched again); teeth black, often with 2 adjacent teeth fused, or black with white margin—not fused (11).
11(10). Central canal occupying 9/10 of stem diam.; stems easily compressed between fingers; sheaths about as wide as long; almost always in standing or slow-moving water . . . . *E. fluviatile*
11. Central canal less than 9/10 of stem diam.; stems not easily compressed between fingers; teeth black or black with narrow to broad white margin; not in standing or slow-moving water (*E. palustre* sometimes in standing water) (12).
12(11). Sheaths longer than wide; first branch internode shorter than adjacent sheath; teeth black with narrow to broad white margins, 2 adjacent teeth not fused; on wet soil but not usually in standing or moving water; often associated with similar unbranched sterile stems that differ only in lack of branches; carinal canals lacking . . . . *E. palustre*
12. Sheaths about as long as wide; first branch internode shorter, or longer than adjacent sheath; teeth with or without white margins, 2 adjacent teeth fused or not; on wet to dry soils; not associated with similar unbranched

Quick Reference Table to Equisetum Species

	Branches	*Teeth*	*Sheaths*	*Stems*	*Ridges*	*Cone tips*	*Canals/area of stem*
E. arvense Sterile stems	Always present, middle and upper stem, ascending, up to 20 regular whorls	4–14, black, stiff, 2 adjacent teeth often coherent	About as long as wide	Usually erect, highly variable, terminating in long tapering tips, 5–90 cm × 0.4–4.5 mm diam.	6–19 ridges	Absent	Centrum 1/4–1/3, 2 to 3 times size of vallecular
E. arvense Fertile stems	Absent	Teeth dark brown with white margins	Sheaths large, loose, ca. 1.25 longer than wide	Light brown, evanescent in spring		Round	
E. fluviatile Sterile and fertile stems	Present midstem, or absent	12–24 black, short, narrow, sharp-pointed, sometimes with thin white border	About as long as wide, green	Easily compressed, erect, tall with long whiplike tips, smooth, 35–113 cm × 2.5–9 mm diam.	12–24 ridges flattened, often obscure	Round and short-stalked, or absent	Centrum 9/10, vallecular usually absent
E. hyemale var. *affine*	Absent (unless stressed or injured)	14–50, twisted, black, often coherent mostly soon shed, leaving a dark crenulated upper sheath rim	Slightly longer than wide, dark girdle above base and at dark wavy upper rim, gray to brown between	Rough, dark bluish-green, long, 20–150 cm × 3–8 mm diam. silica tubercles present	14–50 ridges, ridges not grooved	Sharp-pointed, short-stalked	Centrum 2/3–3/4, vallecular obvious
E. laevigatum	Absent to occasionally slightly branched	Wither early, leaving dark rim on sheath	Twice as long as wide, flared, slightly conelike distally, green with upper black rim	Erect, 20–100 × cm tall, light green, smooth to the touch, silica tubercles absent	10–30 rounded ridges	Round, or with small sharp tip, short-stalked	Centrum ca. 3/4, vallecular small
E. palustre Sterile and fertile stems	Present midstem or absent	5–10, narrow, long, black with white margins	Twice as long as wide, green, deeply grooved, flaring conelike distally	Erect, long, smooth; 4–6 ridges, tips long, thin, tapering	5–10 ridges	Absent, or round on short slender stalk	Centrum 1/6–1/3, vallecular more or less size of centrum

E. pratense Sterile stems	Always present, middle and upper stem, horizontal to drooping	8–18, narrow, dark centers with wide white margins	Slightly longer than wide, clasping	Erect, slender, tips long, thin, tapering	8–18 ridges	Absent	Centrum 1/6–1/3, carinal prominent, vallecular small
E. pratense Fertile stems	Absent	Same as sterile stems	Same as sterile stems	Light brown, turning green and shedding cones in spring	Same as sterile stems	Round	Same as sterile stems
E. scirpoides	Absent	3, dark brown with white margins, sharp-tipped	Longer than wide, 3 segmented, green below, black above	Short, clustered, small, twisted, evergreen	6 ridges	Pointed, small, unstalked	Centrum lacking, vallecular prom-inent
E. sylvaticum Sterile stems	Branches branched again, densely branched regularly from the middle and upper nodes	8–18, reddish-brown, narrow-elongate, pointed, spreading, papery, often fused in groups of 3–4 re-sembling large teeth	Flaring, conelike, or urnlike papery, green at base becoming reddish-brown above	Erect, green, 20–70 cm × 1.5–3 mm	8–18 flat topped	Absent	Centrum 1/2–2/3, clearly larger than prominent vallecu-lar canals
E. sylvaticum Fertile stems	Absent	Same as sterile stems	Same as sterile stems	Light brown, turning green and shedding cones and branching in spring	Same as sterile stems	Round	Same as sterile stems
E. variegatum	Absent	3–12, persistent with dark centers and white margins	Slightly longer than wide, spreading, green with black rim	Stiff, thin	3–12	Sharp-pointed, small, long-stalked	Centrum 1/3 or less, vallecular canals 1/2 diam. of centrum in large stems

sterile stems that differ only in lack of branches; small carinal canals present (13).

13(12). Teeth black, stiff, lacking white margin; adjacent teeth often coherent in pairs; first internode of lowest branches longer than adjacent stem sheath; branches usually ascending, somewhat robust . . . . *E. arvense*

13. Teeth black, with narrow to broad white margins, not stiff; adjacent teeth not adherent; first internode of lowest branches shorter than adjacent stem sheath; branches horizontal, delicate . . . . *E. pratense*

- There are 4 interspecific *Equisetum* hybrids found in Michigan. They are variable in character, often resembling 1 parent species more than the other. Some are evergreen, 1 is partially evergreen, and some are deciduous. You may not see them when they sporulate to see if the spores are green or white.
- If you find a specimen that does not comfortably fit the species in this key, go to the section on *Equisetum* hybrids below.

Equisetum Species Defined By Branching and Winter Stem Persistence

	Present through Summer	*Persistent through Winter*	*Present Only in Spring, (fertile stems)*
Unbranched stems	*E. fluviatile* *E. hyemale* *E. laevigatum* *E. palustre* *E. scirpoides* *E. variegatum*	*E. hyemale* *E. scirpoides* *E. variegatum*	*E. arvense*[a] *E. pratense*[b] *E. sylvaticum*[b]
Branched stems	*E. arvense* *E. fluviatile* *E. palustre* *E. pratense* *E. sylvaticum*	0	Unbranched initially, becoming green and branched *E. pratense* *E. sylvaticum*
Branched and unbranched sterile stems	*E. fluviatile* *E. palustre*	0	0

[a]Evanescent.
[b]Later becoming green and branched.

INFORMAL KEY TO THE MICHIGAN SPECIES OF *EQUISETUM* IN MIDSUMMER

A formal dichotomous key is often not helpful for field identification of *Equisetum* species in the summer (or winter). The observer cannot use keys that include statements as to whether the species is evergreen or deciduous (the previous year's deciduous stems are gone) or whether it is dimorphic or monomorphic (some fertile stems occur only briefly in early spring).

Formal dichotomous keys often do not include the very common hybrids or if they do it is very difficult to key to the right one.

The following key is very informal and irregular and does not follow good rules for dichotomous keys; however it is hoped that this key will help the reader reach the correct identification.

1. Stems without branches (2).
1. Stems with branches (8).
2(1). Stems small, 2.5–26 cm tall; twisted in tangled clusters; unbranched, less than 1 mm diam.; stems with 3 ridges . . . .*E. scirpoides*
2. Stems similar in size or larger; stems straight; not twisted in tangled clusters. (3).
3(4). Stems erect, stiff, straight, dark green, short to long (6–45 cm tall), usually small, slender (0.5–3 mm diam.); 3–12 ridges; teeth persistent with a black center and white margin . . . . *E. variegatum* (If confused here, look under the hybrids *E. ×nelsonii* and *E. ×mackayi.*)
3. Characters otherwise (4).
4(5). Stems smooth; sheaths twice as long as wide, flaring conelike at upper end, green with upper black rim; teeth wither early, leaving a dark rim; cones round tipped (occasionally with small pointed tips) . . . . . *E. laevigatum* (If confused here, look under the common hybrids *E. ×ferrissii* and *E. ×nelsonii.*)
4. Characters otherwise (6).
5(4). Stems rough-surfaced; ridges prominent; sheaths slightly longer than wide with a dark girdle near the base and a dark crenulated upper rim, ashy gray to brown between girdles; teeth usually promptly shed but may persist; cones with sharp-pointed tips; common . . . . *E. hyemale* (If confused here, see also the common hybrid *E. ×ferrissii* and the less common *E. ×mackaii.*)
5. Characters otherwise; unbranched and branched forms (6).
6(5). Central canal filling 90% of the stem diam., easily collapsed when pressed between fingers; smooth; ridges inconspicuous; almost always found in standing or slow-moving water; stems may be branched or unbranched . . . .*fluviatile* (If confused here, see also the hybrid *E. ×litorale.*)
6. Central canal filling less than 90% of stem diam.; not easily collapsed when pressed between fingers; variable habitats, not usually in standing or slow-moving water (7).
7(6). First internodes of lowest whorl of branches shorter than adjacent stem sheaths; teeth with black centers and thin white margins, not fused with adjacent teeth; branches with 5–7 ridges; unbranched stems are found in sunny areas, branched in shade; central canal 1/6–1/3 stem diam.. . . . . *E. palustre*
7. First internodes of lowest whorl of branches longer, or shorter, than ad-

jacent stem sheaths; teeth with black centers and thin white margins, not fused with adjacent teeth, or teeth black usually with 2 adjacent teeth fused; branches with 3 ridges (8).

8(1). Stems with branches that always branch again; 8–18 persistent reddish-brown teeth often fused in groups of 3–4, resembling large teeth .... *sylvaticum*

8. Stems with unbranched branches (*E. arvense* occasionally has some branches that branch again); teeth absent, or black and stiff, or black and softer with white margins (9).

9(8). Teeth entirely black, stiff, 2 adjacent teeth often fused together; branches ascending early, horizontal later; first internode longer than adjacent sheath; very common .... *E. arvense*
(If confused here, see also the rare hybrid *E. ×litorale.*)

9. Teeth with black centers and white margins, soft, not fused together; branches horizontal, delicate, first branch internode shorter or equal to the adjacent stem sheath; uncommon .... *E. pratense*

***Equisetum arvense* Linnaeus**
Equisetum boreale Bongard
Field Horsetail, Common Horsetail

Etymology. Latin *arvensis*, pertaining to *arvum*, arable or cultivated fields, in reference to this species often being found in fields. The vernacular name is a translation of the botanical Latin in reference to this species' preferred habitat.

Plants. Extremely dimorphic; vegetative stems dying back in fall or winter; fertile stems short-lived in spring.

Underground stems. Similar to the aerial stems except they are not hollow; branched and wide-creeping; deep-penetrating, to 2 m deep; occasional tubers present.

Sterile stems. 5–90 cm tall, 1–4.5 mm diam.; usually erect; highly variable; terminating in long tapering tips; internodes 1.4–4.5 cm long; 6–19 ridges: valleys channeled with rows of 2–4 stomata on each side; silica rosettes regularly arranged on ridges; bearing up to 20 regular whorls of branches arising from middle and upper stem.

Canals. Central canal 1/4–1/3 stem diam.; 2–3 times larger than vallecular canals; small carinal canals present.

Branches. Ascending and curving upward when young, spreading when older, 3–4 ridges, **solid** (not hollow), **rarely rebranching, first internode of lowest branch longer than adjacent stem internode (including sheath and teeth)**; branch teeth narrow with long,

often dark tips; valleys channeled with bands 2–3 stomata wide on each side; silica rosettes regularly arranged.

Sheaths. 2–5 × 2–5 mm; about as long as wide, sometimes 1.25× as long as wide.

Teeth. 6–18, 1–3.5 mm long; **dark;** narrow; stiff; **adjacent teeth often coherent in pairs.**

Fertile stems. 5–30 cm tall; **present in early spring, mostly present in April–May** (the green, sterile stems appear later, usually after the fertile shoots have wilted); stalked **cones with rounded tips borne on unbranched, succulent, smooth, ephemeral stems** with 2–5 internodes; **light brown to pale pinkish, never becoming green; shorter than sterile stems; sheaths large, loose,** ca. 1.25 longer than wide;

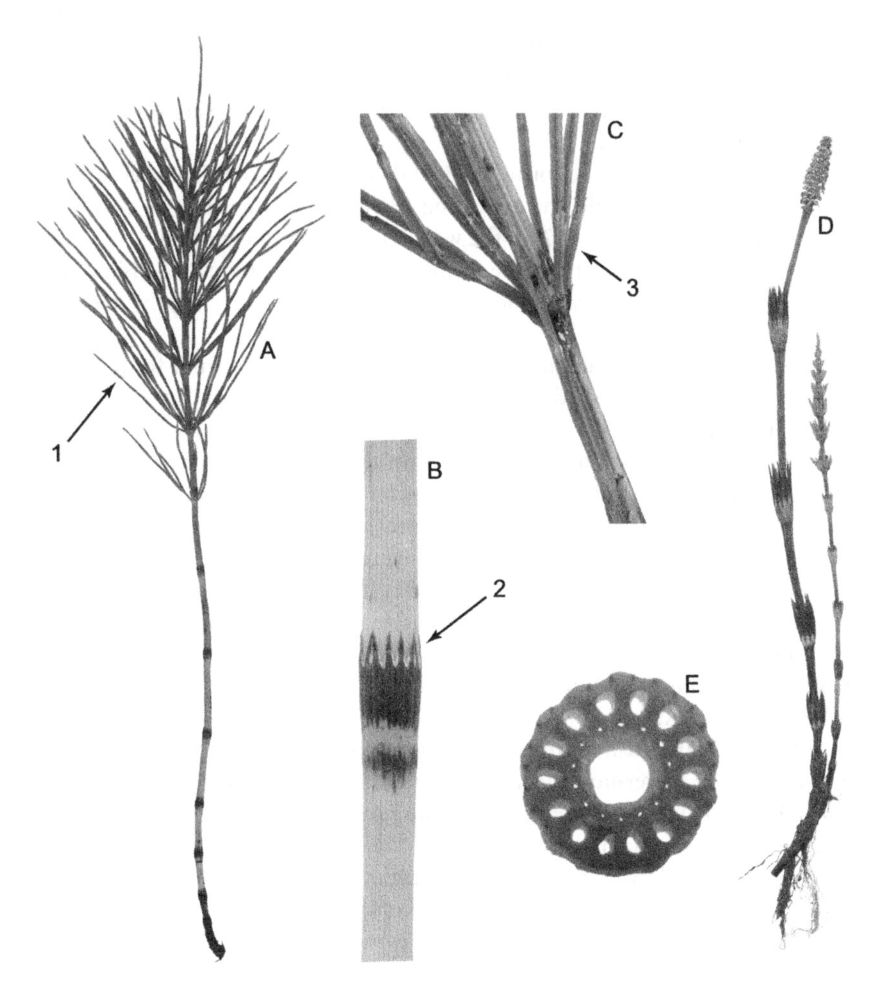

Equisetum arvense: Field horsetail: A. Sterile stem. B. Portion of lower stem. C. First branch internode. D. Fertile stem and early sterile stem. E. Cross section of stem.
Informative arrows: 1. Branches ascending. 2. Stiff black teeth, some coherent. 3. First branch internode longer than stem sheath.

teeth dark brown with white margins; cone tips round (fertile stems rarely persistent and imitate fertile stems).

Habitat. Thrives in many environments but prefers damp, sandy, partially shaded locations, open fields, woods, marshes, swampy areas, riverbanks, disturbed areas such as roadsides, borrow pits, and railroad rights of way.

Distribution. Eurasia, throughout Canada and the continental United States (except in the Southeast). *Equisetum arvense* and *E. hyemale* are by far the most common species in Michigan.

The most widespread species of the genus, *Equisetum arvense* is extremely variable. In exposed sites the plants may be prostrate with injured stems growing back in odd shapes (sometimes with cones on green branches). Plants in favorable sheltered sites may be quite large and have branches that branch again. Dozens of specific and subspecific names have been created to cover this variation. None deserves species recognition.

The rhizomes of *E. arvense* form horizontal layers at descending intervals of about 30 cm and extend to depths exceeding 180 cm. When it grows in colonies the rhizomes may make up to 50% of the weight of the top 25 cm of soil.

The fertile stems appear early in the spring and shed their spores and disappear about the time the sterile stems are coming out. When present they clearly identify *E. arvense.*

E. arvense may accumulate up to 4 1/2 ounces of gold per ton of fresh plant material. Its value in this regard is primarily as an indicator plant rather than a commercial source of gold.

Native Americans made a diuretic tea from the plant, and it was used as a cough medicine for horses and a source of dyes for clothing and porcupine quills. The young shoots were eaten cooked or raw. Constituents for leaf odor were incorporated in perfumes in the 1970s but are not used now. *Equisetum arvense* powder is available by the pound to "repair bone tissue." It is also used as a diuretic and astringent, to alleviate hemorrhaging and menstrual bleeding, and to treat ulcers, cystitis and other infections, prostates, kidney stones, and so on.

E. arvense hybridizes with *E. fluviatile* to form *E. ×litorale*, a sterile hybrid that spreads by rhizome extension.

E. arvense **is distinguished from** *E. palustre* **and** *E. fluviatile* **by its first branch internodes on the lowermost branches, which are longer than the adjacent sheaths. The branches of** *E. arvense* **usually do not branch again, while those of** *E. sylvaticum* **always branch again.**

E. arvense **has stiff, dark teeth with 2 adjacent teeth often fusing, while the teeth of** *E. pratense* **are soft, stand alone, have a dark center with wide white margins, and are not coherent.** *E. pratense* **has more delicate, horizontally arranged branches than** *E. arvense* **with its thicker and ascending branches.**

The branches of *E. arvense* **are coarse and unbranched (unless the plant is injured, stressed, or very robust), while branches of** *E. sylvaticum* **are delicately branched again.** *E. arvense* **has black teeth, while** *E. sylvaticum* **has reddish-brown teeth.**

E. arvense **has a central canal occupying 1/3 to 2/3 of the stem cross section whereas the canal occupies 90% of the cross section in** *E. fluviatile.*

Equisetum arvense **may be easily distinguished from other branched** *Equisetum* **species when its evanescent fertile stems are present. It is the only branched species with unbranched fertile stems that are transient and dry up in the spring before the sterile stems appear. The fertile stems of** *E. pratense* **and** *E. sylvaticum* **are unbranched, persist, and become branched and green when the cone is shed in spring.**

Equisetum fluviatile Linnaeus

Equisetum eburneum Roth; *Equisetum limosum* Linnaeus; *E. maximum* Lamarck

River Horsetail, Pipes, Water Horsetail

Etymology. Latin *fluviatilis* from *flumen*, river, + *-atilis*, a suffix meaning place of growth—"pertaining to rivers"—in reference to one of this species' preferred habitats. The vernacular name describes a preferred habitat and is a translation of the botanical Latin. *Pipes* refers to the hollow stem and easily separated internodes.

Plants. Deciduous; dimorphic (in the sense that stems may be branched or unbranched).

Rhizomes. Same size as stems, wide-creeping, smooth, hollow, light brown with long black clustered roots often shaped like horsetails (the "horses tails" of Dioscorides, who created the name *Equisetum*: Latin *equis* = horse, *setum* = bristle); tubers absent.

Stems. Erect, tall; 35–113 cm long, 2.5–9 mm diam., with long whip-like tips; internodes 2.5–5 cm long, **smooth, 12–24 flattened ridges,** often obscure; **branches present or absent, when present arising from middle nodes only**.

Canals. Central canal 9/10 of the stem diam.; vallecular canals absent or sometimes present and small.

Sheaths. About as long as wide, 4.5–9 × 4–10 mm; green.

Teeth. 12–24, 2–3 mm long, with short, narrow, sharp-pointed, black tips, sometimes with thin white border.

Cones. 10–25 mm long; blunt-tipped; borne on short stalks; sometimes borne on tips of branches.

Branches. Present or absent, when present highly variable in number, **only from midstem nodes,** often with some branches missing; whorled, smooth, slender, spreading to ascending, length variable, very thin-walled, central cavity of stem about 4/5 of the stem diam.; first branch internode at the stem noticeably shorter than the remaining segments.

Habitat. Unbranched forms

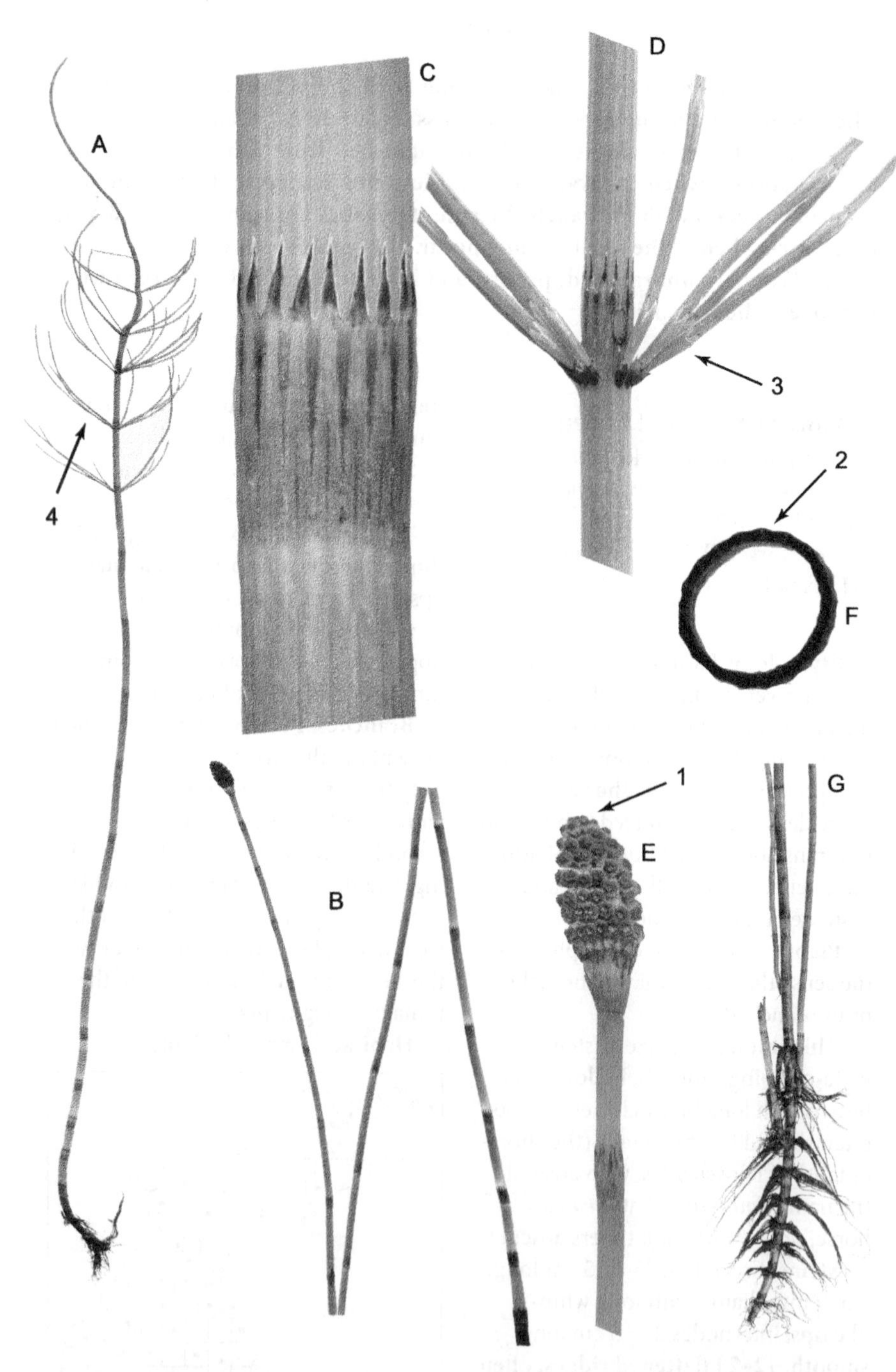

Equisetum fluviatile: Water horsetail: A. Branched stem. B. Unbranched stem. C. Sheath with teeth. D. Stem with branches. E. Round-tipped cone. F. Cross section of stem. G. Roots (horsetails).

Informative arrows: 1. Rounded cone tip. 2. Large central canal. 3. First internodes shorter than stem sheath. 4. Branches, if present, only from midstem nodes.

develop more commonly in open sunny sites, while branched forms are found in shady areas. This variation is demonstrated in some large clones in which the stems extend from sun exposed to shady shores. Unbranched stems may become branched later in the summer. The name *E. limosum* was once applied to unbranched forms. Stem diameter seems to be correlated with nutrient availability.

E. fluviatile can grow in 1.5 m of water. This may be accomplished by its large central canal and hollow rhizomes, allowing air circulation and probably allowing roots to function in an anaerobic environment. It is probably most closely related to *E. arvense*, with which it hybridizes.

Distribution. Eurasia south to northern Italy, China, Korea, Japan. Canada: all provinces. USA: Alaska, including the Aleutian Islands, Washington and Oregon across upper 2 or 3 tiers of states to Maine and Virginia.

Equisetum fluviatile **has the largest central canal of all** *Equisetum* **species, filling about 90% of the stem diam. (***E. palustre* **has central canals occupying 1/6 to 1/3 of stem diam.) It has 12 to 24 stem ridges; small black sheath teeth; and a smooth, thin-walled stem that is easily collapsed when pressed between the fingers. In contrast,** *E. palustre* **rarely grows in water and has 5 to 10 stem ridges and somewhat larger black teeth with white margins. It is distinguished by its aquatic habitat and smooth stem. Its stems may be branched or unbranched.**

Unbranched stems are frequently found when the plant is growing in deep water early in the growing season.

Equisetum hyemale Linnaeus subsp. *affine* (Engelmann) Calder & Roy Taylor

Equisetum affine Engelman; *E. hyemale* Linneaus var. *affine* (Engelmann) A. A. Eaton; *E. affine* (Engelman) Rydberg; *E. hyemale* var. *robustum* (A. Braun) A. A. Eaton; *E. prealtum* Rafinesque var. *affine* (Engelman) M. Broun; *Equisetum robustum* A. Braun var. *affine* Engelmann; *Hippochaete hyemalis* (L.) Bruhin subsp. *affinis* (Engelman) W. A. Weber

Rough Scouring Rush, Rough Horsetail, Snake-grass, Pipes

Etymology. *Hyemale* is derived from the Latin *heims*, winter, or the Greek *xeimos* (χειμών), winter: the stems of this species remain green during the winter. The vernacular name, scouring rush, refers to its early use for cleaning pots made possible by its silica tubercles.

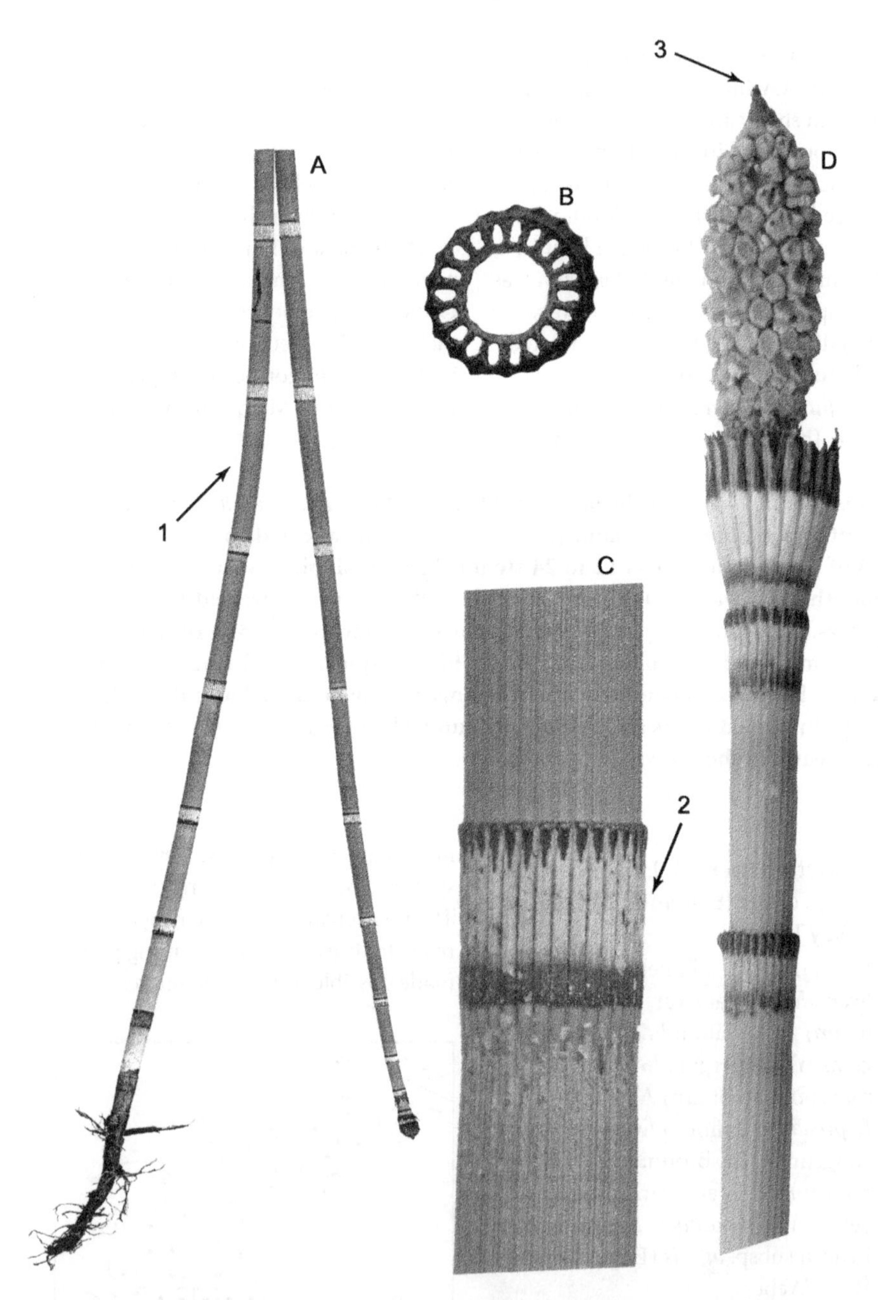

Equisetum hyemale subsp. *affine:* Rough scouring rush: A. Stem with roots. B. Cross section of stem. C. Classic sheath. D. Stem tip with apiculate cone.

Informative arrows: 1. Rough stem. 2. Characteristic sheath with gray band and dark bands above and below. 3. Apiculate cone.

Snake-grass refers to the plant's banded snakelike appearance, or because snakes are thought to live in colonies of it, and pipes possibly because the hollow stems' internodes are connected by means of pipelike joints.

Plants. Evergreen; monomorphic.

Rhizomes. Branching and wide-spreading with multiple horizontal branches (to a depth of 5 m at one roadside cut), center hollow, dull black, glabrous; tubers present.

Stems. 20–150 cm long, 3–8 mm diam., **unbranched,** occasionally branched when injured or stressed, **rough,** 14–50 ridges, ridges not grooved; all the same except for cones on fertile stems, dark bluish-green, single row of silica tubercles along ridgetops, stomata in single lines on each side of ridges.

Canals. Central canal large, 2/3–3/4 of stem diam., vallecular canals obvious.

Sheaths. 4.5–17 × 3.5–18 mm; appressed to stem; **dark girdle above base and at dark wavy upper rim, gray to brown between,** slightly longer than wide; single row of silica tubercles on tops of ridges.

Teeth. Usually shed promptly, leaving a dark crenulated upper rim on the sheath (except for a whorl of teeth at stem apex clasping base of cone), 14–50; long, narrow, 2.5–5 × 0.5–0.6 mm, firm and stiff to thin and paperlike, with long, twisted, tapering pointed tips and narrow to wide frayed white margins; often coherent in groups of 3–4, sometimes persistent.

Branches. Absent unless injured or stressed, sometimes appearing on old stems and producing small cones in the spring.

Cones. 1.2–2.5 cm long, borne on short stalks, **sharply pointed at tips;** cones and spore shedding April through August.

Habitat. Diverse, ditches, moist roadsides, railroad rights of way, woods, fields, sandy areas, and low wet places in woods.

Distribution. Greenland. Canada and USA: Canada from Prince Edward Island to Alaska, south to Mexico and Guatemala. (*E. hyemale* subsp. *hyemale* is found in Europe and Asia to North-west China.)

Equisetum hyemale subsp. *affine* is the most common scouring rush in North America. It often forms large colonies on slopes in moist, moderately shady forests with moist soils.

Except for stem size, the plant is not variable.

Equisetum hyemale subsp. *hyemale* differs from *E. hyemale* subsp. *hyemale* in a few minor characteristics. It is larger and more variable and differs in the number of stem ridges and slightly longer sheath lengths.

E. hyemale earned its vernacular name, scouring rush, from the rows of silica tubercles on the stems that act as a gentle but effective polish. In the past the plant was used to give wood, ivory, silver, pewter, brass, wooden kitchen utensils, arrows, and combs a fine finish. Bunches of the rush were used to scour milking pails and scrub pots and pans in the kitchen. It was imported into England from Holland for those purposes. Even now it could be very useful to campers. It is currently used as "Dutch rush" to prepare reeds for musical instruments.

Equisetum hyemale has a very extensively branching subterranean rhizome system. While most of the rhizome is contained in the top 25 cm of soil, there have been reports of much deeper rhizomes (one as deep as 5 m) in a roadside cut. Rhizomes grow rapidly. A transplanted *E. hyemale* plant may affect 1 hectare of land in 6 years. Large colonies of *E. hyemale*, sometimes 30 m in diameter, probably arising from a single spore, are often seen.

The common belief that *E. hyemale*, with its extensive tangled rhizomes, has been used to help stabilize Holland's dikes is probably erroneous.

It is very difficult to eradicate *E. hyemale* from any location. A persistent small piece of rhizome or stem allows it to regrow. Often the best advice for someone who is contemplating its eradication is "forget about it."

Rounded starch containing tubers, 15–20 mm in diam., are produced at rhizome branchings. These serve both reproductive and starch storage functions and are usually found, often in very large numbers, at depths greater than 50 cm. They store carbohydrates from spring to fall, and detached tubers readily give rise to new plants.

Detached stems, and parts of stems, of *E. hyemale* readily become rooted in wet soils and even form roots and upright shoots while floating in water. These plants are able to form new colonies.

Equisetum hyemale **is distinguished from similar, unbranched** *Equisetum* **species by its rough-surfaced, evergreen stems, cones with pointed tips, and sheaths with dark bands above and below with a wide ashy gray band between them.**

Equisetum laevigatum **has smooth-surfaced stems, cones with rounded tips (sometimes with a small pointed tip), and a green sheath with a black line at the top. Its stems die back in winter.**

The stems of *E. variegatum* **are much narrower and smaller than those of** *E. hyemale* **and have persistent teeth with black centers and white margins.**

***Equisetum laevigatum* A. Braun**
Equisetum funstonii A. A. Eaton; *E. kansanum* J. H. Schaffner; *Hippochaete laevigata* (A. Braun) Farwell
Smooth Scouring Rush, Smooth Horsetail

Etymology. Latin *laevigatus*, smooth, alluding to the plant's smooth low ridges lacking silica tubercles compared to the other scouring rushes.

Plants. Stems monomorphic, **dying back in winter**.

Rhizomes. Branched, wide-spreading, tubers absent.

Stems. 20–100 (–150) cm tall × 3–9 mm wide, erect, light green, 10–30

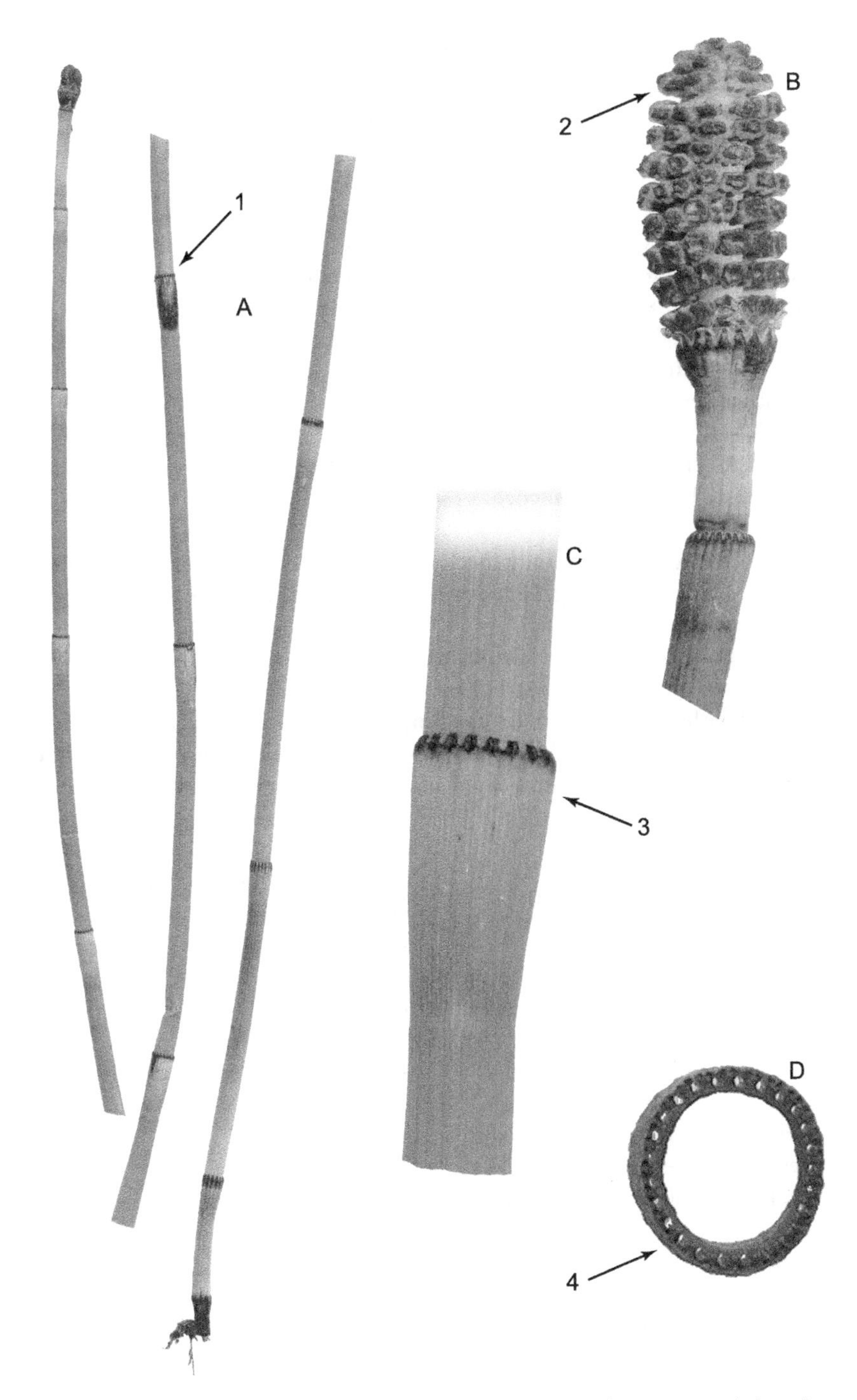

Equisetum laevigatum: Smooth scouring rush: A. Stem in 3 sections. B. Round-tipped cone. C. Sheath. D. Cross section of stem.

Informative arrows: 1. Smooth stem. 2. Rounded tip (sometimes with a small pointed tip). 3. Flared green sheath with dark band at top. 4. Large central canal and small vallecular canals.

rounded ridges, **smooth to the touch, silica tubercles absent,** stomata in a single row on either side of grooves, ridges not grooved.

Canals. Large central canal, ca. 75% of stem diam.; vallecular canals small.

Sheaths. 7–15 × 3–9 mm, **twice as long as wide, slightly flared distally, green with an upper black rim, silica tubercles absent.**

Teeth. 10–30, tiny, **wither early, leaving dark rim on sheath.**

Branches. Usually unbranched.

Cones. Tips round or small and sharp, 1.3–2.5 cm long, short-stemmed.

Habitat. Open, moist, usually sandy sites, lakeshores, riverbanks, roadside ditches, prairies, and pastures.

Distribution. North America only. Canada and USA: Michigan and Ontario west to British Columbia, south to Ohio, Indiana, Missouri, Oklahoma, Texas, northern Mexico, and Baja California, becoming more common in the western United States.

Equisetum laevigatum is the only species of *Equisetum* endemic to North America. It is the only species in subg. *Hippochaete* that has rounded, nonpointed cone tips and dies back in winter.

Distinguished from *Equisetum hyemale* **and** *E. variegatum* **by its smooth stems and usually round-tipped cones.**

The sheaths of *E. laevigatum* **are longer than wide, are slightly flared distally, and have a dark rim.** *E. hyemale* **has sheaths about as long as wide with an ashy gray band with dark lines above and below and a rough surface.**

Equisetum palustre Linnaeus

Equisetum palustre var. *americanum* Victorin

Marsh Horsetail

Etymology. From the Latin, *paluster*, marshy, swampy, or boggy, in reference to this plant's usual habitat in wet places. The vernacular name is a translation of the botanical Latin name and a description of a preferred habitat.

Plants. Stems dimorphic (in that the sterile stems may be branched or unbranched), dying back in winter.

Rhizomes. Creeping and branching, solid, slender, shiny, black to dark brown; tubers often present.

Stems. 20–80 cm × 2–3 mm, erect, **branched, unbranched, or minimally branched, branches only at mid-stem nodes, 5–10 prominent ridges; smooth,** internodes 2.5–5.8 cm long; sterile fronds with long, thin tapering tips.

Canals. Central canal 1/6–1/3 stem diam., vallecular canals about

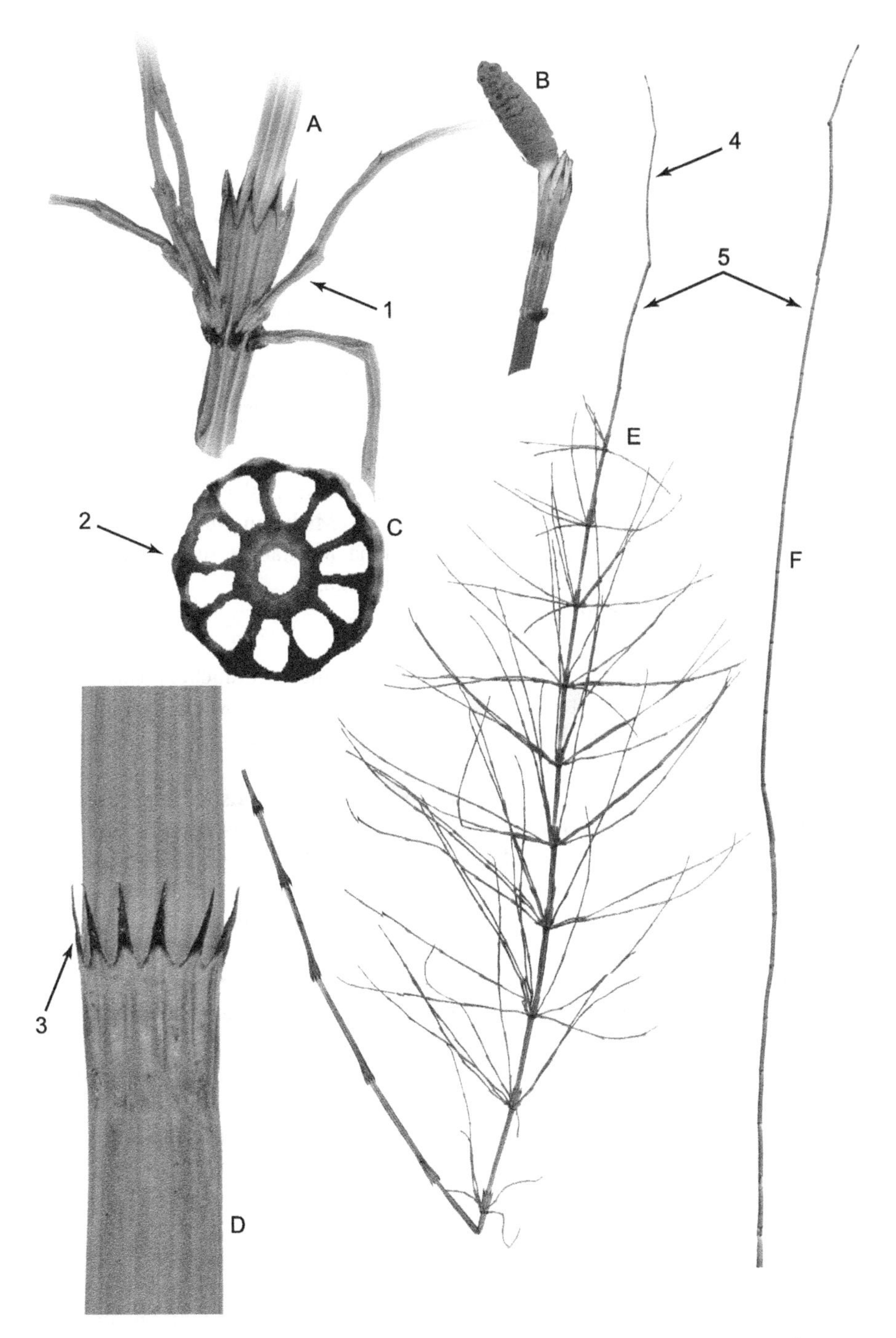

Equisetum palustre: Marsh horsetail: A. Stem internode showing first internodes of branches. B. Cone. C. Cross section of stem. D. Sheath of unbranched stem with teeth. E. Branched form. F. Unbranched form.

Informative arrows: 1. First branch internodes shorter than adjacent sheaths. 2. Central canal and vallecular canals about the same size. 3. Teeth black with white margins. 4. Long whiplike sterile tip. 5. Branched and unbranched forms.

the same size as central canal, carinal canals lacking.

Sheaths. 4–9 × 2–5 mm, twice as long as wide, deeply grooved, flaring somewhat conelike distally, green.

Teeth. 5–10, 2–5 mm long, narrow, black with narrow to broad white margins, teeth on lower sheaths have lower central parts the same color as the sheath.

Fertile stems. Extending 2–3 segments above uppermost branches, with cone on tips.

Branches. Present or absent; may be numerous, spreading in regular whorls from the middle nodes only, first branch internode at the stem noticeably shorter than the remaining segments, smooth, unbranched; firm, central cavity less than 4/5 of the stem diam., 4–6 blunt ridges, lengths variable; ascending to spreading.

Cones. Borne on otherwise typical vegetative stems, 15–32 mm long, on short slender stems, **tips blunt,** sometimes borne on branch tips; appear May through summer.

Habitat. Wet places, swamps, marshes, streams, ponds, lakeshores, usually not growing in water.

Distribution. Eurasia. Japan. Canada and USA: Alaska to Newfoundland, south to Michigan, Maryland, Iowa, Indiana, Montana, Oregon, Idaho, and California.

Equisetum palustre stems growing in the shade are mostly branched, while those in more open, sunny sites tend to be unbranched. This variation is seen in some large clones that cover both habitats. Its hollow rhizomes, allowing air circulation, probably allow roots to function in an anaerobic environment.

E. palustre **resembles** *E. arvense*, **but may be distinguished by its sterile fronds, which are identical to its fertile ones.** *E. arvense* **has very distinct and separate, short-lived fertile fronds. The branches of** *E. palustre* **are found on the middle nodes only, while those of** *E. arvense* **extend to the tips of the stems. The first internode of the lowest branch of** *E. palustre* **is shorter than the adjacent sheath, while on** *E. arvense* **it is longer than the sheath. The teeth of** *E. palustre* **are single (not coherent) with black centers and thin white margins, while the teeth of** *E. arvense* **are dark and stiff with 2 adjacent teeth often coherent.** *Equisetum palustre* **has hollow branches versus solid branches in** *E. arvense*.

The sheaths of *E. palustre* **(sometimes unbranched) retain their teeth while** *E. laevigatum* **sheds its teeth, leaving a black upper rim.**

See the discussion under *E. fluviatile* **for features distinguishing it from** *E. palustre*.

Equisetum pratense Ehrhart
Equisetum umbrosum J. G. F. Meyer
Meadow Horsetail, Shade Horsetail

Etymology. Latin *pratus*, meadow, + *-ense*, a suffix indicating place of growth, growing in meadows, in reference to this species' supposed habitat in meadows. The vernacular name, meadow horsetail, is a translation of the botanical Latin and describes a habitat but not necessarily a favored one.

Plants. Stems semidimorphic, dying back in winter, delicate.

Rhizomes. Deep subterranean, slender, creeping, branching, dull, black; tubers absent; roots wiry.

Sterile stems. 16–50 cm long, erect, slender, 1–3 mm diam., branching regularly from middle and upper nodes, internodes 1.8–3.8 cm long, 8–18 ridges—those on upper branched internodes with long, thin, sharply spiked silica rods, light green, tips long, thin, tapering.

Canals. Central about 1/3 diameter of the stem, vallecular canals many, small; carinal canals many, prominent.

Sheaths. 3–5 × 2–4.5 mm, slightly longer than wide, clasping, pale.

Teeth. 8–18, 1.5–4 mm long, **narrow dark centers and wide white margins.**

Branches. Borne in regular whorls, circa 12.5 cm long, **thin and delicate;** straight, unbranched, **horizontal to drooping**; solid (not hollow); **first branch internode shorter than or equal in length to adjacent stem sheath and never longer,** ridges 3; valleys channeled with stomata in single row on either side; teeth deltoid, slightly incurving, with thin white margins.

Fertile stems. Pink to brown; initially unbranched, persisting and becoming branched and green after spores are discharged and cones wither and fall off in early spring; sheaths 2–2.5× longer than wide, including teeth, teeth dark brown with white margins; appearing in early spring amid emerging sterile stems, scattered but uncommon.

Cones. Round-tipped, 2–2.5 cm long, maturing in late spring, soon falling off.

Habitat. Moist woods, meadows, partial shade to full sun.

Distribution. Circumpolar. Eurasia. Canada and USA: Alaska to Newfoundland, south to British Columbia, Idaho, Montana, North Dakota, Minnesota, Iowa, Illinois, New York, and New Jersey.

Equisetum pratense **resembles a delicate** *E. arvense*. **It may be distinguished from it by its teeth with dark centers and wide white margins that stand separately, while the teeth of** *E. arvense* **are dark and stiff, and two adjacent teeth are often fused. Its finer branches tend to be horizontal rather than ascending as in** *E. arvense*. **The first branch internodes of** *E. pratense* **are shorter than the adjacent sheath, while those of** *E. arvense* **are longer.**

E. pratense **has brown fertile stems that persist, becoming green and branched when the cone is shed, while** *E. arvense* **has separate fertile stems that are unbranched, brown, and evanescent and usually dry up about the time sterile stem appears.**

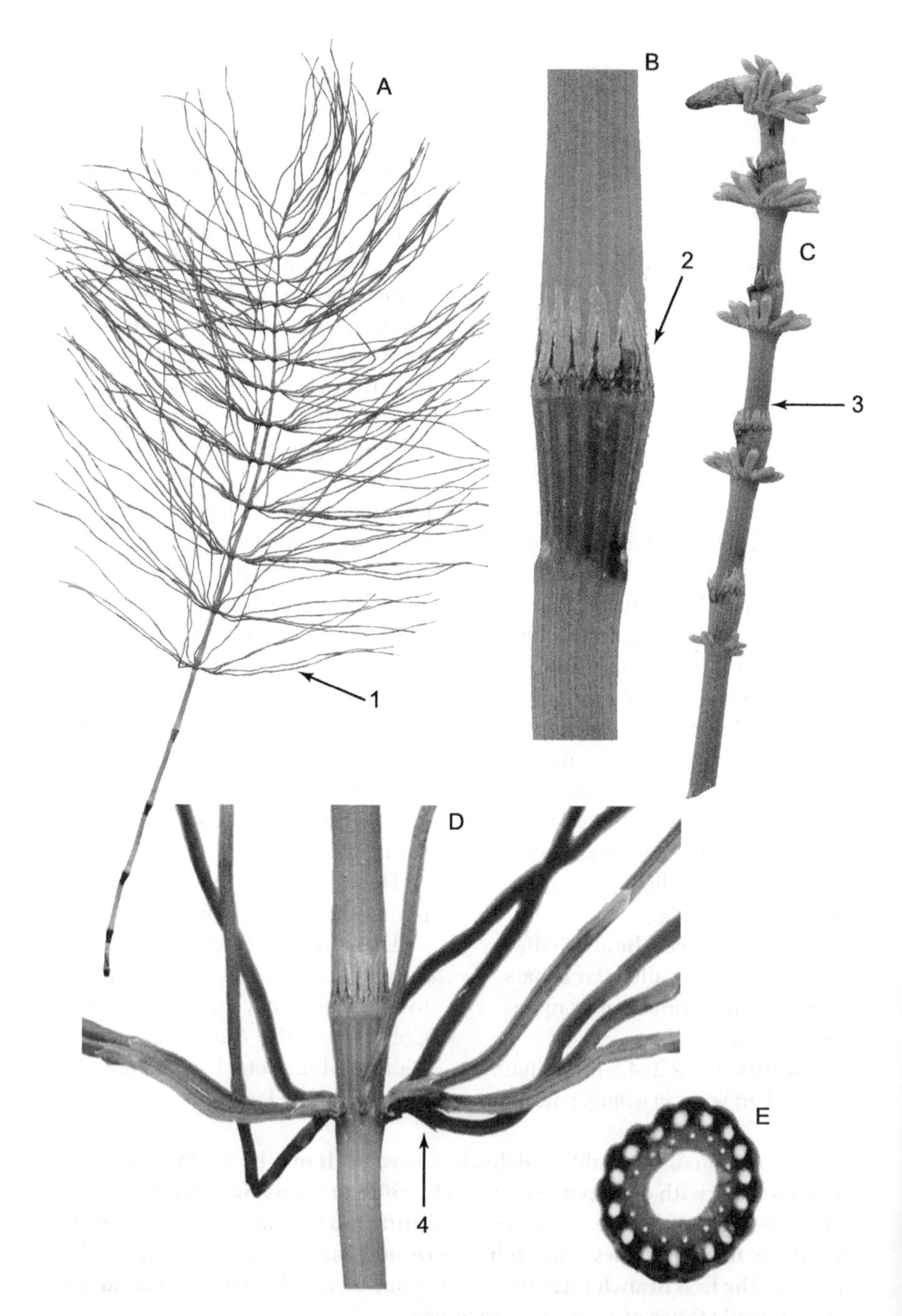

Equisetum pratense: Meadow horsetail: A. Sterile stem with branches spreading horizontally. B. Portion of stem with sheath and teeth. C. Fertile stem (early spring) becoming branched (cone falling off). D. Sheath with teeth and branches showing first internode. E. Cross section of stem.

Informative arrows. 1. Branches horizontal, delicate (*E. arvense* has ascending thicker branches). 2. Sheath teeth dark with wide white margin. 3. Young branches developing on fertile stem. 4. First branch internode shorter than sheath.

It is distinguished from *E. sylvaticum* **by branches that do not branch again and very different sheath teeth. Distinguished from** *E. palustre* **by its smaller size and delicate, horizontal to drooping solid branches rather than thick, ascending, hollow branches.**

Distinguished from *E. fluviatile* **by its much smaller central canal and terrestrial habitat.**

Equisetum scirpoides Michaux

Hippochaete scirpoides (Michaux) Farwell

Dwarf Scouring Rush, Dwarf Horsetail, Sedge Horsetail

Etymology. Latin *scirpoides, scirpus,* + *-oides* a suffix implying resemblance, in reference to its resemblance to a species in the sedge genus *Eleocharis,* which was first called *Scirpus.*

Plants. Stems monomorphic; evergreen.

Rhizomes. Creeping; freely branching; wide-spreading; lustrous black; tubers absent.

Stems. 2.5–26 cm tall, **0.5–1 mm diam., small, slender, twisted, tortuous, clustered,** internodes 1.5–4 cm long, always with 6 ridges (but only 3 sheath teeth), each topped with a single shallow furrow; dark green, stomata in single lines.

Canals. Centrum lacking (except in very large stems), 3 prominent vallecular canals near stem center.

Sheaths. 1–2.5 × 0.75–1.5 mm longer than wide, 3 segmented, broadly and deeply furrowed; often expanding distally; green below, black above.

Teeth. 3, triangular; dark brown with white margins, sharp-tipped; not jointed at contact with sheath.

Branches. Absent.

Cones. Small, 2–3 mm long, **not stalked,** tips sharply pointed, with 3 whorls of sporangiophores; mature April–August, if in late summer may overwinter shedding spores in the spring.

Habitat. Cool, moist shady woods, wet mossy banks, and rotten logs covered with moss.

Distribution. Circumpolar. Northern Eurasia. Canada and USA: Alaska to Labrador and Greenland, south to Washington, Idaho, Montana, South Dakota, Minnesota, Iowa, Illinois, New York, and New England.

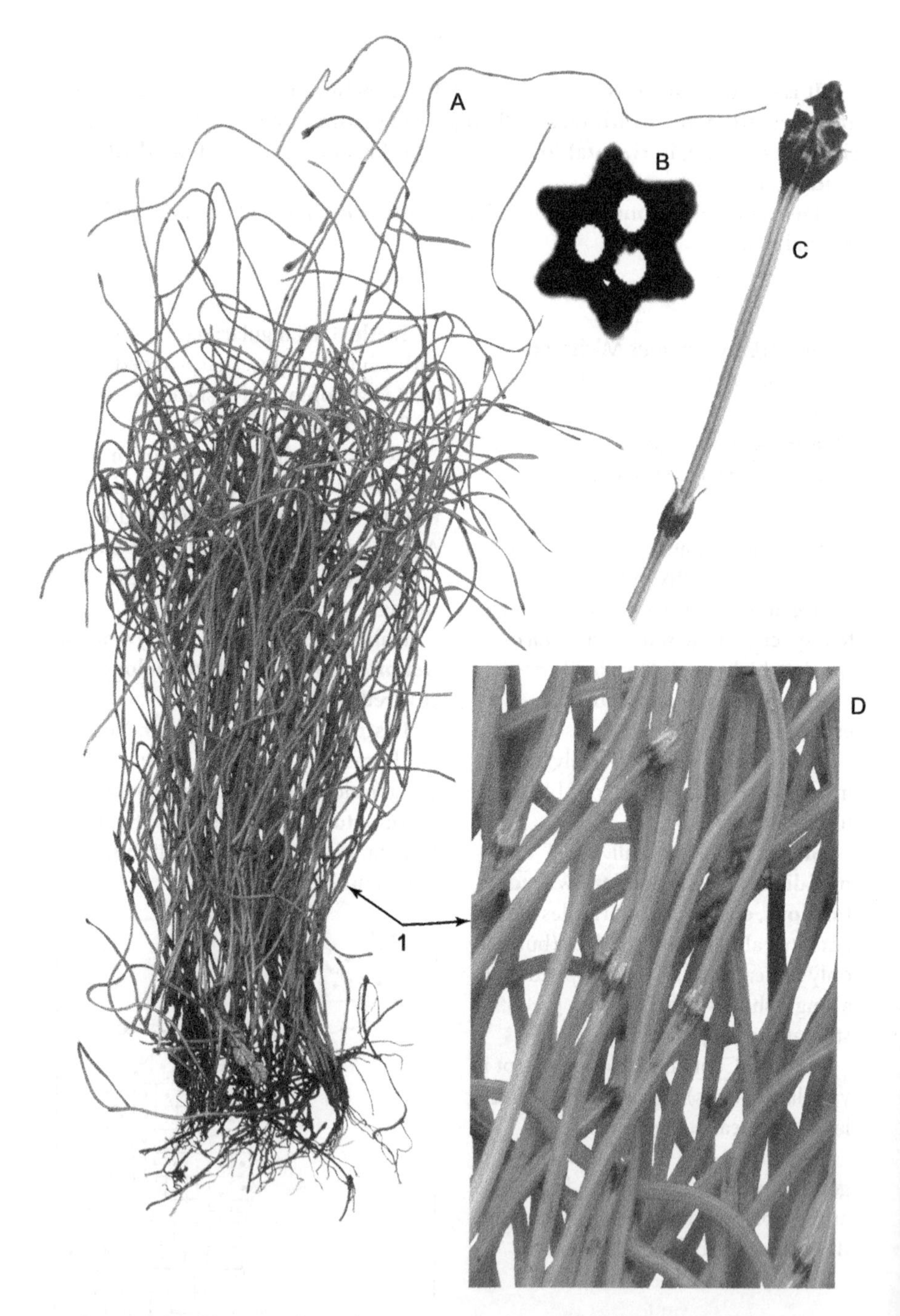

Equisetum scirpoides: Dwarf scouring rush: A. Cluster of stems. B. Cross section of stem. C. Cone and upper stem with sheath. D. Enlargement of tangled stems showing sheaths and teeth.

Equisetum scirpoides **is the smallest species in the genus and is easily identified by its tight clusters of small, slender, tangled curly stems.**

Equisetum sylvaticum Linnaeus

Equisetum capillare Hoffmann; *E. sylvaticum* Linnaeus var. *multiramosum* (Fernald) Wherry

Woodland Horsetail, Sylvan Horsetail

Etymology. *Sylvaticus* is derived from the Latin *silva*, woods, forest, + *-aticus*, a suffix meaning place of growth—"belonging to the woods"—in reference to this species' usual habitat in moist woods. Silvanus was a god of the woods. The vernacular name is a translation of the botanical Latin name and describes its preferred habitat.

Plants. Stems dimorphic, deciduous.

Rhizomes. Subterranean, wide-creeping, shiny, light brown, smooth; tubers present, sometimes large.

Sterile stems. 25–70 cm tall, 1.5–3 mm diam., densely branched regularly from the middle and upper nodes, erect, internodes 2.3–6.5 cm long, 8–18 flat-topped ridges with 2 rows of silica tubercles, valleys channeled with stomata in a single row on either side; green.

Canals. Central 1/2–2/3 stem diam., clearly larger than the vallecular canals, vallecular canals prominent.

Sheaths. 3–6 × 2.5–6 mm excluding teeth, 3–10 ridges, flaring and conical or spreading then narrowing below teeth (urnlike), papery, green at base, becoming reddish-brown above.

Teeth. 1.5–4 mm long, 8–18; persistent, narrow-elongate, pointed, spreading, papery; **often fused in groups of 3–4, resembling large teeth; reddish-brown.**

Branches. In regular whorls, **branches branched again,** solid, delicate, arching, spreading to descending; first internodes longer than adjacent sheaths; 3–4 ridges; valleys channeled.

Fertile stems. Pale brown, initially unbranched, persisting and becoming green and branched after spores are discharged and cones wither and are shed, remaining somewhat smaller and with larger sheaths than sterile stems; sheaths 3× longer than wide, including teeth; teeth reddish-brown with white margins, often fused; cones 1.5–3 cm long, borne on stalks 2–6.5 cm long; scattered among emerging sterile stems, found only in early spring.

Cones. Tips rounded 1.5 × 3 cm long, borne on short stalks, maturing in late spring, withering and falling off after spores shed.

Habitat. Moist woods, swamps, wet banks, wet meadows, stream banks, and mixed conifer–hardwood forests.

Distribution. Circumboreal. Eurasia south to the Mediterranean. Canada and USA: Canada south to Washington, northern Idaho, northwestern Montana, South Dakota, Nebraska, Iowa, Ohio, Kentucky, and Virginia.

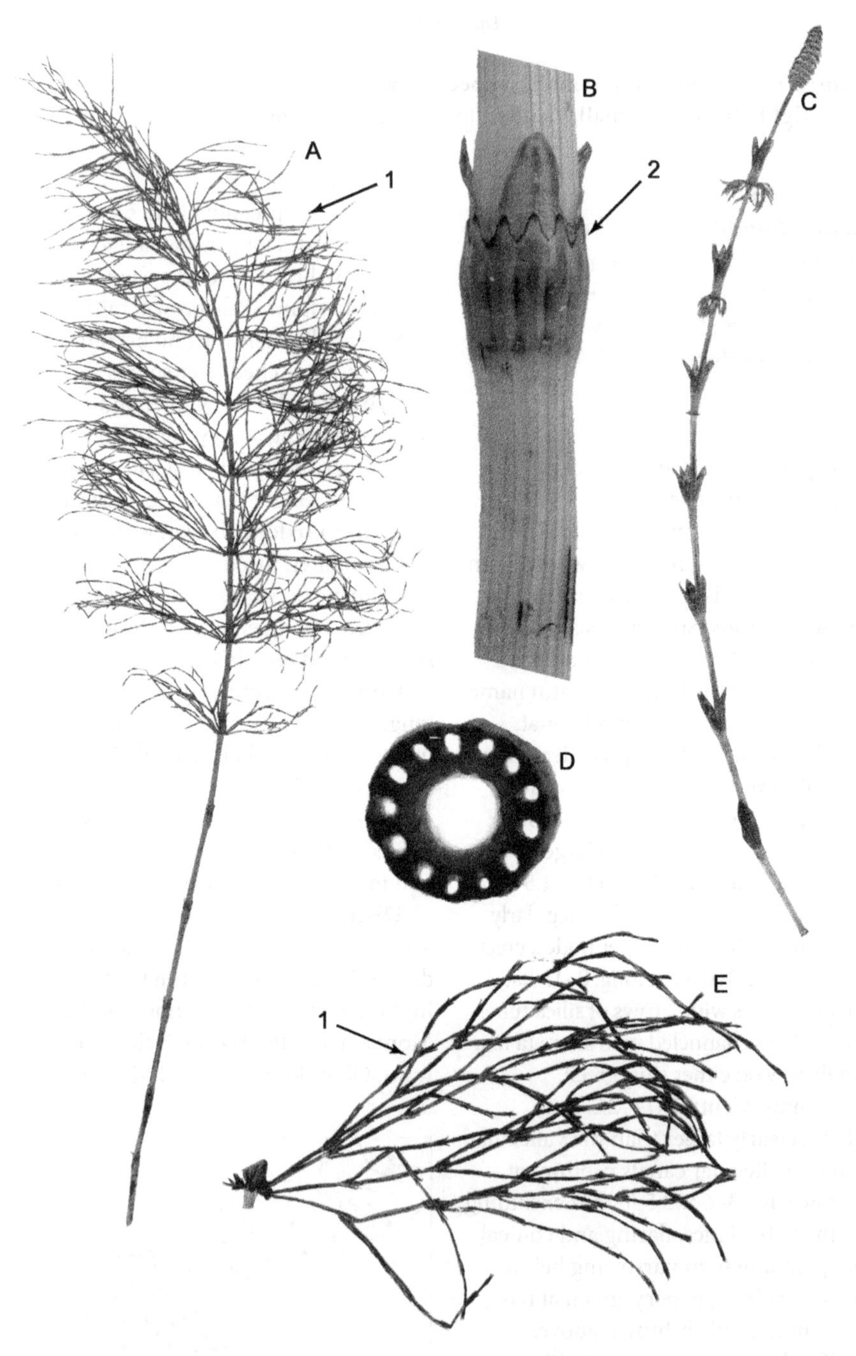

Equisetum sylvaticum, Woodland horsetail: A. Sterile stem. B. Sheath and teeth. C. Fertile stem with round-tipped cone just starting to branch (early spring). D. Cross section of stem. E. Enlarged branch showing repeated branching.

Informative arrows: 1. Branches branching again. 2. Reddish-brown papery teeth fused in groups of 3 or 4.

Equisetum sylvaticum sometimes forms large, beautiful, misty colonies of thin-branched, light green branching stems.

Equisetum sylvaticum **is easily recognized by its delicate, slender branches, which branch again. While other** *Equisetums,* **such as** *E. arvense,* **may occasionally have a few coarse branches that branch again, this is unusual and the branches are thicker. It is the only** *Equisetum* **with reddish-brown sheaths and teeth.**

Equisetum variegatum Schleicher & D. Mohr subsp. *variegatum*

Hippochaete variegata (Schleicher) Bruhin

Variegated Scouring Rush, Variegated Horsetail

Etymology. Latin *variegatus*, of different sorts, particularly colors, in reference to this species' multicolored green sheath with black teeth with white margins.

Plants. Stems monomorphic, evergreen.

Rhizomes. 0.5–1.5 mm diam., creeping near surface, often semiexposed; slender, often hollow, dark, smooth, glabrous, tubers absent.

Stems. 7–30 (–45) cm tall, widest stems (measured between the sheaths) 0.8–1.8 (–2.2) mm in diam.; **small, thin,** erect, straight, **stiff;** clustered; unbranched; 3–12 ridges with a deep central furrow with 2 rows of silica tubercles (1 on each side of furrows); **dark green;** stomata in single lines on either side of ridges.

Canals. Central 1/3 or less the diam. of stem, vallecular canals 1/2 diam. of central canal in large stems and absent in narrow stems.

Sheaths. 1–6 × 1–5 mm, slightly longer than wide, **somewhat funnel-shaped with spreading teeth, sheath nearest cone expanded, partially enclosing young cone, green with a black rim at base of teeth.**

Teeth. 3–12, 1–2 × 0.5–0.75 mm, erect, **persistent with dark centers and white margins.**

Branches. Absent.

Cones. Less than 5–7 mm long, stalked, small, sharply pointed at tips, most cones mature April to August; cones appearing in late summer overwinter and shed spores in the spring.

Habitat. Sandy lakeshores, cool wet woods, marshes, ditches, riverbanks, and meadows.

Distribution. Circumpolar. Eurasia. Iceland. Greenland. Canada and USA: throughout Canada, Alaska, south to Oregon, Utah, Colorado, Minnesota, Illinois, northern Indiana, New York, and Connecticut.

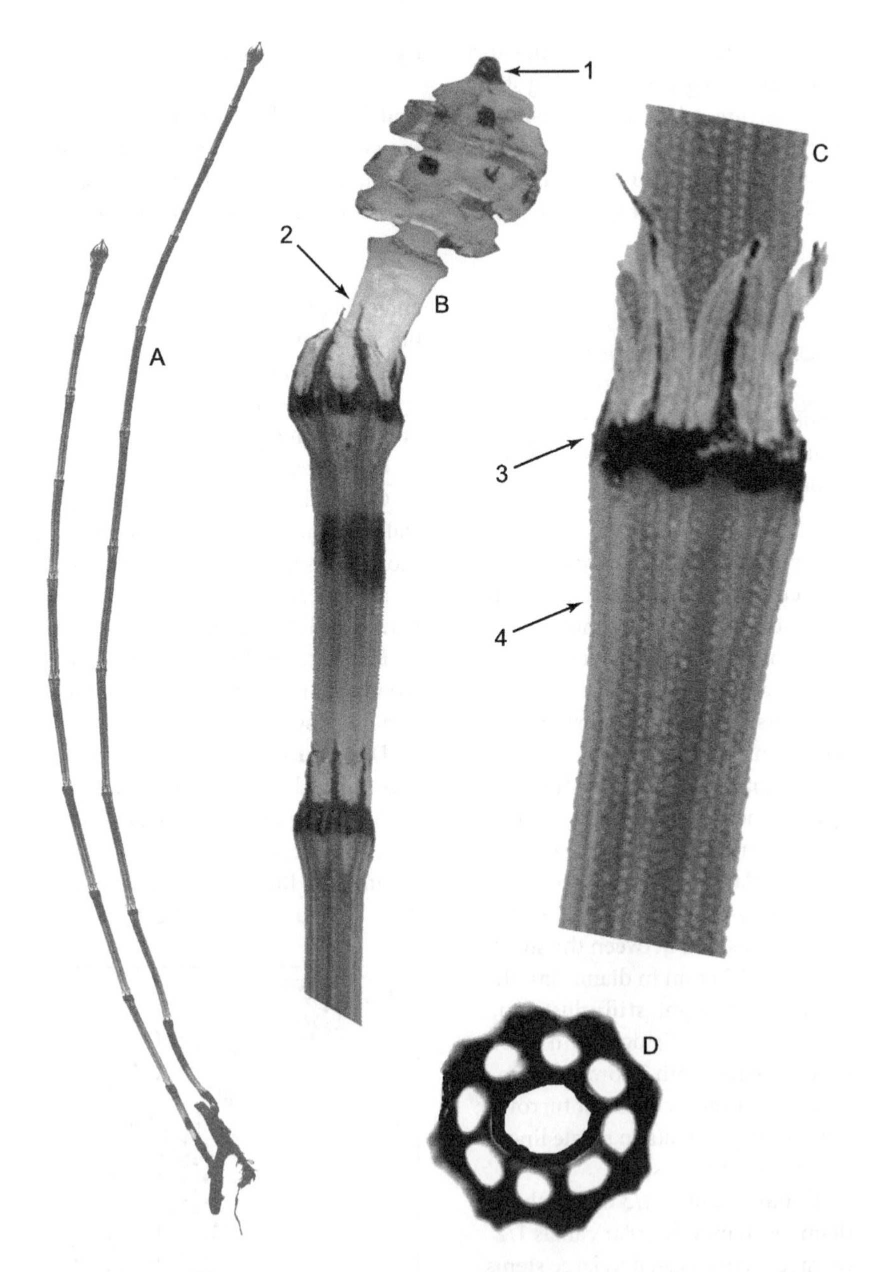

***Equisetum variegatum*, Variegated horsetail: A. Stems and root. B. Upper stem with cone. C. Sheath with teeth. D. Cross section of stem.**

Informative arrows: 1. Small cone with pointed tip. 2. Papery teeth with black center and wide white margins. 3. Green sheath with upper black rim. 4. Somewhat funnel-shaped sheath.

Equisetum variegatum subsp. *alaskanum* (A. A. Eaton) Hultén is limited to the Aleutian Islands, Alaska, Yukon, and British Columbia. It is separated from subsp. *variegatum* by teeth that are incurved, usually all black but sometimes with obscure white margins.

E. variegatum subsp. *variegatum* is an inconspicuous *Equisetum*. Its dark green, slender, and unbranched stems resemble those of rushes and sedges, and it is found in the same locations. It is highly variable with regard to the number of sheath teeth, stem ridges, and stem thickness.

Equisetum variegatum subsp. *variegatum* **may be recognized by stiff stems that are distinctly smaller, more slender, and more delicate appearing than those of** *E. hyemale* **and** *E. laevigatum,* **and by its persistent teeth with black centers and white margins.** *E. hyemale* **has distinctive ashy sheaths with black bands near the bottom and at the top (usually without teeth) while those of** *E. variegatum* **are green with a black rim and persistent teeth.**

E. laevigatum subsp. *variegatum* **has somewhat flaring green sheaths with a dark band on the upper margin at the base of the teeth.**

EQUISETUM HYBRIDS

The *Equisetum* hybrids are being treated more thoroughly than other fern hybrids because they are quite common and often form large clones. Even though the hybrids may occur only rarely, and are sterile, they have the ability to reproduce by fragmentation of their stems and to spread via their extensive rhizomes. Fragments may be carried by water to a new location where they establish a new clone. Some sterile hybrids are found far removed from one of their parental species.

They are recognized by morphologies intermediate between parental species, misshapen spores, and often malformed cones.

It is often difficult to identify an *Equisetum* interspecific hybrid because some clones will resemble one parental species more strongly than the other. A hybrid between a tall and robust parent from one species and a short and weak parent from another will more strongly resemble the former, and a hybrid between two small or weak parents will be different from that produced by robust parents.

Sometimes features other than morphology are helpful. *Equisetum ×nelsonii* at times may resemble *E. variegatum* more than *E. laevigatum*, but it will always be winter deciduous, a character of *E. laevigatum*.

Sometimes it is helpful to observe a hybrid clone throughout the year to more fully recognize all its characters. In *Equisetum ×ferrissii*, the upper part of the stem is clearly deciduous in winter and spring.

Of course it is easier if the clone can be examined during the brief period when the spores are being shed. The spores of the hybrids are white and misshapen—not green and round as in the fertile species.

SUBGENUS *EQUISETUM* HYBRID

Equisetum arvense × E. fluviatile
Equisetum ×litorale Kühlewein
Shore Horsetail

Etymology. Latin *litoralis*, of the seashore, alluding to the habitat of this plant near lakes, ponds, rivers, and streams.

Plants. Deciduous, dimorphic.

Stems. 2–70 cm long, branched, sometimes unbranched, **tips elongate and whiplike,** sterile and cone-bearing stems similar.

Canals. Centrum 1/2 to 3/4 diam. of stem, vallecular canals present.

Branches. When present, mostly from midstem nodes, ascending to spreading, sometimes absent, solid, proximal whorls with first internode of each branch equal in length to adjacent stem sheath, distal whorls with first internode of each branch longer than adjacent stem sheath.

Sheaths. 3.5–8 × 2.5–6 mm, somewhat elongate, black on lower stem, green on upper stem.

Teeth. 8–14,1–3 mm long, narrow, black on lower stem, often black with green center at base and thin white margins on higher stem; not adherent in pairs.

Cones. Tips rounded; present in late summer but misshapen.

Spores. Misshapen; white.

Habitat. Wet meadows, woodlands, ditches, and pond and stream banks.

Distribution. Throughout Canada to Alaska, south to northern tiers of the United States, south to Virginia in the east.

Equisetum ×litorale is an uncommon hybrid between *E. arvense* and *E. fluviatile* and might be expected where the parents coexist.

E. ×litorale **may be distinguished from** *E. fluviatile* **by its sheath teeth with white margins, the smaller central cavity of its stem, its solid branches, and its long first branch internode. It may be distinguished from** *E. arvense,* **which has a still smaller stem central cavity and dark teeth, some of which are coherent.**

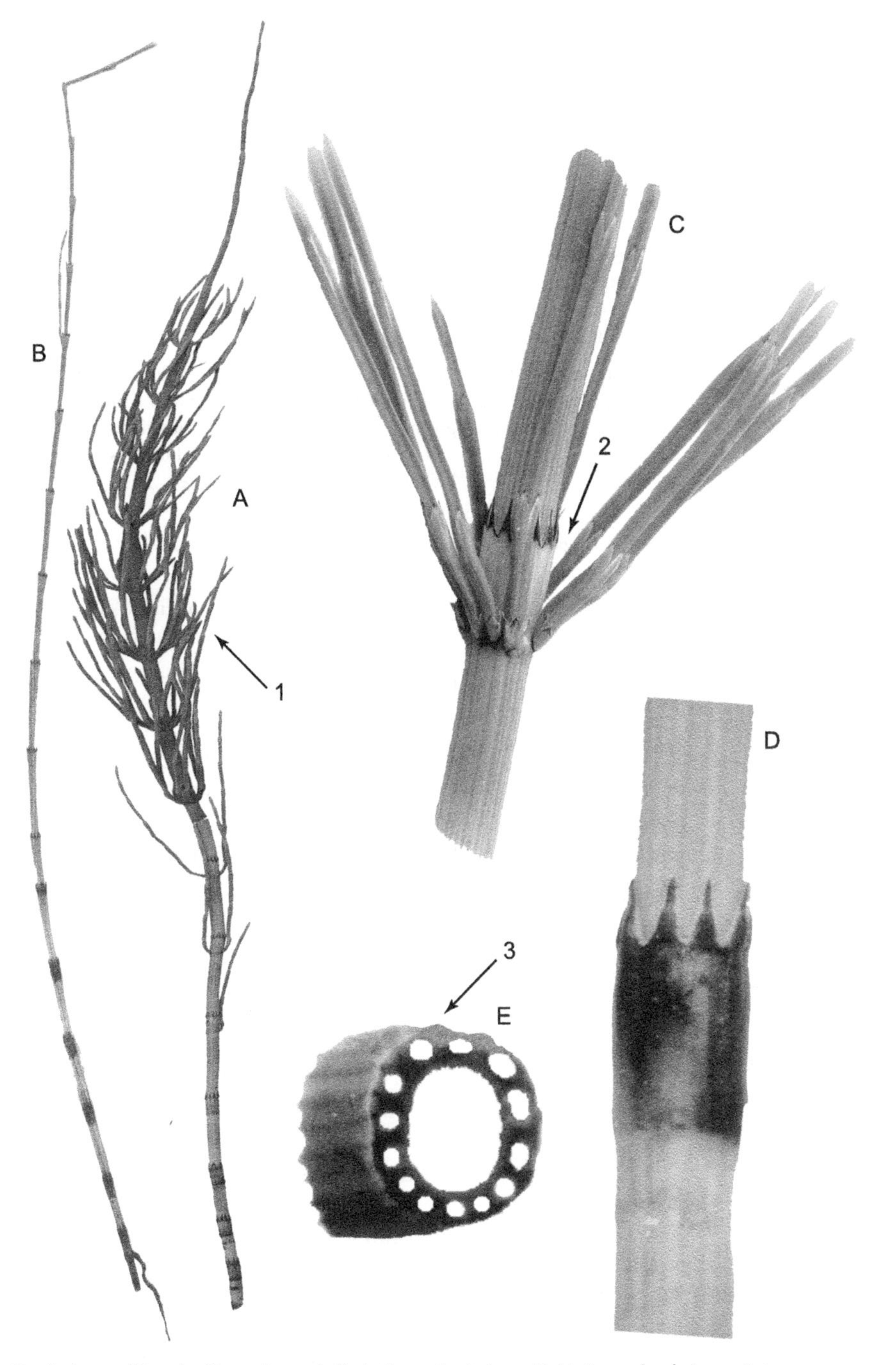

Equisetum ×litorale: **Shore horsetail: A. Branched stem. B. Unbranched stem. C. Lowermost branched internode. D. Sheath with teeth. E. Cross section of stem.**

Informative arrows: 1. Branches ascending, mostly from midstem. 2. First internode equal in length to adjacent sheath. 3. Cross section ridged with vallecular canals and smaller central canal than in *E. fluviatile*.

Quick Reference Table to *Equisetum arvense* × *E. fluviatile*

	E. arvense	*E.* ×*litorale*	*E. fluviatile*
Stems	Dimorphic; short; sterile stems green and branched; fertile stems colorless or brown, unbranched, ephemeral; no whip tip	Monomorphic; medium height; sterile and fertile stems similar; whip tip	Monomorphic; tall sterile and fertile stems similar; long whip tip
Teeth	4–14, dark, narrow, 1–3.5 mm long, adjacent teeth often coherent in pairs	12–14, black on lower stem, often black with green center at base and thin white margin higher, 1.5–3 mm long, not coherent in pairs	12–24, black, short, narrow, 2–3 mm long, tips sometimes with thin white border, sharp-pointed
Stem surface texture	Rough, corrugated	Intermediate	Smooth
Branches	1st internode longer than adjacent sheath, 3–4 ridges	1st internode longer than adjacent sheaths at upper nodes, usually same length at lower nodes, 4–5 ridges	1st internode shorter than adjacent sheath, 4–6 ridges
Central canals	1/4 diam. of stem	1/2–3/4 diam. of stem	About 9/10 of stem diam.
Vallecular canals	Present	Present	Usually absent
Cones	Present only on evanescent fertile stems; 17–36 mm long	Rarely produced; ca. 8–12 mm; remaining tightly closed	Commonly produced; 8–25 mm long, sometimes borne on tips of branches

SUBGENUS HIPPOCHAETE HYBRIDS

Equisetum. hyemale subsp. *affine* × *E. laevigatum*

Equisetum ×*ferrissii* Clute
E. *hyemale* var. *intermedium* A. A. Eaton; *E. intermedium* (A. A. Eaton) Rydberg; *Hippochaete* ×*ferrissii* (Clute) Holub
Ferriss' Scouring Rush

Etymology. Name honors James H. Ferriss (1849–1926), a businessman and civic leader from Joliet, Illinois, who conducted extensive botanical and biological explorations in the arid southwestern United States.

Plants. Evergreen lower stem (like *E. hyemale*), upper stem dying back in winter (like *E. laevigatum*).

Stems. 20–190 cm tall, 3–11 mm diam., unbranched, 14–32 ridges, **rougher than** *E. laevigatum*, **smoother than** *E. hyemale,* silica tubercles on ridges smaller than *E. hyemale*; lines of stomates single.

Canals. Central canal up to 80% of stem diam., carinal and vallecular canals present.

Sheaths. 12–25 mm long, 1 1/2 times as long as wide, lower stem sheaths appressed to stem, dark bands near base and at top, gray to

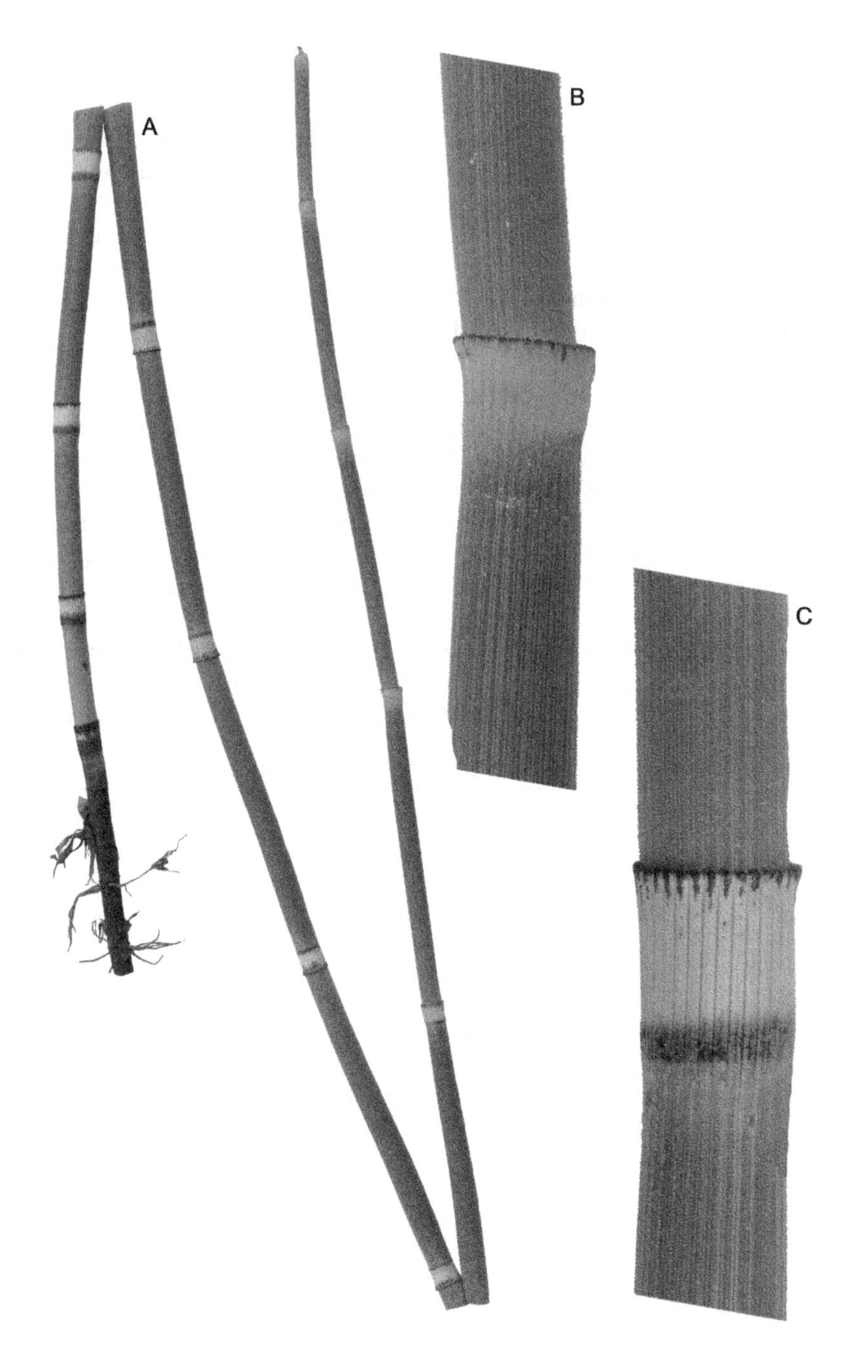

Equisetum ×ferrissii, Ferriss' scouring rush: A. Stem and roots. B. Sheath on upper stem (resembling *E. laevigatum*). C. Sheath on lower stem (resembling *E. hyemale*).

brown between (like *E. hyemale*)**, upper stem sheaths spreading, green with a black rim (like** *E. laevigatum*)**.**

Teeth. Deciduous or partly persistent, leaving a dark rim on the tip of the sheath; 14–32 teeth, jointed to sheath and promptly shed or persistent.

Cones. Somewhat misshapen, apiculum inconspicuous to pointed, mature in late spring or early summer but spores not shed.

Spores. Misshapen, white.

Habitat. Diverse: lakeshores, sand dunes, riverbanks, roadsides, and fields.

Distribution. Southern Canada to Mexico. United States except southeastern states.

Equisetum ×*ferrissii*, a very common hybrid, is unusual in the way the parent species are clearly represented on different parts of the stem. The upper stems resemble *E. laevigatum* (including dying back in winter), and the lower stems resemble *E. hyemale* subsp. *affine* (including persisting over the winter). There is much variability in the proportion of the stem reflecting the characters of *E. hyemale* as compared to *E. laevigatum*. While both taxa are often represented in an equal length on the stems, some will have a much larger portion of their stems resembling one species.

This sterile hybrid is quite common and probably spreads by rhizome fragments breaking off and floating along lakeshores and canals. It is found in some areas of the northeastern United States far beyond the range of *E. laevigatum.*

Quick Reference Table to *Equisetum hyemale* subsp. *affine* × *E. laevigatum*

	E. hyemale	*E.* × *ferrissii*	*E. laevigatum*
Cones	Tips apiculate	Inconspicuously to prominently apiculate	Tips rounded, sometimes finely apiculate
Stem	Evergreen, rough	Upper portion deciduous, rougher than *laevigatum* smoother than *hyemale* in texture	Deciduous, smooth
Stem size	20–220 cm tall, 3–8 mm wide	20–190 cm tall, 3–11 mm wide	20–150 cm tall, 3–9 mm wide
Silica spicules	Prominent single row of silica spicules along ridge	Rougher than *laevigatum*, smoother than *hyemale*	No silica spicules
Stem ridges #	14–50	14–32	10–32
Sheath	Gray with black rim and black band near base	Upper sheaths like *laevigatum*, lower sheaths like *hyemale*	Green with black rim
Sheath shape	About as long as wide	About 1 1/2× as long as wide	About 2× as long as wide
Silica tubercles	Single rows along ridgetops	Intermediate	No tubercles
Spores	Green, round	White, malformed	Green, round

Equisetum hyemale* subsp. *affine* × *E. variegatum* subsp. *variegatum

Equisetum ×*mackayi* (Newman) Brichan

Equisetum hyemale Linnaeus var. *mackaii* Newman; *E. hyemale* subsp. *trachyodon* A. Braun; *E.* ×*trachyodon* (A. Braun) W. D. J. Koch

Mackay's Scouring Rush

Etymology. Name honors Dr. Alexander Howard Mackay (1848–1929), a Nova Scotia teacher who studied fungi, lichens, and pteridophytes among many other interests.

Plants. Monomorphic, evergreen.

Stems. 20–90 cm tall, 2–5 mm diam., erect to decumbent, unbranched; rough; (7–) 10–16 ridges with central furrow and silica spicules on each side; single line of stomata; grayish-green.

Canals. Central canals ca. 1/3 cm diam. of stems.

Branches. Absent unless stressed or injured.

Sheaths. 3.5–8 × 2–5.5 mm, longer than wide, appressed, on lower stem black band at base and white in wide center area with dark rim at base of teeth (resembling *E. hyemale*), sheaths on upper stem green with long teeth with black center and wide white margins (resembling *E. variegatum*), many sheaths intermediate between these 2, each ridge with central furrow and a row of silica tubercles on top of each side of the ridges.

Teeth. Persistent; 7–16, up to 3 mm long, center brown, margins white, tips usually brown, acute, sometimes long and tapered.

Cones. 4–5 mm long, tips pointed, misshapen, usually half remaining within the uppermost sheath, mature in late summer but abortive spores not shed.

Spores. Misshapen, white.

Habitat. Lakeshores, sandy beaches, riverbanks, and marshes.

Distribution. Northern Europe; Greenland, eastern Canada, scattered populations in mid- and western Canada, northeastern United States west to Minnesota, scattered in northwestern United States.

E. ×*mackayi* **is sometimes mistaken for small forms of** *E. hyemale*. **The persistent teeth on the sheaths and the usually 10–16 stem ridges are helpful in distinguishing it from its parental species.**

Equisetum ×mackayi: Mackay's scouring rush: A. Stems and root. B. Sheath with teeth from upper stem (somewhat resembling *E. variegatum*). C. Sheath with teeth from lower stem (sheath somewhat resembling *E. hyemale*).

Quick Reference Table to *Equisetum hyemale* subsp. *affine* × *E. variegatum* subsp. *variegatum*

	E. hyemale	*E.* ×*mackayi*	*E. variegatum*
Cone	Apiculate, tips sharply pointed; 7–15 mm long, partially concealed by teeth of uppermost sheath	Tips apiculate; 4–5 mm long, half within uppermost sheath	Apiculate at tips, 5–7 mm long, stem protruding from uppermost sheath
Stem	30–150 cm × 3–8 mm	20–90 cm × 2–5 mm	10–50 cm × 1–3 mm
Centrum size	2/3–3/4 diam. of stem.	1/2 diam. of stem	ca. 1/3 diam. of stem.
Silica	Ridges without central furrow, 1 row of tubercles on ridge	Ridgetops furrowed, with 1 row of tubercles on each side of furrow	Ridgetops furrowed, with 1 row of tubercles on each side of furrow
Stem ridges #	14–50	(7–) 10–16	3–12
Sheath	Gray or brown with black rim and black band near base, about as long as wide	Black base with ashy gray band between black rim on lower stem, becoming green with black rim at base of teeth on upper stem, almost totally black in between, longer than wide	Green with black rim at base of teeth, longer than wide
Teeth	Usually promptly shed, leaving dark crenulated upper margins	Persistent, finely attenuate tips, with dark centers and dry white margins	Persistent, broadly triangular, with dark centers and broad white margins
Sheath shape	About as long as wide (0.8–1.7×)	Longer than wide (1.3–2.4×)	Mostly longer than wide (0.8–2.2×)
Spores	Green, round	White, malformed	Green, round

Equisetum laevigatum × *E. variegatum* subsp. *variegatum*

Equisetum ×*nelsonii* (A. A. Eaton) J. H. Schaffner

Equisetum variegatum Schleicher & D. Mohr var. *nelsonii* A. A. Eaton

Nelson's Scouring Rush

Etymology. Name honors N. L. T. Nelson (1862–1932), a bryologist who served on the faculties of several midwestern and southern universities and first collected this plant.

Plants. Monomorphic; deciduous (sometimes lowest parts persist over the winter).

Stems. 20–60 cm long, 2–3 mm wide; unbranched; 6–14 ridges, single line of stomata; single row of silica tubercles along ridgetops.

Sheaths. Longer than wide, 3.5–7.5 × 2–4 mm, green, with wide black rim and persistent teeth.

Teeth. 6–14; prominent; tips long filiform; centers dark brown; margins white.

Cones. Tips slightly pointed; mature in early summer; spores not shed.

Spores. Misshapen; white.

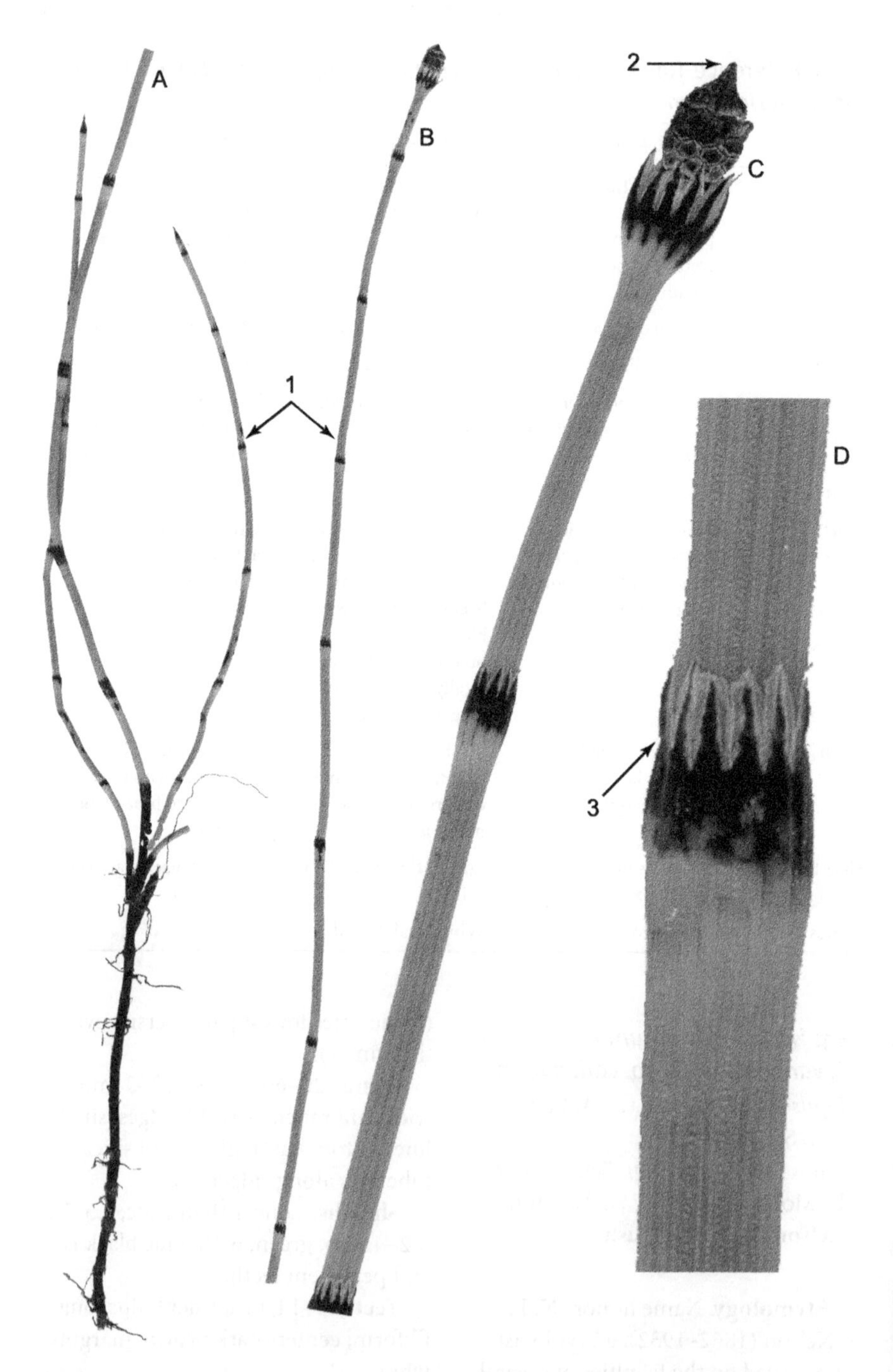

Equisetum ×nelsonii: Nelson's scouring rush: A. Stems and root. B. Fertile stem. C. Upper stem and cone. D. Sheath and teeth.

Informative arrows: 1. Slender stems with sheaths longer than wide. 2. Pointed cone tip. 3. Persistent sheath teeth with dark centers and white margins.

Quick Reference Table to *Equisetum laevigatum* × *E. variegatum* subsp. *variegatum*

	E. laevigatum	*E. ×nelsonii*	*E. variegatum*
Cones	Tips rounded, sometimes with fine-pointed tips	Tips slightly pointed	Tips pointed
Stem	Deciduous, 20–150 cm tall × 2–7 mm	Deciduous, 20–60 cm tall × 2–3 mm wide	Evergreen, 10–50 cm tall × 1–3 mm wide
Sheath	Green with black rim, teeth soon deciduous	Black rim with teeth like *variegatum*, persistent	Green with black rim, teeth persistent
Sheath ridges #	10–32	6–14	3–12
Silica	Ridges not furrowed, no silica tubercles	Ridges not furrowed, with 1 row of tubercles on ridgetop	Ridges furrowed, with 1 row of tubercles on each side of furrow
Teeth	Shed early, leaving black rim	Persistent, dark center with white margins	Persistent, dark center with white margins
Silica tubercles	Ridges not furrowed, silica tubercles absent	Ridges not furrowed, single row along ridgetops	Ridges furrowed, 2 rows of tubercles on furrow tops
Spores	Green, round	White, malformed	Green, round

Habitat. Lakeshores, dunes, and riverbanks.

Distribution. Canada: Ontario and Quebec. USA: New York, Illinois, Indiana, Michigan; scattered in Minnesota, Montana, Wyoming, and Oregon.

Equisetum ×nelsonii **has pointed cone tips, sheaths, and teeth resembling those of** *E. laevigatum*. **The stems die back in winter, a characteristic of** *E. laevigatum* (*E. variegatum* **is evergreen). It is fairly common in its preferred habitats.**

PART 2

Ferns (Polypodiopsida)

Key to the Genera of Leptosporangiate and Eusporangiate Ferns (Polypodiopsida) as Found in Michigan

1\. Plants aquatic or partially so; small and free-floating or clover-shaped and rooted in shallow water or mud (2).

1\. Plants terrestrial on soil, rock, or rooted in moist or wet soil; not free-floating and not ordinarily rooted in shallow water or mud (3).

2(1). Plants free-floating on quiet water surfaces; fronds 2-lobed, imbricate, tiny—about 0.5–1.5 mm long; whole plant resembles a cedar branchlet; sori borne in leaf axils on undersurface of plants. . . . . *Azolla*

2\. Plants growing in water or rooted in mud, sometimes becoming stranded; blades resembling a 4-leaf clover, with 4 equal pinnae at tips of stipes, each less than 2 cm long; sori borne in hardened sporocarps at bases of stipes . . . . *Marsilea*

3(1). Fronds long, climbing, vinelike, and twining; pinnae palmately lobed, fertile and sterile pinnae very different (dimorphic); sporangia borne in tight clusters at tips of pinna lobes. . . . . *Lygodium*

3\. Fronds of various shapes but not vinelike; fertile and sterile blades the same or different; sporangia may or may not be borne in tight cluster of pinna lobes (4).

4(3). Fronds simple, linear, strap-shaped, or undivided long triangular hastate; sori long-linear along veins . . . . . *Asplenium* (in part) (*A. rhizophyllum, A. scolopendrium*)

4\. Fronds various shapes from ternate to 1-pinnate to 3-pinnate; sori of various types (5).

5(4). Fronds fan-shaped; rachises divided into 2 equal parts curving away from each other; pinnae or pinnules trapezoidal or dimidiate (unequally sided, the basiscopic side narrower and lower margin closer to midvein); sori linear under false indusia formed by leaf margins folding . . . . . *Adiantum*

5\. Fronds of various shapes but not fan-shaped; rachises not divided; pinnae and pinnules mostly not trapezoidal; sori various in size, shape, and location (6).

6(5). Sporangia borne on very different, separate erect spikes arising separately from sterile fronds arising at ground level or on very different flaccid fertile pinnae in the middle or at tips of fronds; fertile regions green in early spring, turning brown and wilting in early summer or becoming dark, hardened, and persistent over winter (7).

6\. Sporangia borne in sori on undersurface of blades, pinnae, or pinnules, not on separate spikes or separate pinnae; sterile and fertile blades alike or fertile blades with narrower pinnae and pinnules; fertile and sterile blades share same seasonal variation in color and wilting (9).

7(6). Sterile fronds 1-pinnate-pinnatifid, fertile fronds 30–60 cm long with all pinnae shorter, narrower, and flaccid, green in early spring, soon turning brown, dying, and wilting; or sterile fronds and lower parts of fertile fronds fully 2-pinnate with several much smaller, darker fertile pinnae found at tips of fronds; or sterile fronds and upper and lower divisions of fertile frond 1-pinnate-pinnatifid with much smaller, darker fertile pinnae between fertile parts not persisting through winter. . . . . *Osmunda*

7. Sterile blades pinnatifid to 1-pinnate-pinnatifid; sterile and fertile pinnae arising separately from ground; fertile blades shorter, hard, dark, and persisting through winter (8).

8(7). Sterile blades 1-pinnate-pinnatid throughout, rachises not winged, veins free. . . . . *Matteuccia*

8. Sterile blades pinnatifid, rachises winged, the lowest pinnae sometimes free, veins netted . . . . *Onoclea*

9(6). Sori linear or slightly curved along veins, or parallel to costae, or in continuous bands along margins of pinnae under a false indusium (10).

9. Sori round or nearly so, along veins from costae toward margin or near margins of pinnae or pinnules (17).

10(6). Sori long-linear along veins; cross section of stipe bases reveals 2 back to back, C-shaped vascular bundles uniting to form an X-shape higher up stipe; scales clathrate. . . . . *Asplenium* (in part) (*A. platyneuron, A. ruta-muraria, A. trichomanes, A. viride*)

10. Sori short- or long-linear along veins, pinna, or pinnule margins or along costae of pinna or pinnule midribs; cross section at stipe bases reveals 2 straplike, parallel vascular bundles; scales nonclathrate (11).

11(10). Sori linear or j-shaped along veins that extend from midribs toward margins (12).

11. Sori linear along margins of pinnules or along areolar veins near and parallel to midrib (14). . . . . . . *Asplenium, Athyrium, Cryptogramma, Deparia, Homalosorus, Pellaea, Pteridium*

12(11). Blades 1-pinnate; pinna margins entire; fertile fronds narrower and with much narrower pinnae. . . . . *Homalosorus* (*H. pycnocarpos*)

12. Blades 1-pinnate-pinnatifid to 2-pinnate-pinnatisect; sterile and fertile fronds the same shape or nearly so (13).

13(12). Grooves of pinna costae shallow, not continuous with grooves of rachises; blades 1-pinnate pinnatifid; sori linear. . . . . *Deparia* (*D. acrostichoides*)

13. Grooves of pinna costae deep, continuous with grooves of rachises; blades 1-pinnate-pinnatisect to 2-pinnate-pinnatifid; sori linear or J-shaped hooked over vein tips . . . . *Athyrium* (*A. filix-femina*)

14(11). Sori short-linear in single chainlike rows on anastomosing areolar veins on each side of, and parallel to, costae, and sometimes along midribs; veins joining to form a single row of areoles near and on both sides of midribs, then free and branching to margins . . . . *Woodwardia*

14. Sori short- to long-linear; along pinna or pinnule margins; veins free, not forming areolae near costae (15).

15(14). Fronds 35–180 cm long, 20–50 cm wide; well separated, growing in large colonies on long, creeping underground rhizomes; sterile and fertile fronds the same; blades broadly triangular; growing in loose soil in open areas; stipes straw-colored to green or brown. . . . . *Pteridium*

15. Sterile fronds 3–50 cm long, 2–20 cm wide; fronds clustered in small tufts

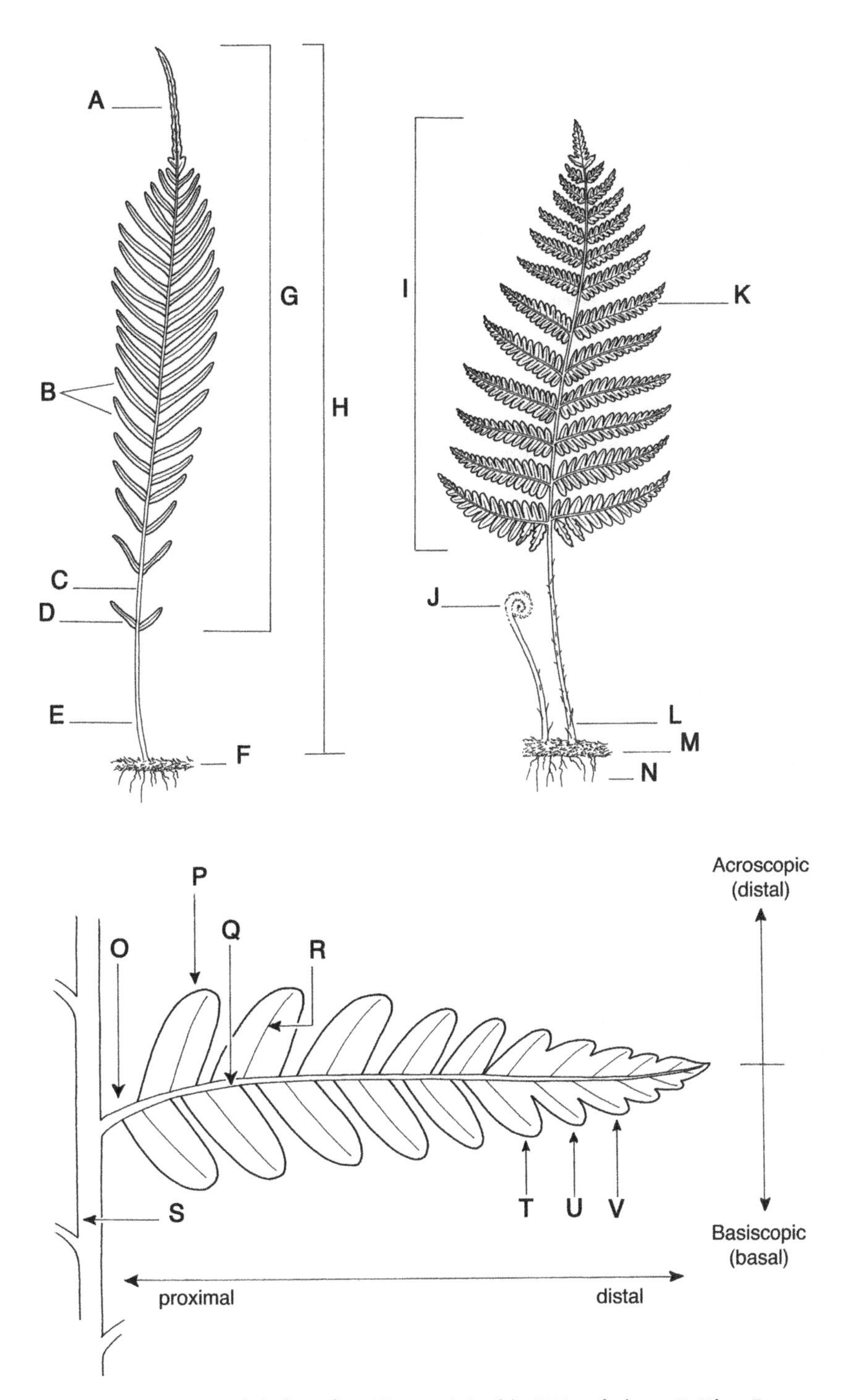

Fronds Anatomy: A. Frond tip (apex). B. Pinnae. C. Rachis. D. Basal pinna. E. Stipe. F. Rhizome. G. Blade. H. Frond. I. Blade. J. Crosier. K. Pinna. L. Stipe. M. Rhizome. N. Roots. O. Pinna stalk. P. Pinnule. Q. Costa. R. Costule. S. Rachis. T–V: Ultimate segments.

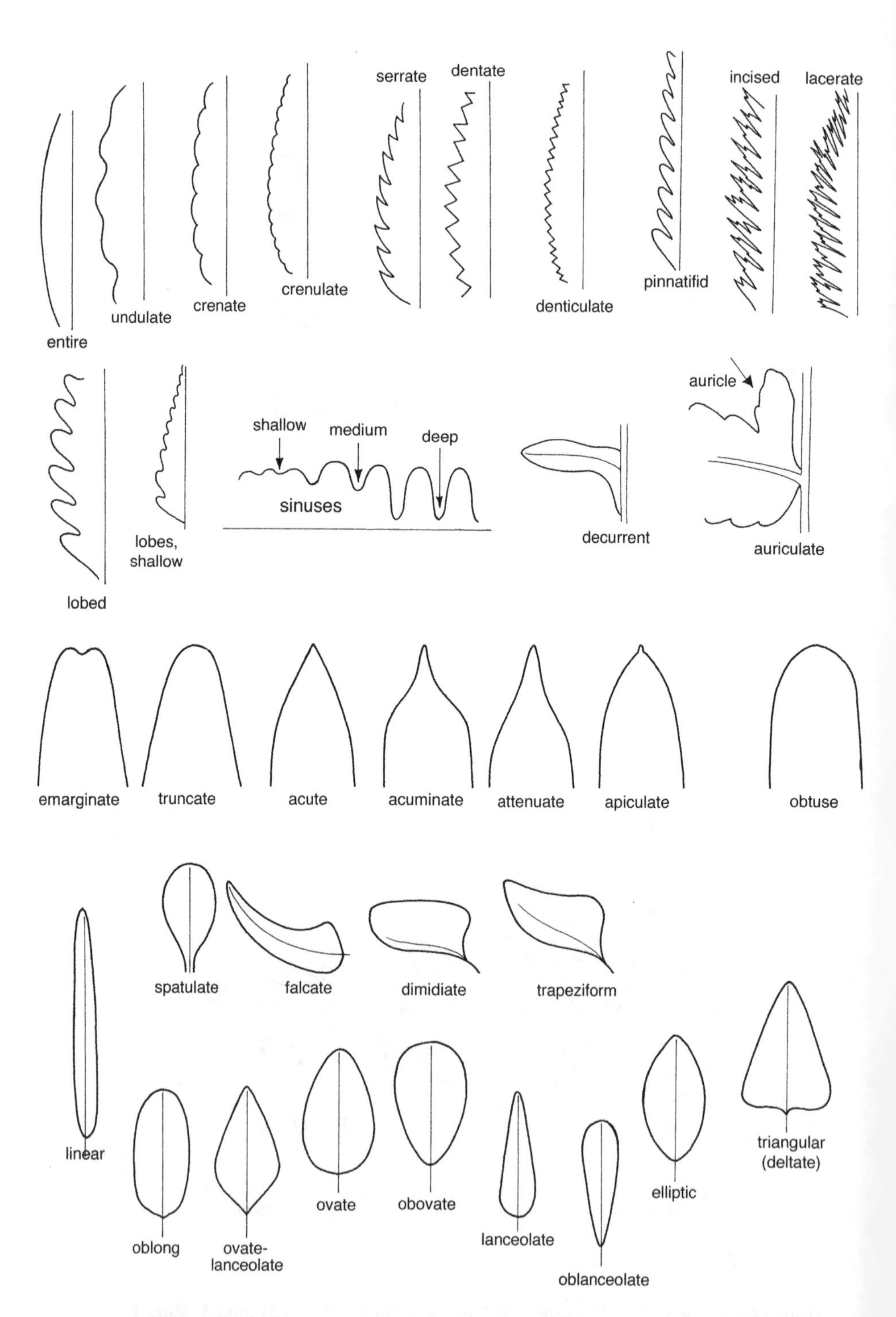

Pinna Forms and Margins

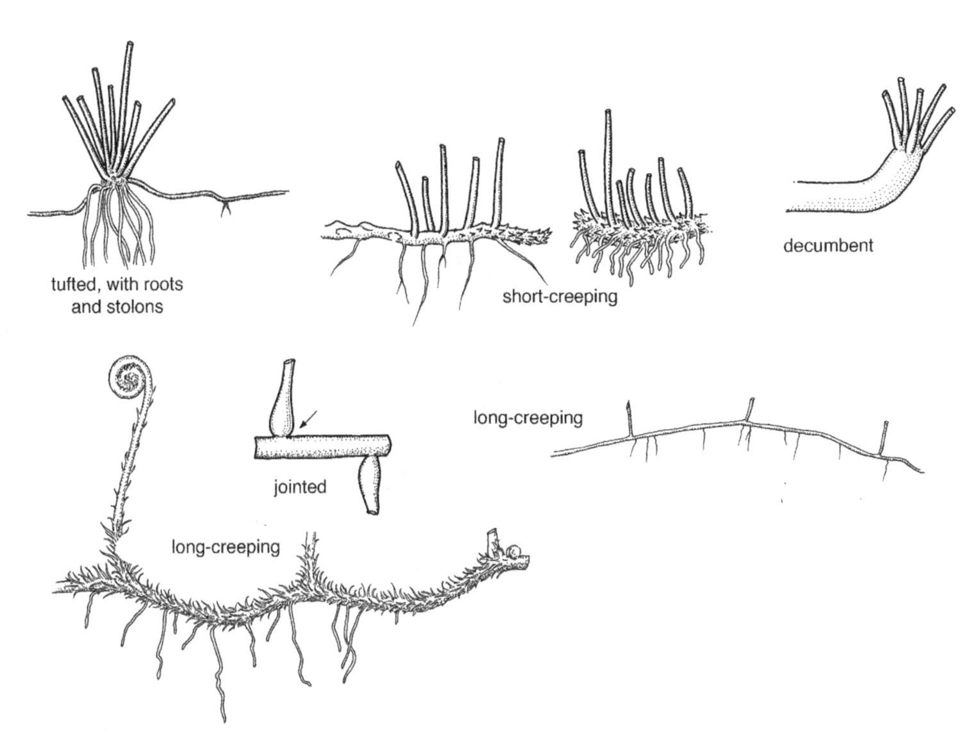

Rhizomes

on short ascending to erect rhizomes; sterile blades different; usually growing on rock; stipes dark purple to black or stipes brown at base, green or straw-colored distally (16).

16(15). Stipes dark, shiny, brown to dark purple to black; fronds somewhat dimorphic or not; sterile and fertile pinnae linear-oblong to oblong-lanceolate . ... *Pellaea*

16. Stipes brown at base, green or straw-colored distally; fronds strongly dimorphic; fertile fronds obviously longer than the sterile, narrow, elongate, usually with revolute pinnules; sterile pinnules oblong- to fan-shaped, fertile pinnules narrow, elongate .... *Cryptogramma*

17(9). Blades deeply pinnatifid; rhizome long-creeping; indusia lacking........ .... *Polypodium*

17. Fronds 1-pinnate to 3-pinnate; rhizomes erect; short- or long-creeping; indusia present or absent (18).

18(17). Blades 1-pinnate, pinna margins sharply toothed with long, stiff, bristlelike projections; indusia peltate; pinnae eared at base; sterile and fertile pinnae same size or fertile pinnae near frond tips much smaller ..... *Polystichum*

18. Blades 1-pinnate-pinnatifid to 3-pinnate-pinnatifid; pinna margins with or without fine teeth; indusia of various types; sterile and fertile fronds same size or different, fertile pinnae near frond tips not smaller (19).

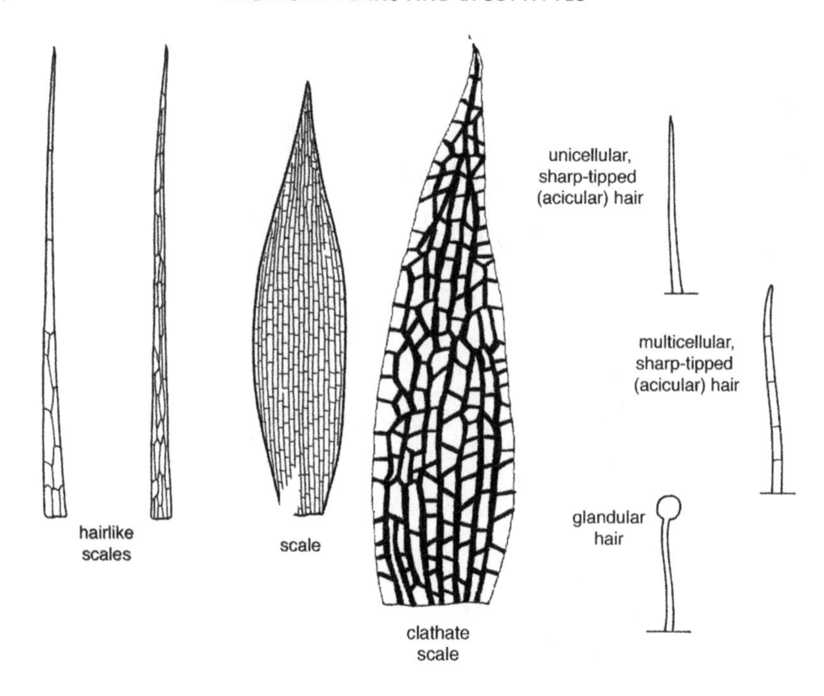

Scales and Hairs

19(18). Indusia absent; fronds ternate or broadly triangular with winged rachises (20).

19. Indusia of various types present; fronds not ternate or broadly triangular; rachises not winged (21).

20(19). Fronds ternate; rachises not winged; stipes 1.5 to 3 times length of blade; costae grooved adaxially; lacking small, white, sharply pointed (acicular) hairs. . . . . *Gymnocarpium*

20. Fronds broadly triangular; rachises winged; stipes 1 to 1.5 length of blade; costae not grooved adaxially; small, white, sharply pointed (acicular) hairs on undersurface of pinnae and rachis . . . . *Phegopteris*

21(19). Blades 1-pinnate-pinnatifid; fine, white, sharp-tipped, translucent needlike (acicular) hairs present on rachises and costae; cross section at base of stipe shows two slightly curved, linear vascular bundles that unite upward to a U-shape. . . . . *Thelypteris*

21. Blades 1-pinnate to 3-pinnate-pinnatifid; acicular hairs absent; cross section at base of stipe various, not with two linear vascular bundles (22).

22(21). Rhizomes ascending to erect; indusia above sori, round or reniform; 1-pinnate-pinnatifid to 3-pinnate-pinnatifid: glandular or not. . . . . *Dryopteris*

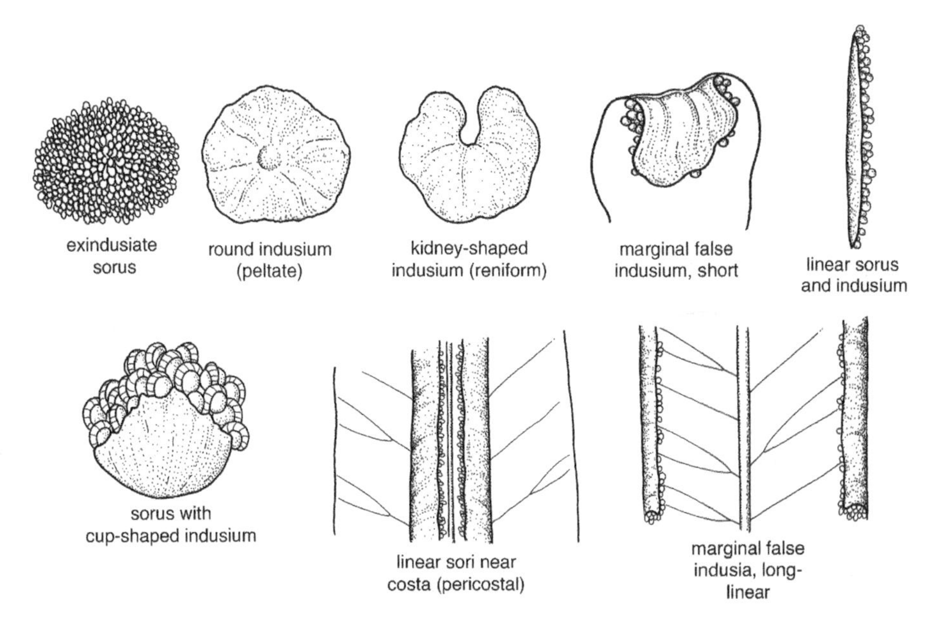

Sori and Indusia

22. Rhizomes ascending to erect or long-creeping subterranean; sori with cup-like or tubular indusia or ribbonlike or filamentous hairs under indusion spreading to cover it. (23).

23(22). Rhizomes long-creeping below soil surface; stipes and rachises hairy, scales lacking; sori marginal at vein tips in sinuses, small (less than 0.5 mm in diam.); cup-shaped circular or slightly 2-valvate cups; fronds 30–90 long (rare, Jackson County). . . . . *Dennstaedtia*

23. Rhizomes decumbent to erect; sori larger than 0.5 mm; stipes and rachises scaly, with or without hairs; sori not marginal; indusia cup-shaped, attached at base, looking outward or composed of many narrow to hairlike segments attached beneath and encircling sori; fronds shorter than 50 cm long (24).

24(23). Indusia hoodlike (soon shriveling and absent), opening outward, attached at bases and partially overlapping sori; old stipe bases not persistent or sparsely persistent on rhizomes; fronds up to 50 cm long (*Cystopteris bulbifera* often longer) . . . . *Cystopteris*

24. Indusia composed of many narrow to hairlike segments attached beneath and encircling sori; old stipe bases persist on rhizome; fronds up to 25 cm long (*Woodsia obtusa* sometimes longer) . . . . *Woodsia*

ADIANTUM Linneaus Pteridaceae

Maidenhair Ferns

See Pteridaceae in appendix I for a description of the family and a key to the Michigan genera.

Etymology. Greek *adiantos*, unwetted; an ancient name alluding to the water-repellent fronds. Maidenhair, a name used for obscure reasons, is applied to ferns with delicate fronds and slender black stipes (see also *Asplenium trichomanes*).

Plants. Monomorphic; deciduous; fronds in loose clusters in clones.

Rhizomes. Creeping.

Fronds. Stipes erect with blade horizontal to ground.

Stipes. Round, usually brown to purple or black, shiny, mostly glabrous except at very base, sometimes sparsely hairy or with fibrils.

Blades. 2-pinnate to 3-pinnate.

Pinnae. Elongate, tips obtuse.

Pinnules. Rhomboidal, trapezoidal, fan-shaped, or dimidiate, sessile to short-stalked, never adnate.

Veins. Free, repeatedly branching dichotomously.

Sori. Less than 4 mm long, short-linear, marginal along several vein tips.

Indusial flaps. Consisting of reflexed margins of pinnae, rectangular or U-shaped.

Sporangia. Distinctively borne on undersurface of the false indusia.

Chromosome #: $x = 29, 30$

Adiantum is a genus of about 200 species widely distributed in tropical and temperate areas, mostly at low- to mid-elevations and in wet forests. Nine species are treated in the *Flora of North America North of Mexico*, vol. 2 (1993). One species is found in Michigan.

Adiantum pedatum Linnaeus
Maidenhair Fern, Northern Maidenhair Fern

Etymology. Latin, *pedatum*, footlike; the frond outline resembles a bird's foot. Linnaeus used this term to describe the fronds of this fern.

Plants. Monomorphic, deciduous.

Rhizomes. Short-creeping, thick, often with attached old stipe bases; scales yellowish, margins entire.

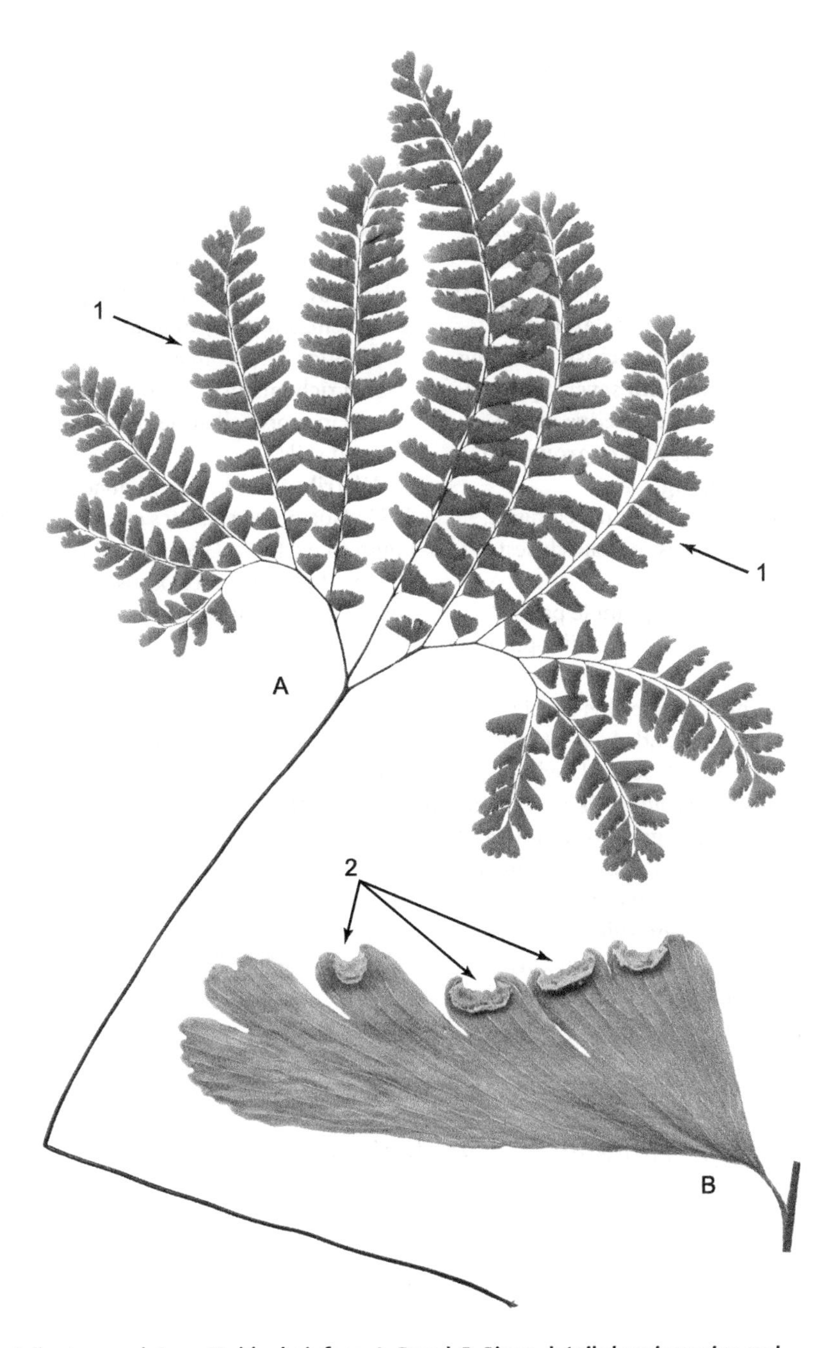

Adiantum pedatum: **Maidenhair fern: A. Frond. B. Pinna detail showing veins and marginal linear sori.**

Informative arrows: 1. Dimidiate pinnules. 2. Marginal linear sori.

Stipes. 1–2 mm diam., erect, round, glabrous, dark brown or purplish-brown to black; stipe tips fork to form 2 rachises spreading in a Y-fashion.

Fronds. 35–75 cm long, delicate; crosiers reddish-brown when young.

Blades. 15–30 × 20–40 cm, 2-pinate to 3-pinnate, **semicircular, fan-shaped; formed by rachises dividing into 2 branches with several pinnae forming on the outer edge of the branches; horizontal to ground to slightly drooping.**

Pinnae. 5–20 cm long, elongate, tips obtuse, longest pinnae closest to stipes.

Pinnules. 8–21 pinnule pairs, 3 times as long as wide, smaller pinnules triangular, alternate, dimidiate, lower margin straight, acroscopic margins lobed, lobes separated by a narrow sinuses 1–2 (–3) mm deep.

Veins. Free, forked repeatedly.

Sori. 1–3 mm long, linear to slightly curved, marginal on tips of lobes, covered with elongate, green to gray, reflexed, marginal false indusia.

Habitat. Grows in a variety of soils, best in rich, mature, deciduous shady woods in moist, alkaline, and humus-covered soils.

Distribution. Eastern Canada. USA: Northeastern states extending south to northern Georgia and Oklahoma. The distribution is essentially that of the deciduous forests of eastern North America.

Some authors divide this species in North America into 3 to 4 subspecies. *Adiantum pedatum* and *A. aleuticum* (which does not occur in Michigan) are here recognized as 2 closely related species in North America.

Adiantum pedatum **is a graceful fern with distinct fronds that is unique within the Michigan flora and easily recognized. Its silhouette alone is diagnostic. Its horizontal or drooping fan-shaped blades are borne on narrow, dark stipes and are divided into 2 equally branched rachises. Its forked rachises radiate pinna on one side only. Its pinnules are on narrow stalks, and the linear sori on their outer edges are attached to the undersurface of an in-rolled margin, forming a false indusium.**

ASPLENIUM Linnaeus Aspleniaceae

Spleenworts

Aspleniaceae, as recognized here, is comprised of a single, huge, extremely diverse genus, *Asplenium*. *Asplenium* contains about 700 species worldwide, mostly in the tropics. The family and generic treatments are identical and found below.

Etymology. Greek, *a,* not, + *splen.* In ancient Greece, the plants were believed to cure spleen diseases. The vernacular name derives from the Greek: spleenworts.

Plants. Small to medium-sized, terrestrial or epipetric.

Fronds. Monomorphic, small to medium-sized; clustered to remote.

Rhizomes. Short- or long-creeping, decumbent or erect, **tips usually covered with clathrate scales.**

Stipes. Clustered to remote, short to long, usually flattened or grooved adaxially, often dark and shiny, often glossy, black, purple, dark brown, or tan to light green, smooth; sometimes winged, often scaly with **clathrate scales** at bases, sometimes with hairs and glands, **cross section at base revealing two C-shaped vascular bundles oriented back to back, which join distally to form a single X-shaped bundle.**

Blades. Extremely variable, simple to 2-pinnate-pinnatifid in Michigan, of diverse sizes and shapes.

Ultimate segments. Variable in shape and size.

Veins. Free, forking, pinnate or anastomosing.

Sori. Linear, long, borne along one side of vein, seldom back to back of veins; indusia linear along veins, with short tapering ends.

Sporangia. Long-stalked, stalk with 1 row of cells.

Habitat. Diverse.

Chromosome #: $x = 36$

Asplenium is one of the world's largest fern genera with more than 700 species. Twenty-eight species are treated in the *Flora of North America North of Mexico*, vol. 2 (1993). Six species, including 1 subspecies, are found Michigan.

Asplenium montanum is not treated in this book because, while it was reported collected once in Keweenaw County in 1888, subsequent studies have found this report to be incorrect.

KEY TO THE SPECIES OF ASPLENIUM IN MICHIGAN

Except for *Asplenium platyneuron*, all *Asplenium* species in Michigan are uncommon (*A. trichomanes* and *A. viride*) or rare and endangered (*A. rhizophyllum, A. ruta-muraria,* and *A. scolopedrium* var. *americanum*).

1. Fronds simple, long-triangular, or strap-shaped (2).
1. Fronds 1-pinnate to 2-pinnate-pinnatifid (3).
2(1). Blades long-triangular with wide-lobed bases, tapering to narrow tips; some tips root and form new plants . . . . *A. rhizophyllum*
2. Blades simple, linear, strap-shaped; tips rounded to tapering; tips not rooting to form new plants . . . . *A. scolopendrium* var. *americanum*
3(1). Blades 1-pinnate to 2-pinnate-pinnatifid; pinnules long-stalked, fan-shaped; stipes green throughout; growing on calcareous rocks . . . . *A. ruta-muraria*
3. Blades 1-pinnate; pinnules lacking; stipes green or dark reddish-brown to black; growing on calcareous rocks or not (4).
4(3). Dimorphic, *fertile fronds* erect, 30–50 cm long, longer than sterile ones; *sterile fronds* spreading horizontally; pinnae with conspicuous basal auri-

cles overlapping rachises; sori submedial; plants mainly in soil. . . . . *A. platyneuron*

4. Monomorphic, fertile and sterile fronds the same, 5–20 cm long; pinnae bases not overlapping rachis; sori medial to supramedial; plants mainly growing on rock (5).

5(4). Stipes and rachises purplish-brown to black throughout; fronds tapering at both ends; stipes 1/6–1/4 length of blade . . . . *A. trichomanes*

5. Stipes and rachises green throughout or dark only at stipe base; fronds not tapering at base; stipes 1/4–1/2 length of blade . . . . *A. viride*

Asplenium platyneuron (Linnaeus) Britton, Sterns, & Poggenburg

Acrostichum platyneuron Linnaeus; *Asplenium ebeneum* Aiton

Ebony Spleenwort

Etymology. Greek *platy*, broad, + *neuron*, nerve or vein. The name arose from an early illustration of the species that erroneously exaggerated the veins. The dark reddish-brown color of the stipes and rachises is the source of the common name. Spleenwort is a name applied to many *Asplenium* species.

Plants. Evergreen, **dimorphic.**

Rhizomes. Very short-creeping to erect, unbranched, trophopods (nutrient-storing stipe bases) persistent, scales linear-deltate, 2–4 × 0.3–0.6 mm, dark brown to black, **clathrate,** margins entire.

Fronds. Tufted, 14–16 per plant.

Fertile fronds. Erect, stiff, 30–50 × 1–4 cm, narrow, tapering at both ends.

Sterile fronds. Shorter than fertile, to 10 cm, spreading horizontally, sometimes touching ground, twice as many as fertile fronds.

Blades. 1-pinnate, linear, widest above the middle, tapering to either end, often with glandular hairs; scales sparse, linear.

Pinnae. 15–45 pairs, 10–25 mm long and 3–5 mm wide at the base, alternate, lower pinnae smaller; *sterile pinnae* lanceolate with more rounded tips, **bases slightly eared acroscopically,** sometimes basiscopically, close together, margins less finely toothed than fertile pinnae; *fertile pinnae* linear-oblong, eared at acroscopic base, margins finely toothed.

Stipes. 1–10 cm long, 1/10–1/3 length of blades, **shiny,** at first green, becoming **dark reddish-brown to black,** brown hairlike clathrate scales at base, glabrous above.

Veins. Free and forked.

Sori. Linear, circa 2 mm, **1–18 per pinna, paired across midvein, cre-**

Asplenium platyneuron: Ebony spleenwort: A. Fertile pinnule. B. Frond cluster. Informative arrows: 1. Linear sori. 2. Erect fertile frond. 3. Small spreading sterile frond.

ating a herringbone pattern; indusia white or translucent.

Habitat. Grows in a wide variety of habitats: second-growth forests, disturbed dry woodlands, open fields, and shady banks. Plants are often widely separated.

Distribution. South Africa (an extreme disjunct). Canada: Ontario to Quebec. USA: Atlantic coast west to Arizona, Colorado, Nebraska, and Minnesota. In Michigan more common in southern counties, gradually becoming less common northward.

Chromosome #: $2n = 72$

Asplenium platyneuron mostly grows on loose soil in contrast to the other Michigan *Asplenium* species, which grow on rocks and in rock crevices. This fern has extended its range northward in the past several decades and is often found in northern counties in Michigan's Lower Peninsula where it was not previously seen. In the past, when found, it usually occurred singly and was considered uncommon to rare. The reasons for its northward spread could include its ability to grow in diverse environments and in competition with other plants, increasing areas of second-growth forestland, and possibly climate warming.

***Asplenium platyneuron* is distinguished from other Michigan *Asplenium* species by its dark reddish-brown stipes and rachises, its dimorphic fronds (the fertile fronds are longer and more erect), its 1-pinnate fronds with basal auricles on its pinnae, and the herringbone pattern of its linear sori.**

***Asplenium rhizophyllum* Linnaeus**
Camptosorus rhizophyllus (Linnaeus) Link
Walking Fern

Etymology. Greek *rhizi-*, root, + *phyllon*, leaf, alluding to the ability of this fern to root at the tips of the fronds and "walk." The vernacular name forthe plantlets at the tips of some fronds touching the ground and sprouting new plants is "walking."

Plants. Evergreen.

Rhizomes. Short, erect or ascending, mostly unbranched; scaly, scales clathrate, dark brown, narrowly triangular.

Fronds. Clustered, 2.5–38 cm long, 2.5–5 cm wide at base, fertile fronds usually larger than sterile fronds.

Stipes. Reddish-brown at base, green above, dull, sometimes shiny at base, 0.5–12 cm long, 1/10 to 1 1/2 times length of blade; scales narrowly triangular, dark brown, clathrate, with minute glandular hairs near tips.

Blades. Highly variable in size and shape (even on the same rhizome), 1–30 × 0.5–5 cm, simple, slender,

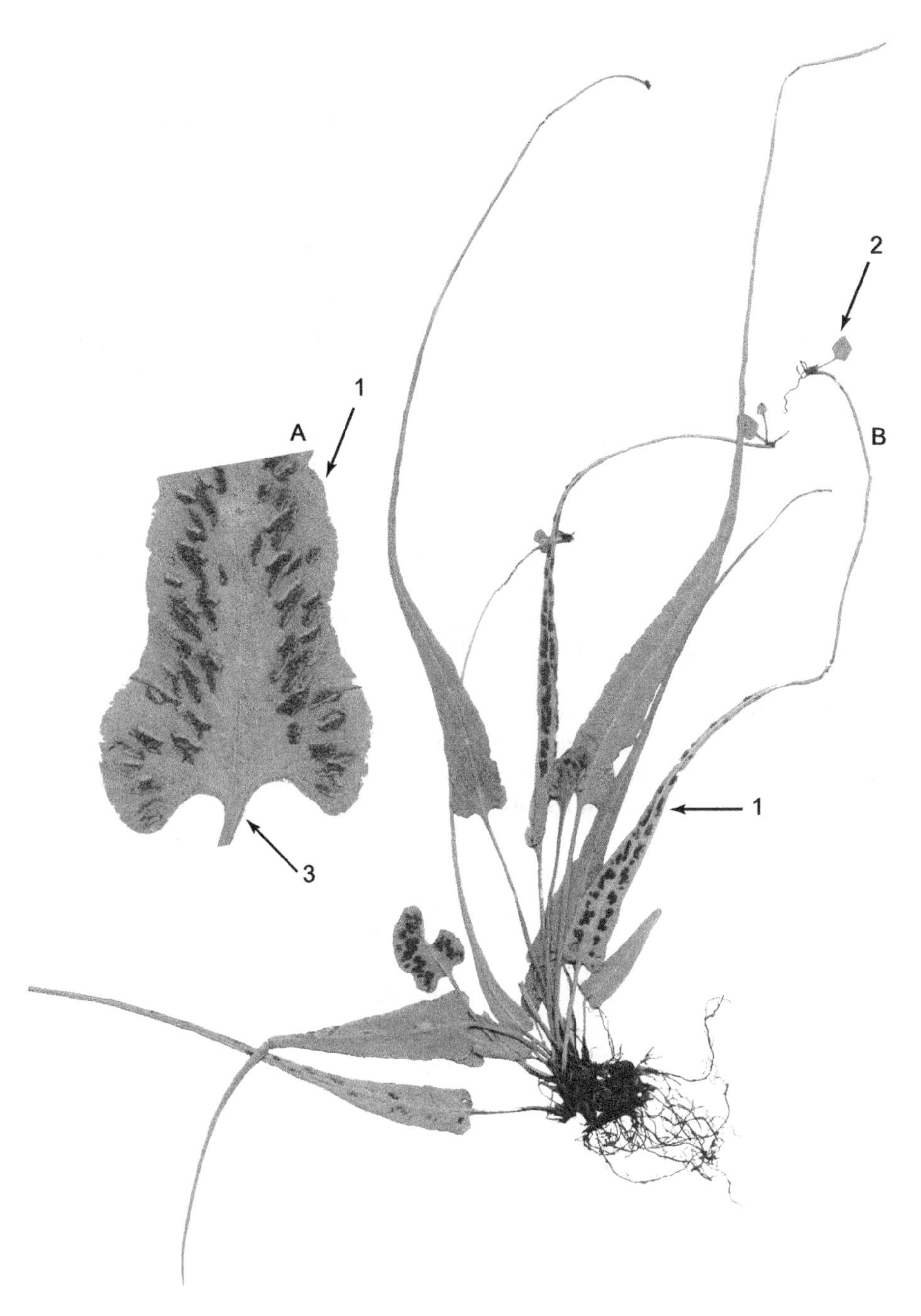

Asplenium rhizophyllum: Walking fern: A. Enlargement of blade base. B. Cluster of fronds.

Informative arrows. 1. Linear sori of various size irregularly arranged. 2. Plantlets at tips long narrow frond tips. 3. Heart-shaped base of blades.

long-tapering, narrowly triangular to lance-shaped, leathery, sparsely hairy, hairs more numerous abaxially, margins entire to sinuate, **tips rounded to very long, tapering gradually, often rooting at tips, bases heart-shaped to strongly lobed or occasionally hastate;** rachises green, dull, nearly glabrous.

Veins. Obscure, netlike, forming areolae near midribs.

Sori. Scattered irregularly over netted veins, numerous, linear or curved, often joined where veins meet.

Habitat. In Michigan primarily shaded, moist, moss-covered boulders and outcrops of Niagaran limestone and dolomite in second-growth mesic northern forests. On South Manitou Islands it is found on mossy cedar logs in wooded dunes.

Distribution. Eastern North America: ranging from southern Ontario and Quebec in Canada south to Georgia, Alabama, and Mississippi, occurring west to Wisconsin, Iowa, Kansas, and Oklahoma. In Michigan it is rare and local.

Chromosome #: $2n = 72$

Asplenium rhizophyllum is a threatened (legally protected), critically imperiled species in Michigan. It is globally secure.

Previously included in the genus *Camptosorus*, this distinctive fern is very different in appearance from all other Michigan *Asplenium* species. It forms dense clonal colonies by "walking"—proliferating by developing long-tapered, arching frond tips that root and form new plants when touching the ground.

There is no other Michigan fern that looks like *A. rhizophyllum* **with its characteristic shape, netted veins, and long-tapering tips, which sometimes form roots and produce new fronds**

Asplenium ruta-muraria Linnaeus

Asplenium cryptolepis Fernald; *A. ruta-muraria* var. *cryptolepis* (Fernald) Wherry

Wall-rue

Etymology. Latin, *ruta,* the herb rue (*Ruta graveolens,* a southern European plant), + *muralis,* growing on walls, + *-aria,* a suffix indicating connection—the "wall-rue." The vernacular name refers to its favored habitat and is a translation of the Latin name.

Plants. Monomorphic, evergreen.

Fronds. 7.5–19 cm long.

Rhizomes. Short-creeping to erect, often branched, dark brown, scaly; scales dark brown, clathrate, narrowly triangular, 1–3 × 0.1–0.25 mm, marginal teeth widely spaced.

Asplenium ruta-muraria: Wall-rue: A. Detail of pinnae and pinnules with sori. B. Plant with rhizome and cluster of fronds.

Informative arrows: 1. Pinnae well separated. 2. Fan-shaped pinnules. 3. Long-stalked pinnules.

Stipes. Reddish-brown at base, green upward, dull, 1–2 times length of blade, scales dark brown, narrowly triangular at base, tapering to hairlike tips.

Blades. 2-pinnate to 2-pinnate-pinnatifid; deltate-ovate, 2–6 × 1–4 cm, somewhat thick, glabrous, base obtuse, tips rounded to acute, bluish-green to olive-green: rachises green, dull, glabrous or with scattered, minute hairs.

Pinnae. 2–4 pairs, 7–30 × 5–20 mm, widely spaced, alternate, deltate-ovate, lowest pinnae largest.

Pinnules. Widely spaced, **mostly fan-shaped, stalks distinct,** tips rounded to pointed, finely lobed or toothed.

Veins. Free, **branching 1–4 times fanlike.**

Sori. Linear, usually 1–5 per pinnule, on both sides of midrib.

Habitat. Calcareous shale, cliffs and boulders, stonework.

Distribution. Europe. East Asia. Canada: New Brunswick, Ontario, Quebec. USA: New England to Georgia, west to Arkansas, north to Michigan. Found only on Drummond Island in Chippewa County in Michigan.

Asplenium ruta-muraria is a Michigan endangered species. It is critically imperiled in Michigan but globally secure.

Chromosome #: 2*n* =144

Asplenium ruta-muraria, **a very rare fern in Michigan, is quite distinct and unlikely to be confused with any other fern growing in the state. It may be recognized by its 2-pinnate to 2-pinnate-pinnatifid blade and fan-shaped pinnules and by its preference for calcareous substrates.**

Asplenium scolopendrium Linnaeus var. *americanum* (Fernald) Kartesz & Gandhi

Phylitis scolopendrium (Linnaeus) Newman var. *americanum* Fernald; *P. fernaldiana* Å. Löve; *Scolopendrium vulgare* J. E. Smith

American Hart's Tongue

Etymology. Latin, *skolopendra*, an old Greek name for an unknown plant resembling a centipede, + *-ium*, a suffix suggesting resemblance. It alludes to the 2 rows of sori that suggested a centipede's legs to Linnaeus. The common name suggests the frond's resemblance to a deer's tongue.

Plants. Monomorphic, evergreen.

Fronds. Clustered, **strap-shaped,** 10–45 × 2.5–6 cm, 10–40 per rhizome, elongate.

Rhizomes. Erect, short, unbranched, covered with cinnamon-colored clathrate lanceolate scales, 3–6 × 1–1.5 mm, margins entire.

Asplenium scolopendrium var. *americanum:* American hart's tongue: A. Single frond. B. Cluster of fronds on rhizome.

Informative arrows: 1. Tongue-shaped blade with wavy margin and heart-shaped base. 2. Long-linear sori of irregular length.

Stipes. Short, 3–10 cm long, 1/8–1/3 length of blade, green to purplish-brown to straw-colored, eventually becoming green and dull in midrib; scales on young stipes brown, narrowly lanceolate, sometimes absent on older stipes.

Blades. Linear 6–40 × 2.5–5 cm, simple, strap-shaped or lanceolate, tips rounded or tapering to a point, base heart-shaped, margins entire becoming undulate with age, glossy, nearly glabrous, bright green, somewhat leathery, scales silvery, soon falling; rachises brown at base, straw-colored upward, dull, glabrous.

Veins. Free, forking, not reaching margin, obscure.

Sori. Linear, along veins, prominent, appearing as linear brown stripes, usually restricted to upper 1/2 of blade, often of variable lengths; indusia thin, whitish, flaplike, opening toward the vein.

Habitat. North- and east-facing cool, moist, talus slopes; **shaded, moist boulders; ledges of calcareous rocks; sinkholes; and rocky, young, sugar maple forests, always in deep shade.** All sites in Michigan are dominated by relatively young hardwood forests of sugar maple.

Distribution. Rare and spotty. Mexico. Canada: Ontario's Bruce Peninsula. USA: Alabama, Michigan, New York, and Tennessee.

Possibly Michigan's rarest fern, it is considered endangered in Michigan and threatened federally.

Chromosome #: $2n = 144$

Asplenium scolopendrium var. *americanum* is distinguished with difficulty from European plants (var. or subsp. *scolopendrium*) on the basis several morphologic features. The best-defined distinguishing feature is that var. *americanum* is tetraploid while var. *scolopendrium* is diploid.

While the *A. scolopendrium* var. *americanum* is rare and limited to North America, var. *scolopendrium* is common in Europe and is found in Eurasia and northwestern Africa.

A. scolopendrium **var.** *americanum*, **with its linear, strap-shaped blades and conspicuous linear veins and sori, is not likely to be confused with any other Michigan fern species.**

Asplenium trichomanes Linnaeus

Maidenhair Spleenwort

(Including *Asplenium trichomanes* **subsp.** *trichomanes* **and** *Asplenium trichomanes* **Linnaeus subsp.** *quadrivalens* **D. E. Meyer; both subspecies are found in Michigan, and their descriptions are combined because of their similar appearance.)**

Etymology. *Trichomanes* is an ancient Greek name used by botanists before Linnaeus. Its etymology is not clear. It may be an allusion to the way this fern's delicate texture resembles that of the filmy ferns of the genus *Trichomanes*. The subspecies name *quadrivalens* comes from the Latin *quadri-*, four, + *valens*, strong, vigorous, probably referring to the fact that the fern is tetraploid. *Maidenhair* is a term applied to ferns with delicate fronds and slender black stipes (see also *Adiantum pedatum*).

Plants. Monomorphic, evergreen.

Fronds. 7–20 × 1.2–2 cm, growing in rosettes, narrow, tapering at both ends, fertile fronds erect, sterile fronds often flattened to the ground, fronds a mixture of new fronds with all pinnae present and old fronds from which half the pinnae are gone.

Rhizomes. Short-creeping to erect, often branched, scales black or black with brown borders, clathrate, lanceolate, 2–5 × 0.2–0.5 mm, margins entire to toothed.

Stipes. Reddish-brown or blackish-brown throughout, lustrous, brittle, wiry, 1/6 –1/4 length of blade; scales black, linear-lanceolate or filiform at base or lacking.

Blades. 1-pinnate, 3–22 × 0.5–1.5 cm, glabrous, tapering at both ends, dark green, linear; *rachis* **dark purplish-brown throughout, shiny**, mostly glabrous.

Pinnae. 15–35 pairs, opposite to subopposite, more widely spaced near base, oblong to oval, pinnae in middle of frond 2.5–5 (–8) × 2.5–4 mm, becoming smaller at base and tips, bases broadly wedge-shaped, margins

shallowly crenate to small-toothed or entire, tips rounded.

Veins. Free, mostly 1–2 forked.

Sori. 2–5 pairs per pinna on both sides of costa, short.

Habitat. Crevices in moss-covered rocks and moist woods. *A. trichomanes* **subsp.** *trichomanes*, **grows mainly on acidic to neutral, noncalcareous rocks** (rarely on calcareous rocks), while **subsp.** *quadrivalens* **grows on calcareous rocks.**

Distribution. The species *Asplenium trichomanes* has a worldwide distribution.

Chromosome #: *A. t.* subsp. *trichomanes*: $2n = 72$

A. t. subsp. *quadrivalens*: $2n = 144$

Asplenium trichomanes is divided into 4 subspecies worldwide. The 2 subspecies found in Michigan are similar in appearance but prefer different habitats and grow on different substrates. The subspecies, while very similar, may be distinguished by their spore size: subsp. *trichomanes* (diploid) is 27–33 μm in size, and subsp. *quadrivalens* (tetraploid) is 37–43 μm (larger to hold twice as many chromosomes). Some would recognize these 2 subspecies as distinct and separate species. They cannot interbreed to produce fertile offspring.

A. t. subsp. *trichomanes* is somewhat more delicate with a slender stipe and rachis and fewer pinnae. The fronds of subsp. *trichomanes* tend to arch away from the rock substrate. *A. t.* subsp. *trichomanes* is found in large areas of North America and worldwide.

A. t. subsp. *quadrivalens* is somewhat stouter with somewhat squarer pinnae

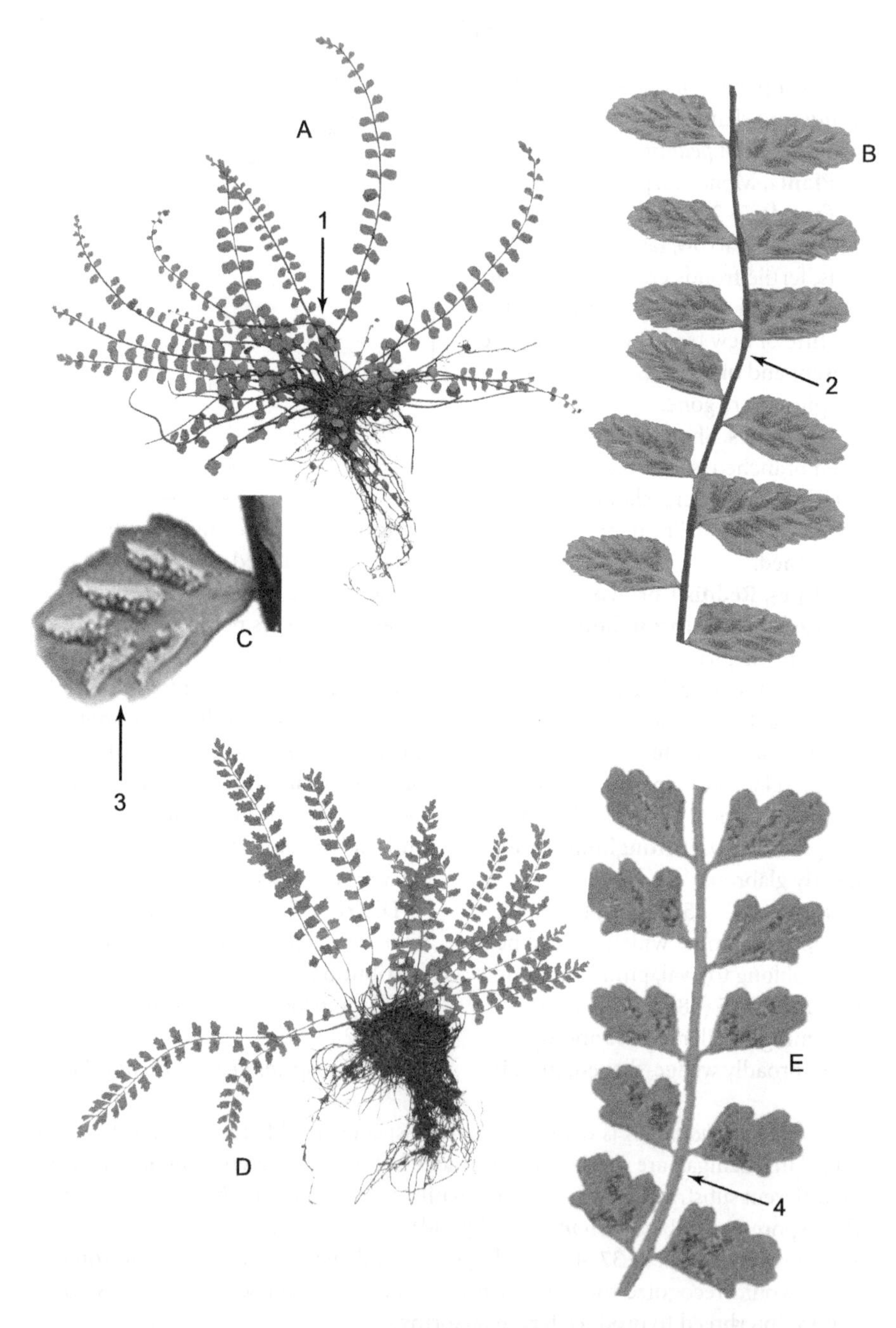

Asplenium trichomanes: Maidenhair spleenwort: A. Cluster of fronds. B. Portion of pinna. C. Pinnule. D–E. *A. viride:* Green spleenwort: D. Cluster of fronds. E. Portion of pinna showing green rachis.

Informative arrows: 1. Short stipes. 2. Black shiny rachis. 3. Linear sori. 4. Dull green rachis.

with shorter stalks. The fronds of susp. *quadrivalens* tend to grow closer to the rock substrate. *A. t.* subsp. *quadrivalens* is also found worldwide. Triploid hybrids between the above 2 subspecies have been found.

The 1-pinnate blades of *Asplenium trichomanes* **subspecies may be distinguished from** *A. viride* **by their shorter, dark, shiny, reddish-brown to black stipes and rachises. Those of** *A. viride* **are green from above the lower stipes through the rachis, and the stipes are generally longer.**

Asplenium viride Hudson

Asplenium trichomanes-ramosum Linnaeus as *Asplenium Trich. ramosum* Linnaeus

Green Spleenwort

The name *A. trichomanes-ramosum* **is commonly used for this species, including in** *Flora of North America North of Mexico*, **vol. 2 (1993). As the first name applied to this fern, it would usually have priority. However,** *A. viride* **is the correct name. Linnaeus described this fern in 1753 as** *Asplenium Trich. ramosum.* **Under the rules of the International Code of Botanical Nomenclature, phrase names such as** *Asplenium Trich. ramosum* **are to be treated as orthographic errors. (The intercalation of** *Trich.* **for** *trichomanes* **is to be disregarded.) Linnaeus's name was rejected in 1993 in favor of Hudson's later name,** *Asplenium viride* **(published in 1762).**

Etymology. Latin *viride*, green, in reference to the green stipes and rachises of this species. The vernacular name, green, refers to the green stipe and rachis of this species.

Plants. Monomorphic, evergreen.

Fronds. 5–18 cm long, growing in tufts, slender, delicate, semierect.

Rhizomes. Short-creeping, becoming erect, frequently branching.

Stipes. Reddish-brown at base, green upward, shiny, 1–5 cm long, 1/8–1/2 the length of blade; scales dark reddish-brown to black, narrowly deltate, clathrate, 2–4 × 0.2–0.4 mm, margins entire to undulate or with widely spaced shallow teeth, glandular hairs present.

Blades. 1-pinnate, linear, 2–13 × 0.6–1.2 cm, lower and middle pinnae same size and widely spaced, glabrous or with sparse minute hairs acute; *rachises* **dull green,** flattened, glabrous or with scattered hairs.

Pinnae. 6–20 pairs, subopposite, basal pinnae same size as middle pinnae, oblong, oval, round, or wedge-shaped, margins entire or slightly to deeply scalloped, upper and lower sides often unequal, tips rounded to acute, pinnae in middle of frond 5–6 × 4–5 mm.

Veins. Free, apparent.

Sori. 2–4 pairs per pinna, on both sides of vein.

Habitat. Cool, shaded crevices, calcareous cliffs, and talus slopes.

Distribution. Circumboreal. Europe. Asia. Greenland. In North America it has a western and eastern distribution with a large gap between. In the west, Alaska to California east to the Northwest Territories, Alberta, Montana, South Dakota, Wyoming, and Colorado. In the east, Quebec to the Atlantic, Wisconsin to New York, Vermont, and Maine.

Chromosome #: $2n = 72$

Asplenium viride is a Michigan threatened species and legally protected but apparently globally secure.

Asplenium viride **is smaller and more delicate than** *A. trichomanes* **and is easily distinguished by its longer stipes and green, dull stipes and rachises above the base. The rachises of** *A. trichomanes* **are dark, shiny, reddish-brown to black throughout, and its dark-colored stipes are generally shorter.**

ATHYRIUM Roth Athyriaceae

Lady Ferns

See Athyriaceae in appendix I for a description of the family and a key to the Michigan genera.

Etymology. Greek *athyros*, doorless, possibly because in some species the indusia opens slowly and tardily.

Plants. Medium-sized, terrestrial.

Rhizomes. Short-creeping, decumbent to erect.

Fronds. Clustered at tips of rhizomes.

Stipes. Adaxially grooved, usually V-shaped in cross section. **Cross-sections at bases reveal ends of two ribbonlike vascular bundles (appearing as linear and parallel lines).**

Blades. 1-pinnate-pinnatisect to 2-pinnate-pinnatifid, triangular, herbaceous.

Rachises. Grooved, grooves continuous with pinnae costae and costules, hairs absent on costae.

Sori. Linear or J-shaped along veins.

Indusia. Present.

Chromosome #: $x = 40$

Athyrium is a genus of about 180 species with worldwide distribution. Two species are treated in the *Flora of North America North of Mexico*, vol. 2 (1993). It is represented in Michigan by a single species.

In the past the genus *Athyrium* included *Deparia* and *Homalosorus* (until recently recognized as *Diplazium*) and was included in the family Athyriaceae. These are now recognized as separate genera, and *Homalosorus* has been placed in the new family Diplaziopsidaceae.

Athyrium filix-femina (Linnaeus) Roth var. angustum (Willdenow) G. Lawson
Polypodium filix-femina Linnaeus;
Aspidium angustum Willdenow;
Athryium angustum (Willdenow) C. Presl
Lady Fern, Northern Lady Fern

Etymology. Latin *filix*, fern, + *femina*, female: lady fern. Lady fern was the medieval name for this and other ferns with graceful, finely divided, delicate leaves, especially compared to those of *Dryopteris filix-mas*, the male fern.

Plants. Deciduous, monomorphic, usually in clonal clusters.

Fronds. 40–110 × 10–36 cm, yellow-green, clustered in a somewhat shuttlecock arrangement, new fronds appearing all summer.

Rhizomes. Short-creeping to ascending, often branching, forming an asymmetric clump, stout, 20–50 mm diam. including stipe bases, scaly, many old dead stipes attached.

Stipes. Straw-colored with dark brown or black, (in some forms very red) swollen bases, 1/2 to same length as blade; **scales scattered, plentiful—especially at base—light to dark brown,** linear- to ovate-lanceolate, 2–8 × 1–5 mm.

Blades. Elliptic, 1-pinnate-pinnatisect to 2-pinnate-pinnatifid; herbaceous, 30–60 × 10–20 (–35) cm, widest in the middle at about the fifth or sixth pinna pair from base, tips acuminate, yellow-green to bright green; rachises straw-colored (in some forms red), often with extensive glandular hairs, sometimes glabrous.

Pinnae. (12–) 15–35 alternate pairs, 3–20 × 1–4 cm, sessile or short-stalked, oblong-lanceolate, with narrow pointed tips, **3–4 lowermost pinnae gradually becoming smaller, lower pinnae often bending forward out of plane of blade,** deep V-shaped grooves on costae continuous with grooves of rachises, costae often with glandular hairs.

Veins. Free, forked, mostly not touching margins.

Pinnules. Decurrent into winged costae or sessile, obtuse to acute, **lobes cut up to halfway to costules; pinnules on both side of costae of equal size;** margins serrate or lobed, tips and segments blunt or acute with small teeth.

Sori. 3–10 per pinnule, **J-shaped and curved around a vein at far end**

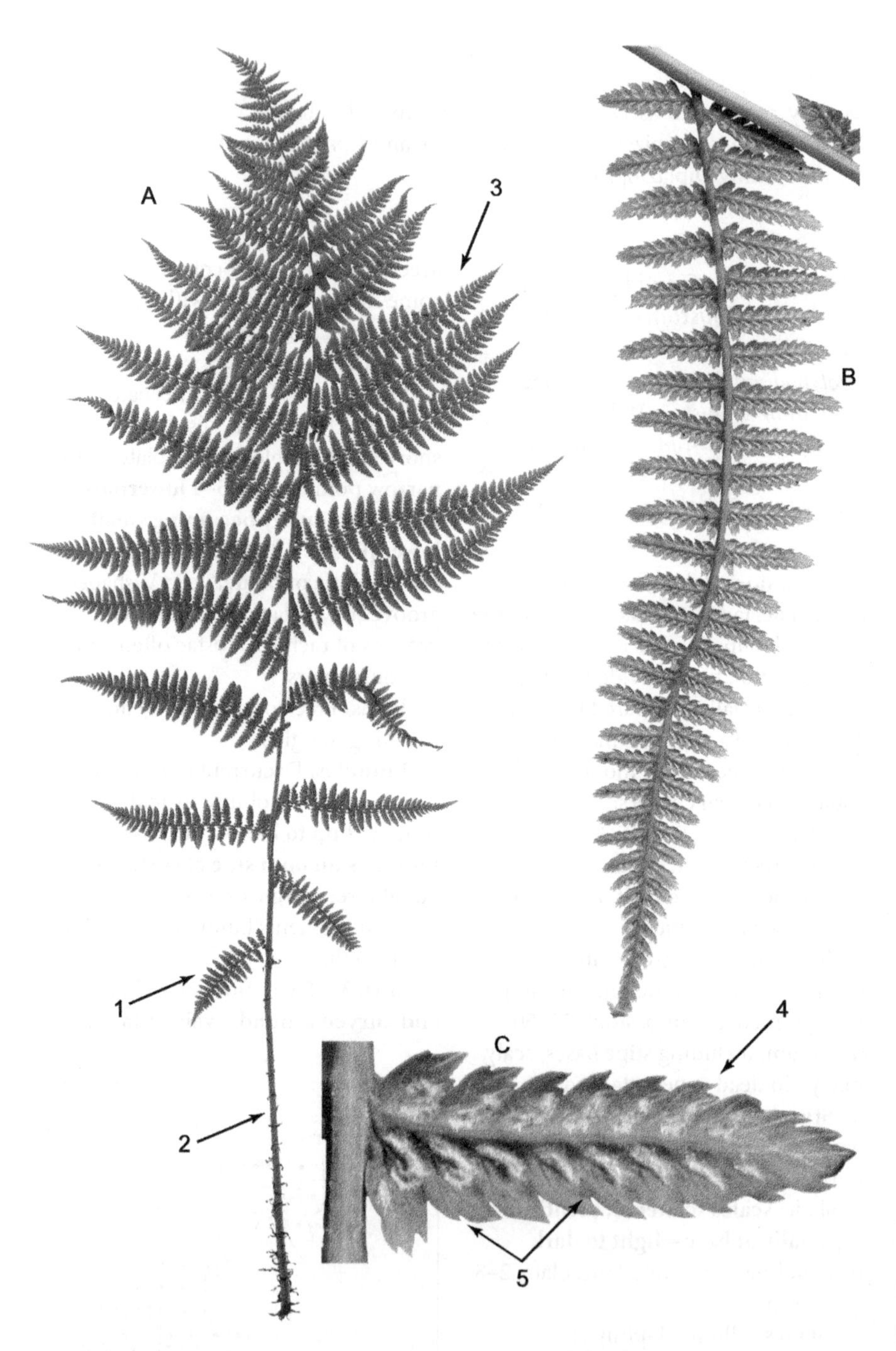

Athyrium filix-femina: Northern lady fern: A. Frond. B. Pinna. C. Pinnule.
Informative arrows: 1. Basal pinnae smaller. 2. Stipe smooth with scattered dark brown scales. 3. 2-pinnate blade. 4. Toothed pinnule margins. 5. Sori J-shaped and straight.

or linear and straight; indusia elongate, toothed.

Habitat. Moist woods, swamps, and meadows.

Distribution. Greenland. Canada and USA: Manitoba to Atlantic coast, south to North Carolina, west to Nebraska and North Dakota.

Chromosome #: $2n = 80$

Athyrium filix-femina is circumboreal in distribution and is found as far south as Central and South America. Five varieties are recognized worldwide. Four of these are treated in the *Flora of North America North of Mexico*, vol. 2 (1993). As far as is known, var. *angustum* is the only one found in the Great Lakes region. It is sometimes recognized at the rank of species as *A. angustum*.

Athyrium filix-femina is one of the most variable ferns, as demonstrated by its several named varieties and about 200 horticultural forms (mostly derived from the European var. *filix-femina*) all of which seem to intergrade. Occasionally plants in nature have red stipes and reddish fronds (*A. filix-femina* var. *angustum* f. *rubellum*). A horticultural variety with slightly smaller elliptic, lacy, light green fronds with burgundy-red stipes and rachises is widely sold as Lady in Red by nurseries.

Athryium filix-femina **var.** *angustum* **may be confused with** *Thelypteris noveboracensis* **but may be distinguished by the its J-shaped or linear sori (vs. round on** *T. novaboacensis*)**, somewhat reduced basal pinnae, and pinnules with toothed margins (vs. very much smaller basal pinnae with smooth margins on** *T. novaboacensis*)**.** *Thelypteris palustris* **is distinguished by its round sori and basal pinnae, which are mostly not reduced in size. Both** *Thelypteris* **species have acicular hairs that are lacking in** *A. filix-femina* **var.** *angustum*.

Dryopteris carthusiana **and** *D. intermedia* **may be distinguished by pinnules on the basiscopic sides of the pinnae that are longer than those on the acroscopic sides (vs. equal length in** *A. filix-femina*)**, are more deeply cut, and have round sori.**

AZOLLA Lamarck Azollaceae

Azollaceae is a family with a single genus, *Azolla*. The treatment of the family is identical to the treatment of that genus below.

The species of *Azolla* are difficult to identify because most specimens lack sori (which are necessary for identification). A scanning electron microscope is needed to see megaspore ornamentation, and a light microscope is needed to examine the hairs on the upper leaf lobe. At least 40× magnification is required to see the hairs.

When fertile, *Azolla* produces numerous sporocarps on the undersides of branches. Male sporocarps are greenish or reddish, resemble insect egg masses, and produce up to 130 microsporangia, each containing 32 or 64 microspores. The microspore forms a male gametophyte with a single *antheridium,* which produces 8 swimming sperm.

Female sporocarps are less numerous and contain only 1 sporangium with 1 megaspore (a large female spore). The megaspore produces a female gametophyte that protrudes from the megaspore and bears a small number of archegonia, each containing a single egg.

Azolla species have a symbiotic relationship with *Anabaena azollae*, a nitrogen-fixing, blue-green cyanobacterium that is found at the stem tips, under indusia, and in cavities of the upper leaf lobes. *Anabaena* colonies are nitrogen fixers, and other, more tropical species of *Azolla* are planted with rice crops where they decompose and release green fertilizer.

Azolla species with more southern distributions form thick mats on still waters that are presumed to smother mosquito larvae, leading to the common name mosquito ferns.

Azolla caroliniana Willdenow

Evrard and Van Hove (2004) suggest that the correct name for the plant is *A. cristata* Kaulf.

Duckweed Fern, Water Fern, Carolina Pond Fern, Mosquito Fern

Etymology. *Carolinana* is the latinized name for Carolina. One common name, water fern, describes its preferred habitat in the American Carolinas. Another name, mosquito fern, derives from its dense growth, which is said to cover the water so thickly that mosquito larvae are excluded. Duckweed fern derives from its resemblance to duckweed, which ducks are said to eat.

Plants. Minute, free-floating, forming mats.

Fronds. 1.25–2.5 cm long, dark green or with margins of bright crimson or whole plants dark red, free-floating, forming multilayer mats 2–4 cm thick in good environments on still waters; seldom fertile.

Stems. Prostrate, 0.5–1 cm long, thin, branching pinnately frequently, brittle, pale brown; roots minute.

Pinnae. 2-lobed, oval, in 2 rows, overlapping, upper emergent lobe dark green or with margins of bright crimson (red in full sun, green in shade) and a submerged, colorless lower lobe, tiny, 5–6 mm diam., largest hairs on upper lobe near stem with 2 or more cells, broad hair-stalk cell often 1/2 or more height of hair, apical cell curved, with tips nearly parallel to leaf surface.

Sori. Infrequent, megasporocarps and microsporocarps rarely collected; when found borne in pairs at base of pinnules, 1 large round sporocarp contains masses of microspores and 1 smaller acorn-shaped sporocarp contains one megaspore; megaspores without raised angular bumps or pits, consistently covered with tangled filaments.

Habitat. Floating on still water in ponds, on mud, or on muddy or

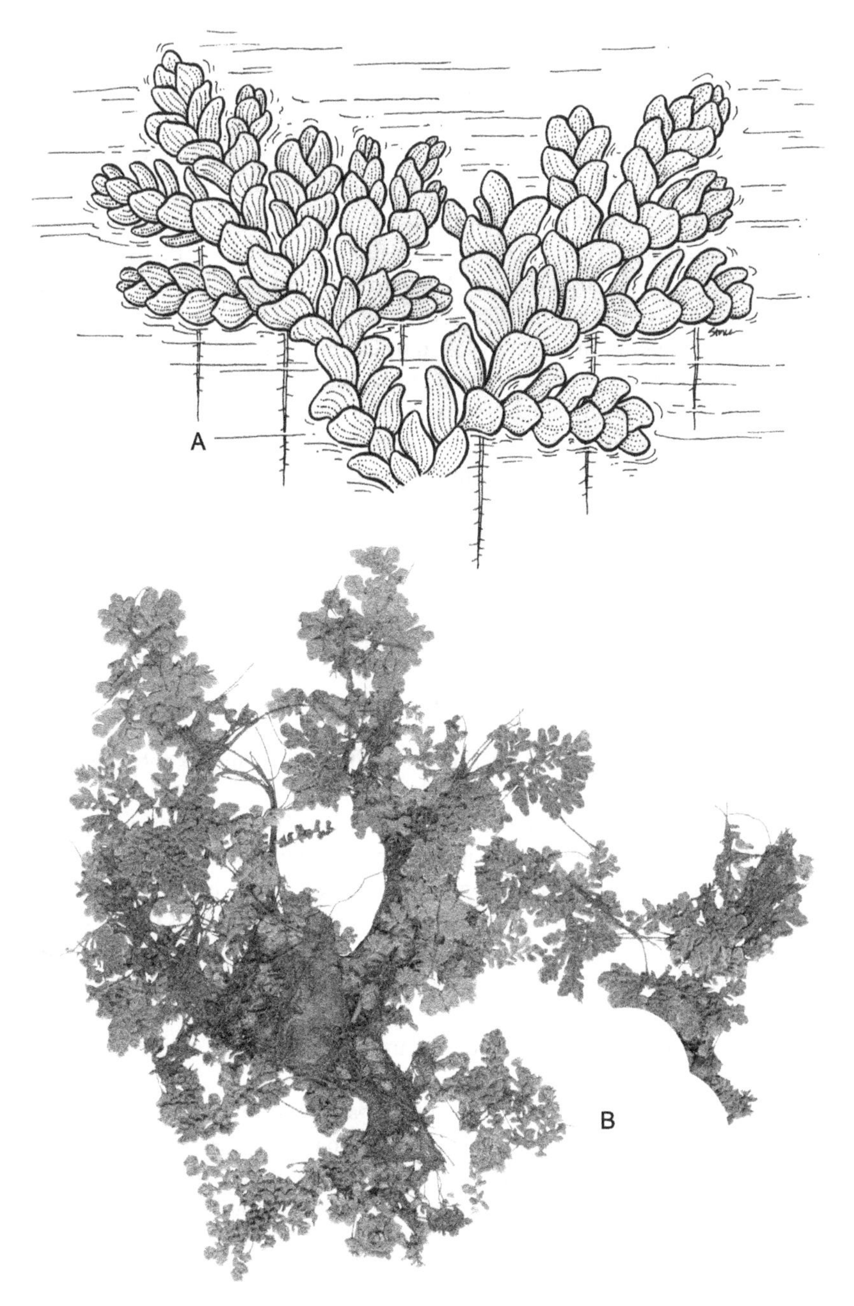

Azolla caroliniana: Duckweed fern: A. Generic *Azolla* line drawing of portion of floating frond. B. Cluster of dried fronds. (Drawing by Anna Stone.)

moist mossy banks (where it can grow and creep up to several inches above water level).

Distribution. Eurasia. South America. Canada: British Columbia and Ontario. USA: Massachusetts south to Florida, west to Texas, north to South Dakota and Michigan. *Azolla caroliniana* reaches the northern limits of its range in Berrien and Macomb counties in southern Michigan.

Chromosome #: $2n = 48$

Azolla caroliniana resembles a small floating moss bearing slender, pendulous roots on its underside. It may produce reddish anthocyanin in the fronds when growing in full sunlight, especially in late summer and fall. The colors varying from dark green to red create fall-like colors in the floating patches. It commonly overwinters in temperate areas but is killed by extreme cold.

A. caroliniana has tiny branches that are so fragile that water turbulence can break off the ends, causing small plants to float free and disperse. These small plants may be spread from 1 pond to another by birds.

A. caroliniana **is a tiny, free-floating, aquatic fern whose floating plants resemble the tip of a eastern white cedar leaf.**

CRYPTOGRAMMA R. Brown Pteridaceae

Parsley Ferns, Rock Brakes, Cliff Brakes

See Pteridaceae in appendix I for a description of the family and a key to the Michigan genera.

Etymology. Greek *kryptos*, hidden, + *gramme*, line (a word often used for fern sori) in reference to the linear sori at the pinnule margins, which are partially hidden by revolute margins, forming a false indusium. Brake is a name often applied to ferns in the family Pteridaceae. Parsley derives from fronds that resemble parsley.

Plants. Fronds green over winter, persistent after withering, strongly dimorphic.

Fronds. Scattered or densely tufted, stiff and erect, fertile fronds obviously longer; sterile fronds shorter, spreading.

Rhizomes. Creeping, ascending to erect, sparsely to much branched; scales pale brown, bicolored, ovate, or linear-lanceolate, margins entire.

Stipes. Dark brown at base, light brown to green above, grooved on upper surface, scaly, single round vascular bundle in cross section at base.

Blades. 1-pinnate-pinnatifid to 2-pinnate-pinnatifid, deltate to ovate-lanceolate, somewhat leathery to herbaceous, glabrous to sparsely hairy, dull to somewhat lustrous; sterile blades shorter and finely dissected.

Sterile pinnules. Short-stalked or sessile, ovate, lanceolate, ovate-lanceolate, or fan-shaped, dull green, margins entire, crenate or toothed, often somewhat more deeply incised.

Fertile pinnules. Very different from sterile pinnules, lanceolate to linear; 2–4 mm wide often less than 2 mm wide, with narrow, elongate ultimate segments.

Veins. Free, pinnately branched, usually obscure.

Sori. Linear along entire length of margins on marginal veins, margins reflexed to form false indusia, greenish to brown, broad, not modified in texture, at first covering and usually concealing young sporangia, often becoming flat at maturity.

Habitat. Usually on rock.

Distribution. Eurasia and North America, with disjunct populations in South America.

Chromosome #: $x = 30$

A genus of 8 to 11 species of temperate areas of North America, South America, and Eurasia. Four species are included in *Flora of North America North of Mexico*, vol. 2 (1993). The two species found in Michigan are rare in the state but have a wider distribution elsewhere.

KEY TO THE SPECIES OF *CRYPTOGRAMMA* FOUND IN MICHIGAN

1. Fronds densely clustered; stipes dark brown in proximal 1/8 or less, greenish distally; rhizomes stout, 4–20 mm diam., decumbent to erect; much branched; mostly on noncalcareous rock .... *C. acrostichoides*
1. Fronds scattered along rhizomes; stipes mostly dark brown in proximal 1/2–2/3, greenish distally; rhizomes thin, 1–1.5 mm diam., creeping, little branched; mostly on calcareous rock........................ *C. stelleri*

Cryptogramma acrostichoides R. Brown

Cryptogramma crispa (Linnaeus) R. Brown subsp. *acrostichoides* (R. Brown) Hultén

American Parsley Fern, American Rock Brake

Etymology. *Acrostichum*, a tropical fern genus with sori completely covering the undersurface of its pinnae, + *-oides*, a Greek suffix indicating resemblance, alluding to the undersurface of the fertile fronds, which appear to be covered with sori. The vernacular name alludes to its resemblance to parsley. Brake is a name often applied to ferns in the family Pteridaceae.

Plants. Fronds green over winter, dimorphic; sterile fronds green over

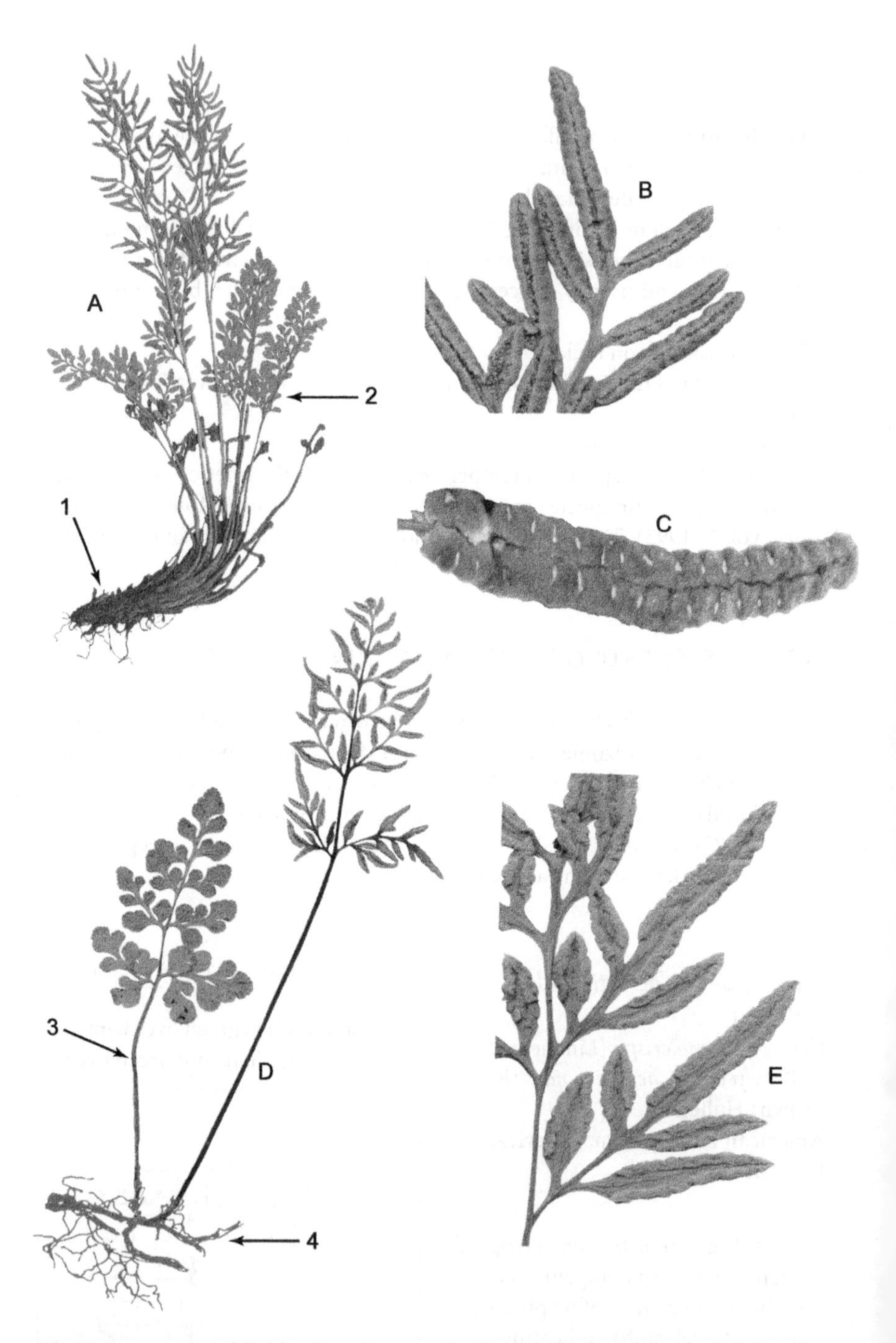

Cryptogramma acrostichoides: American parsley fern: A. Cluster of fronds. B. Enlargement of fertile pinna. C. Enlargement of young fertile pinnule.

Cryptogramma stelleri: Slender rock brake: D. Sterile and fertile fronds on creeping rhizome. E. Fertile pinnae showing marginal sori.

Informative arrows: 1. Stout decumbent to erect rhizome. 2. Fronds densely clustered. 3. Stipes dark brown at base, green distally. 4. Narrow creeping rhizome.

winter with shriveled fronds persistent the following spring.

Fronds. Forming dense clusters, persistent after withering.

Rhizomes. Short-creeping, decumbent to erect, **much-branched,** covered with old, stiff, sharply cut off stipe bases of more or less the same length, **stout, 10–20 mm diam., including stiff old stipe bases;** scales dense, bicolored, to 6 × 2 mm, linear to broadly lanceolate.

Stipes. Long, up to 1–2.5 × as long as blade, 1–2 mm diam., stiff, green to straw-colored, dark brown in proximal 1/8 or less; fine hairs scattered along grooves; grooves of rachis continuing on to the costa on abaxial surface; scales at base becoming sparse above, bicolored or somewhat concolored.

Sterile fronds. Spreading, 3–17 cm long, much shorter than fertile fronds.

Sterile blades. 2–3-pinnate, deltate to ovate-lanceolate, somewhat leathery, smooth, bright green, hydathodes sunken below leaf, fine appressed hairs scattered along grooves of stipes and along costae and costules of adaxial blade surface.

Sterile pinnae. Oblong to fan-shaped, short-stalked, folded into 3 dimensions, sometimes overlapping, margins toothed.

Sterile pinnules. 5–6 pairs, oblong to ovate-lanceolate, bases cuneate, 1.5–3 cm long, delicate, almost translucent, distal 2/3–1/2 of segments, crenate to dentate, every second tooth often somewhat more deeply incised.

Fertile fronds. Erect, long and narrow, 8–25 × 2–5 cm.

Fertile blades. Much narrower than sterile blades, 2-pinnate.

Fertile pinnae. Horizontal to ascending, linear, contracted, 3–12 × 1–2 mm; margins entire.

Fertile pinnules. Much narrower than sterile pinnules, lanceolate, pointed; margins of fertile segments reflexed covering sporangia.

Veins. Free.

Sori. Linear, submarginal, along reflexed margins forming false indusia; coalescing at maturity.

Habitat in Michigan. Noncalcareous cliff crevices; rock outcrops and talus; usually relatively dry habitats with cool summers.

Distribution. Asia. Reported in Mexico in Baja California. Canada and USA: Alaska to Ontario, south to Michigan, west along northern tier of United States and scattered in southwest from Colorado to New Mexico to California. A mostly western fern reaching its eastern limits in Michigan and Ontario. The only known Michigan locations are on Isle Royale on dry, sunny bedrock glades near the Lake Superior shoreline.

Chromosome #: $2n = 60$

Cryptogramma acrostichoides is a Michigan endangered species and legally protected. It is globally secure.

Cryptogramma acrostichoides **grows on noncalcareous rocks, while** *C. stelleri* **grows on calcareous rocks. The fronds of** *C. acrostichoides* **are densely tufted on short, much-branched rhizomes, and its fronds are more leathery and usually larger than those of** *C. stelleri.* **The fronds of** *C. stelleri* **are scattered on a more elongate, slender rhizome, and they are smaller and more fragile than those of** *C. acrostichoides.*

Cryptogramma stelleri (S. G. Gmelin) Prantl
Pteris stelleri S. G. Gmelin
Slender Rock Brake, Fragile Rock Brake

Etymology. The name *stelleri* honors Georg Wilhelm Steller (1709–46), a collector with the Bering expedition to Kamchatka and North America in 1741. His name is applied to several northwestern species, including Steller's jay and Steller's sea lion. Common names derive from its delicate form and preferred habitat. Brake is a name often applied to ferns in the family Pteridaceae.

Plants. Fronds deciduous, withering, dimorphic, new growth produced in spring, dying and shed by late summer.

Fronds. Scattered along rhizomes, dying by late summer, soon shed; glabrous.

Rhizomes. Creeping, slender, 0.5–1.5 mm diam., with few branches, soft, fleshy, somewhat translucent when young, hard and wrinkled later; shriveling in second year following frond emergence; scales sparse, inconspicuous, pale brown, ovate, 0.4 × 0.3 mm.

Stipes. 1 mm diam., dark brown at base, becoming greenish distally, glabrous.

Sterile fronds. Erect to low arching, 3–15 cm long, glabrous.

Fertile fronds. Erect, 5–20 cm long, glabrous.

Stipes. Stipes usually longer than blades on both sterile and fertile fronds, circa 1 mm diam. wide when dry; smooth, glabrous, lower 1/3 or less dark brown, becoming green or light brown above.

Sterile blades. 2.5–15 (–20) cm long, ovate to ovate-lanceolate, fragile; 1-pinnate-pinnatifid to 2-pinnate-pinnatifid, 3–8 × 2–6 cm, herbaceous to membranous, thin, hydathodes superficial, often poorly developed or absent; rachises green, glabrous.

Sterile pinnae. 5–6 pairs, opposite and alternate, lowest pair longest.

Sterile pinnules. 1–3 pairs per pinna, 5–15 × 3–10 mm, short-stalked, broadly round, ovate-lanceolate to fan-shaped, distal 1/2–1/3 shallowly lobed, margins entire or toothed, flat, delicate, almost translucent, often crowded, often overlapping; costae and costules glabrous.

Fertile blades. ovate-lanceolate to ovate, Longer than sterile, erect to spreading; 5–20 cm long; fertile fronds larger, 10–20 cm, with narrower leaflets, thin.

Fertile blades. 1-pinnate-pinnatifid to 2-pinnate, lanceolate, herbaceous to membranous.

Fertile pinnae. Horizontal to ascending, often only partially differentiated from sterile pinnae, lanceolate to linear, 8–25 × 2–4 mm, margins

reflexed forming continuous false indusia.

Fertile pinnules. Narrow, lanceolate to linear, tips pointed, stalked, margins reflexed forming false indusia, horizontal to ascending, 8–25 × 2–4 mm.

Veins. Free, forked, not reaching margins.

Sori. Linear, marginal, reflexed unmodified pinnule margins form false indusial coalescing at maturity and often cover entire underside of pinnules.

Habitat. Cool boreal habitats, moist, **calcareous rock ledges and shady calcareous cliff crevices,** outcrops in mesic northern hardwoods, and coniferous forests.

Distribution. Eurasia. Western China to Himalaya. Canada and USA: In western North America from Arizona to Alaska. In eastern North America from Labrador and Newfoundland west to Minnesota, south to West Virginia.

Chromosome #: $2n = 60$

Cryptogramma stelleri is a species of special concern in Michigan, although its frequency and status in the state are unknown. It is globally secure. It may be more common in Michigan than previously thought.

Cryptogramma stelleri, **in contrast to** *C. acrostichoides,* **grows on calcareous rocks and has smaller more delicate fronds. The fronds of** *C. stelleri* **are scattered on elongate, slender rhizomes, while the fronds of** *C. acrostichoides* **are densely tufted on short, stout, much-branched rhizomes.**

CYSTOPTERIS Bernhardi Cystopteridaceae

Bladder Ferns, Brittle Ferns

See Cystopteridaceae in appendix I for a discussion of this family and a key to its Michigan genera.

Etymology. Greek *kystos*, bladder, + Greek *pteris*, fern (from *pteron*, wing, feather)—the bladder fern. The indusium looks like an inflated bladder.

Plants. Monomorphic, deciduous, sending up new fronds during summer; small to medium-sized.

Rhizomes. Short- to long-creeping, decumbent, densely covered with old, persisting, erect, somewhat succulent leaf bases.

Stipes. Slender, grooved, 1/3–3 times length of blades, straw-colored to medium brown, translucent, smooth, mostly glabrous, bases with occasional membranous scales; cross sections at bases reveal 2 round or oblong vascular bundles.

Blades. 1-pinnate-pinnatifid to 2-pinnate-pinnatifid, ovate-lanceolate to triangular to long-triangular, gradually tapering to pinnatifid tips, herbaceous to somewhat leathery, glabrous or slightly scaly.

Pinnae. Proximal pinnae sessile or stalked, not reduced or one pair slightly reduced; hairs and glandular hairs absent or with sparse or plentiful gland-tipped hairs abaxially or multicellular uniseriate hairs in pinnae axils; costae grooved, grooves continuous with rachis grooves.

Pinnules. Margins crenulate, dentate, or serrate.

Veins. Free, simple or forked, reaching margins.

Sori. Round, in 1 row between midribs and margins; indusia hoodlike or cup-shaped when young, small, thin, ovate or oval, attached at bases, free at margins, arching over sorus, opening outward, scalelike and often obscured by mature sporangia, usually ephemeral, often obscure at maturity.

Habitat. Diverse, moist limestone substrates, wet woods, and springy areas with mesic loamy soils.

Chromosome #: $x = 42$

Cystopteris is a genus of about 15 to 20 species widespread in temperate zones and at higher elevations in the tropics. Nine species are treated in *Flora of North America North of Mexico*, vol. 2 (1993). Five species are found in Michigan.

The genus *Cystopteris* contains diploid, tetraploid, pentaploid, and even octoploid taxa. Some species are homopolyploids and some allopolyploids. Interspecific hybrids are common wherever *Cystopteris* species occur together (or even where they don't). Hybridization and doubling of chromosome numbers (allopolyploidy) involving *C. bulbifera* and other North American *Cystopteris* species have generated several allopolyploid species. If a plant is found that has the characters of two or more species there is a possibility that it is a hybrid. Hybrids usually have shriveled and distorted spores. Glandular hairs and deformed bulblets may be present if *C. bulbifera* is one of the parents.

Other than *C. bulbifera, Cystopteris* species are often difficult to differentiate because characteristics distinguishing the species frequently overlap. Their morphology is remarkably plastic, and leaf morphologies can vary greatly depending on environment. Sterile plants, small fertile plants, or plants growing in stressful habitats will be especially difficult to identify. Observing complete fertile specimens is important.

For amateurs, or those only mildly interested, precise diagnosis of the species may require more effort than they may want to expend. For precise species diagnosis it may be necessary to consult an expert. It may be simplest and best to identify *C. bulbifera*, *C. fragilis*, and *C. protrusa* and recognize that the other *Cystopteris* species can be difficult.

Spore size aids in species identification because it reflects the chromosome number: a larger spore is needed to contain the additional chromosomes found in polyploids.

Some *Cystopteris* species have trophopods (persistent swollen stipe bases) densely covering the rhizomes. These serve as storage organs for species that often grow in hostile environments and are subject to adverse conditions. They allow the plants to grow new fronds if the old ones are destroyed by drought.

Cystopteris tennesseensis and *C. montana* have been found on the Ontario shore of Lake Superior and should be looked for in Michigan.

KEY TO THE SPECIES OF *CYSTOPTERIS* FOUND IN MICHIGAN

See the above discussion regarding the difficulty of identifying some *Cystopteris* species.

1. Blades of two types—early season short-triangular sterile blades, and later long-attenuate-triangular fertile blades, widest at base; rachises and costae of larger blades frequently with bulblets; rachises, midribs of pinnules, and indusia densely covered with minute glandular hairs; stipes reddish when young, becoming green or straw-colored in mature specimens. . . . . *C. bulbifera*

1. Blades monomorphic, sterile and fertile fronds the same, narrowly elliptic, mostly widest at or just below the middle of the blade; rachises and costae lacking bulblets (sometimes with usually misshapen bulblets on *C. laurentiana*); rachises, costae, indusia, and midribs of pinnules lacking glandular hairs (or only sparse in *C. laurentiana*); stipes dark brown to straw-colored, or green (2).

2(1). Rhizome tips protrude 1–5 cm beyond fronds; rhizomes with abundant golden-yellow hairs; sori absent from early season blades, subsequent fronds are fertile; pinnules closest to rachises on lowest pinnae cuneate on short stalk circa 0.5–1 mm long. . . . .*C. protrusa*

2. Fronds clustered at rhizome tips; rhizomes nearly hairless; nearly all fronds are fertile; pinnules closest to rachises on lowest pinnae cut square, rounded or wedge-shaped at base, sessile or on very short stalks less than 0.5 mm long (3).

3(2). Pinnae typically at less than a 90° angle to distal rachises, often curving toward blade tips; pinnae along distal 1/6 of blades ovate to narrowly elliptic; pinna margins crenulate or with rounded teeth; basiscopic pinnules of lowest pinnae stalked, wedge-shaped to rounded at base; dark color of stipe base extending to rachises . . . .*C. tenuis*

3. Pinnae typically perpendicular to rachis, not curving toward blade tips; pinnae along distal 1/6 of blades deltate, ovate to oblong; pinna margins with teeth or serrate; basiscopic pinnules of lowest pinnae sessile or nearly so, cut square to rounded at base; stipe base entirely stramineous or dark only at base (4).

4(3). Mostly 30–50 cm long; blade widest just above base; glandular hairs sparse on rachises, pinna costae, and indusia (mostly shed by early summer—except on indusia); margins of pinnae serrate: occasional vestigial bulblets (small hairy bodies) at axils of pinnae . . . . *C. laurentiana*

4. Mostly under 30 cm; blade widest just above middle; glandular hairs lacking; margins of pinnae with sharp teeth; vestigial bulblets always absent. . ... *C. fragilis*

Cystopteris bulbifera (Linnaeus) Bernhardi
Polypodium bulbiferum Linnaeus
Bulblet Fern, Bulblet Bladder Fern

Etymology. Latin *bulbus*, a swelling or bud, + *-fera*, carry, bear, referring to the bulblet-bearing fronds. The vernacular name alludes to the bulblets that form on the underside of the rachises and drop off to grow into new ferns.

Plants. Deciduous, somewhat dimorphic.

Fronds. Clustered in groups of 3–8, 20–80 × 6–15 cm, somewhat flaccid and rarely standing erect.

Rhizomes. Creeping, covered with old stipe bases; scales lanceolate, tan to light brown, somewhat clathrate.

Stipes. Reddish when young, later green or straw-colored, **1/4–1/2 length of blade,** sparsely scaly at base.

Blades. 2-pinnate to 2-pinnate-pinnatifid, 24–65 × 6–15 cm, pale green to yellow-green, **broadly to narrowly triangular, widest at base, gradually tapered to slender tips; earliest fronds shorter and lack sori, subsequent fronds with sori longer, narrower, and arching; rachises, costae, and indusia covered with unicellular glandular hairs (more easily seen with a hand lens),** may be shed during the growing season**; larger more mature fronds often have bulblets on abaxial surface of rachises and in axils of pinnae.**

Pinnae. 18–40 pairs, long-triangular, subopposite to alternate, perpendicular to rachises, sometimes with bulblets.

Pinnules. Sessile to short-stalked, bases truncate to obtuse, tips rounded; margins finely toothed.

Veins. Free, pinnately forked, usually running to sinuses between teeth.

Sori. Medial in pinnule lobes; indusia cup-shaped, tips cut off in a curved or straight line, usually with unicellular, glandular hairs, evanescent.

Habitat. Commonly grows on moist limestone substrates, but in northern Michigan it is found growing on soil in swamps, wet woods, and springy areas.

Distribution. Canada: Ontario to Newfoundland. USA: Eastern United States, Minnesota south to Alabama, Arkansas west to Arizona and Utah, South Dakota, and Nebraska.

Chromosome #: $2n = 84$

Cystopteris bulbifera is recognized by bulblets found on the fern's fronds that are capable of producing a new plant when they touch the ground, still attached or

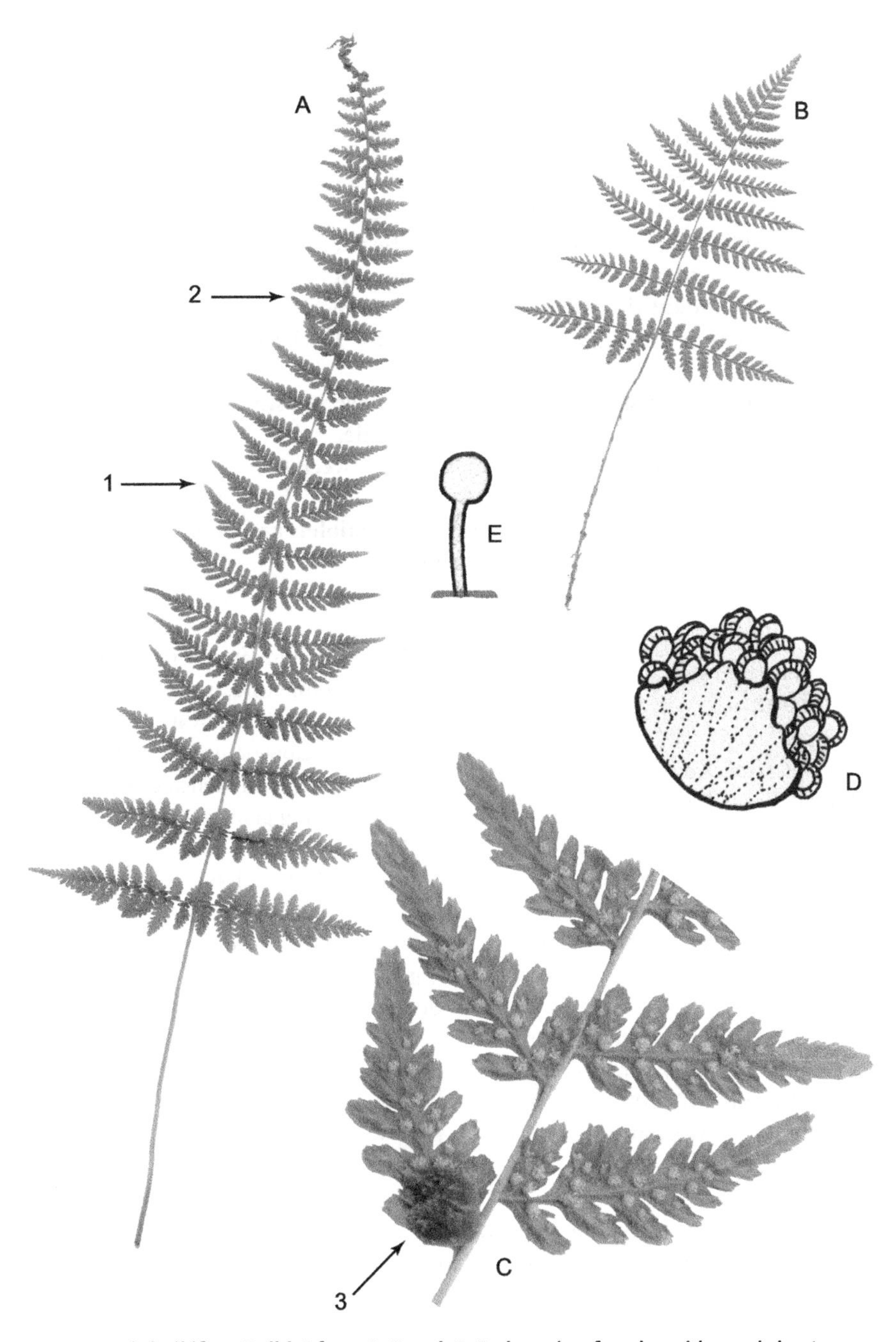

Cystopteris bulbifera: Bulblet fern: A. Frond. B. Early spring frond—wider and shorter. C. Enlargement of pinnae with bulblet and young sori. D. Line drawing of sorus. E. Line drawing of glandular hair found on blades.

Informative arrows: 1. Long narrow triangular frond. 2. Look for glandular hairs. 3. Bulblet at base of pinna.

not. This asexual form of reproduction allows for the formation of clonal colonies with the same genotype as the parent.

Cystopteris bulbifera frequently hybridizes with other *Cystopteris* species, and hybrids are frequently found when two or more species grow together.

Cystopteris bulbifera **is usually easy to recognize and may be distinguished from** *C. fragilis*, *C. protrusa*, **and** *C. tenuis* **by the characteristic bulblets near the tips of some fronds; by its long, narrowly tapering, triangular mature fronds; and by the abundant tiny glandular hairs on its rachises and costae. (Viewing with a 10× lens in bright light is often required. The hairs are very small, but they really are there.)**

Early fronds are usually short-triangular, have red stipes, and usually lack sori. Later fronds have green stipes and are fertile.

C. laurentiana **is a fertile hexaploid hybrid between** *C. bulbifera* **and** *C. fragilis* **(which contributes 2/3 of its chromosomes and which it more closely resembles). It sometimes produces small or abortive bulblets and often has scattered glandular hairs—particularly on the indusia.**

Cystopteris fragilis (Linnaeus) Bernhardi

Polypodium fragile Linnaeus;
Cystopteris dickieana R. Sim
Fragile Fern, Brittle Fern

Etymology. Latin *fragilis*, easily broken, brittle, alluding to the stipes and blades that are easily broken when bent. The vernacular name derives from the Latin name.

Plants. Deciduous, monomorphic.

Fronds. 3–8, loosely clustered at tips of rhizomes, 10–40 cm × 3–8 cm, more than 2.5 times longer than wide, highly variable, erect, prostrate or arching.

Rhizomes. Short-creeping, slender, covered with old stipe bases, hairs absent; roots numerous, black; scales lanceolate, tan to light brown.

Stipes. Shorter than or nearly equaling blade length, reddish-brown at base, green to straw-colored above, brittle, easily broken off particularly near base, scales absent to sparse near base.

Blades. 1- pinnate-pinnatifid to 2-pinnate-pinnatifid, nearly all bearing sori, 10–30 × 3–8 cm; ovate-lanceolate, widest just below or at middle, tips pointed, light green, delicate texture, rachises lacking glandular hairs and bulblets.

Pinnae. 8–15 pairs; ± opposite, ± perpendicular to rachises, lower pinnae widely spaced, nearly all bearing sori, margins serrate to sharply dentate.

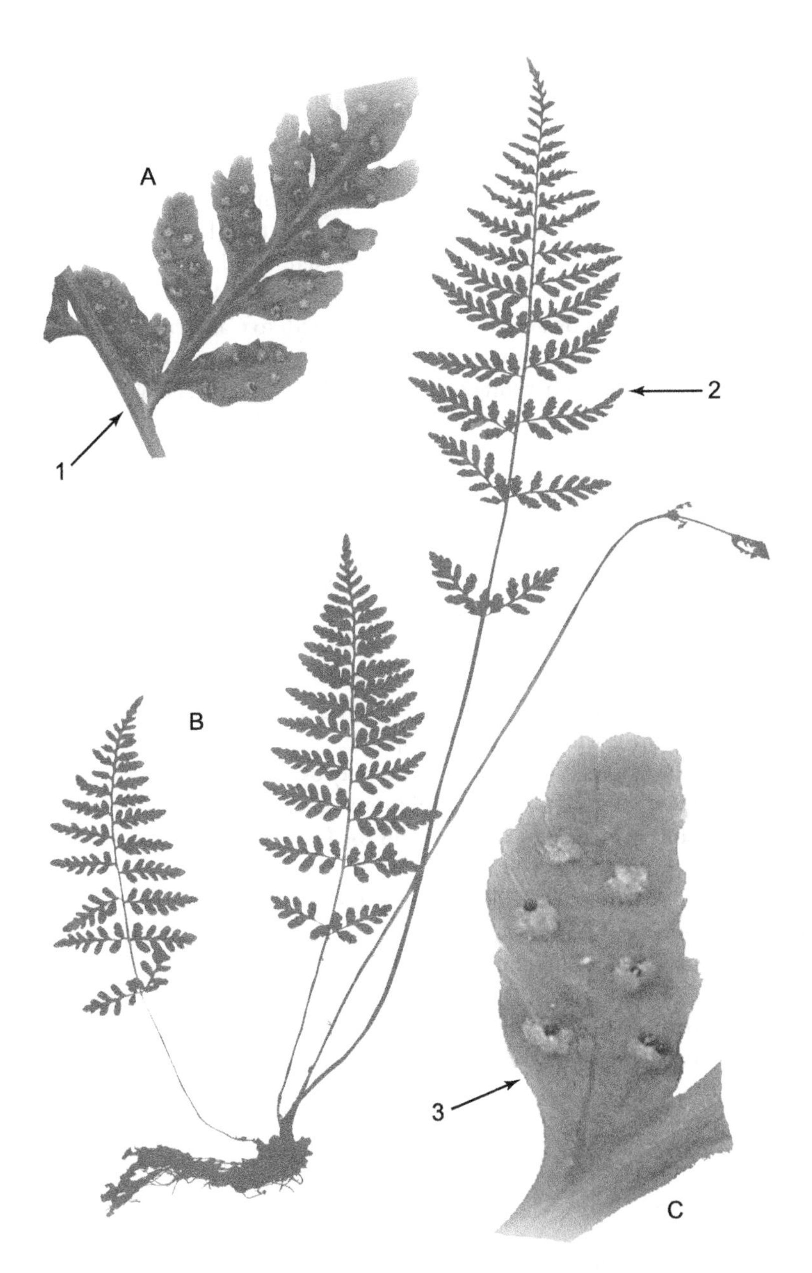

Cystopteris fragilis: **Fragile fern: A. Enlargement of pinna. B. Cluster of fronds. C. Enlargement of pinnule.**

Informative arrows: 1. Downward-facing pinnules of lowest pinnae rounded or cut square at base. 2. Widest at or just below middle of blade. 3. Sori with young cup-shaped indusia.

Veins. Free, forked, mostly ending in marginal teeth or small notches, not in sinuses.

Pinnules. Those nearest rachis with bases rounded or cut square at base, margins smooth to sharply toothed, downward-facing pinnules of lowest pinnae sessile.

Sori. Round; indusia ovate to cup-shaped, forming a hood over the sorus, opening outward, shriveling and nearly invisible with maturity, glandular hairs lacking.

Habitat. Sheltered, mostly calcareous cliff faces, moist woods in humus-rich soil, and in thin soil over rock.

Distribution. Widespread in northern hemisphere. Southern tip of South America. Hawaii. New Zealand. Kerguelen Island. Greenland. Canada and USA: From Alaska to Newfoundland, south to California, Nevada, New Mexico, Texas, Iowa, Illinois, Indiana, Ohio, Pennsylvania, New York, and New England.

Chromosome #: $2n = 168, 252$

Cystopteris fragilis is essentially worldwide in distribution. In warmer climates it is limited to higher elevations in mountains.

Cystopteris fragilis as previously interpreted included species now recognized as *C. laurentiana*, *C. tennesseensis*, *C. protrusa*, and *C. tenuis* as varieties. Subtle morphological characters, and even microscopic examination of the spores are sometimes required to identify this difficult group of species. Confusion is sometimes increased when *C. fragilis* hybridizes with *C. laurentiana* and *C. tenuis*, producing sterile hybrids where they are sympatric.

Cystopteris fragilis **is quite variable and may be confused with** *C. protusa*. *C. protusa* **is somewhat larger and has rhizomes that are covered with abundant golden hairs and protrude 1–5 cm beyond the last frond and blades with more rounded teeth on pinnules (vs. pointed teeth in** *C. fragilis*).

Cystopteris fragilis **has basiscopic pinnules of lowest pinnae sessile, cut square to rounded at base (vs. basiscopic pinnules of lowest pinnae stalked, wedge-shaped to rounded at base in** *C. tenuis*).

Cystopteris fragilis **may be distinguished from** *C. bulbifera* **by the absence of glandular hairs and bulblets and by fronds widest at or just below the middle of the blade (vs. widest at base with long-tapering fronds in** *C. bulbifera*).

Cystopteris laurentiana (Weatherby) Blasdell

C. fragilis (Linnaeus) Bernhardi var. *laurentiana* Weatherby

Laurentian Fragile Fern

Etymology. *Laurentiana*, a Latinized name suggesting connection, referring to the Laurentian Great Lakes on the Canadian-US border and the large plateau extending from the Great Lakes to the Arctic Ocean on which this fern was first recognized. The vernacular name derives from the generic name.

Plants. Monomorphic, deciduous.

Fronds. 20–50 cm long to 12 cm wide, clustered at rhizome tips, nearly all bearing sori, nearly all looking alike.

Rhizomes. Short-creeping, inter-

nodes very short, less than 5 mm, heavily covered with old stipe bases, hairs absent; scales uniformly brown to ± clathrate.

Stipes. 3–8, clustered; stipes usually dark at base, grading to straw-colored distally, shorter than blade—usually 2/3 its length, sparsely scaly at base.

Blades. 2-pinnate to 2-pinnate-pinnatifid, ovate to lanceolate, widest above base, tips short-tapering, occasionally with malformed bulblets, light green or yellowish-green; rachises and costae with sparse ephemeral glandular hairs that may be shed during summer, glands are usually absent after the end of June.

Pinnae. 12–16 pairs, typically perpendicular to rachis, not curving toward blade tips, margins serrate, stalks and costae with sparse ephemeral glandular hairs, distal pinnae ovate to oblong.

Veins. Free, simple or forked, ending in both teeth and notches.

Ultimate segments. Basiscopic pinnules not enlarged, basal basiscopic pinnules sessile to short-stalked, base truncate to obtuse.

Sori. Round, in 1 row between midrib and margin; indusia cup-shaped, with sparse unicellular, glandular hairs.

Habitat. Cracks and ledges on cliffs,

	C. bulbifera	*C. laurentiana*	*C. fragilis*
Frond size	Mostly more than 50 cm	Mostly 30–50 cm	Mostly under 30 cm
Blade shape	Long-triangular, widest at base	Blade ovate to lanceolate, widest above base	Ovate-lanceolate, widest just below or at middle
Pinna pairs	18–40	12–16	8–15
Vein termination	Teeth	Teeth and sinuses	Sinuses
Glands on rachis and costae	Plentiful	Sparse, easily and soon shed	Absent
Bulblets	Frequent	Occasional, malformed	Absent
Stipe color	Reddish when young, later green or straw-colored	Stipes usually dark at base, grading to straw-colored above	Reddish-brown at base, green to straw-colored above
Stipe length	1/4–1/2 length of blade	Usually 2/3 length of blade	Shorter than or nearly equaling blade length
Chromosome #	$2n = 84$, diploid	$2n = 252$f, allohexaploid	$2n = 168$, tetraploid

Cystopteris laurentiana: Laurentian fragile fern: A. Cluster of fronds on rhizome. B. Single frond.

Informative arrows: 1. Pinnae fewer than in *C. bulbifera*, larger and more numerous than in *C. fragilis*. 2. Look for sparse glandular hairs. 3. Look for uncommon or distorted bulblets. 4. Stipes of young fronds not red. 5. Blade widest just above the base. 6. Stipe usually 2/3 length of blade.

often on calcareous substrates.

Distribution. Canada: Ontario to Newfoundland. USA: Minnesota to Michigan, Iowa, Illinois, Connecticut, Massachusetts, Pennsylvania, and Vermont.

Chromosome #: $2n = 252$

Cystopteris laurentiana is a species of special concern in Michigan.

Cystopteris laurentiana is a fertile allohexaploid hybrid between *C. bulbifera* (diploid) and *C. fragilis* (tetraploid). It is intermediate between its parent species in several respects, and it more strongly resembles *C. fragilis*, the contributor of 2/3 of its chromosomes.

Sterile pentaploid hybrids between *C. laurentiana* and *C. fragilis* have been discovered where the 2 species grow together.

Cystopteris laurentiana **is distinguished from** *C. bulbifera* **by being smaller; by having only sparse glandular hairs on the blades; by having bulblets that are uncommon, absent, or distorted; and by having stipes that are not red on young fronds.**

Cystopteris protrusa (Weatherby) Blasdell

Cystopteris fragilis (Linnaeus) Bernhardi var. *protrusa* Weatherby

Southern Bladder Fern, Lowland Brittle Fern

Etymology. Latin *protrusus*, pushed out, exerted, referring to the extension of the rhizome beyond this year's set of fronds. The vernacular name reflects the Latin name and refers to the extension of the rhizome.

Plants. Deciduous, seasonally dimorphic (early fronds sterile and smaller than later-emerging fertile fronds).

Fronds. 15–45 × 5–10 cm, closely spaced along creeping rhizome in dense clusters, narrowly lanceolate to oblong; **emerging 1 to several cm behind rhizome tips.** There is an early spring flush, with many dying back in midsummer, and a second flush in late summer; early fronds smaller and sterile, margins with rounded teeth; later fronds larger and fertile, more finely divided, margins with sharply pointed teeth.

Rhizomes. Long-creeping, **protruding 1–5 cm beyond persistent old stipe bases,** 3–4 mm diam., **hairy with tan to golden hairs**, covered with old persistent stipe bases, roots thick, fibrous; scales tan to light brown, 5–6 mm long, ovate-lanceolate to lanceolate.

Stipes. Usually shorter than or sometimes nearly equaling blade

Cystopteris protrusa: Southern bladder fern: A. Cluster of fronds with rhizome. B. Single frond. C. Rhizome.

Informative arrows: 1. Rhizome protruding 1–5 cm beyond last frond (covered with abundant golden hairs). 2. Pinnules closest to rachises on lowest pinnae wedge-shaped on short stalks.

length, grooved, dark at base, straw-colored to green above, sparsely scaly at base, scales lanceolate.

Blades. 1-pinnate-pinnatifid to 2-pinnate, ovate-lanceolate, widest at or just below middle, tips broadly acute; rachis and costae lacking glandular hairs and bulblets, glabrous.

Pinnae. 6–12 pairs, mostly alternate, **at right angles to the rachis, not curving toward blade tips,** lower pinnae closely spaced, lowest pinna pair lanceolate, widest just below the middle, distal pinnae deltate to ovate, margins dentate to serrate.

Pinnules. Basiscopic pinnules not enlarged, basal basiscopic pinnules stalked, base truncate to obtuse, pinnules closest to rachises on pinnae cuneate on short, circa 0.5–1mm stalks, marginal teeth pointed, spreading, or teeth pointing outward.

Veins. Free, simple or forked, usually ending in teeth.

Sori. Round, in 1 row between midrib and margin; indusia ovate to cup-shaped, forming a hood over the sorus, shriveling with maturity, glandular hairs lacking.

Habitat. Soil of mesic or moist deciduous forests, rich soils of slopes, ravine bottoms, and stream banks, seldom on rocks. Can form substantial colonies.

Distribution. Canada: Southern Ontario. USA: Central to eastern United States from New York to Minnesota, south to Oklahoma, Alabama, and Georgia.

Chromosome #: $2n = 84$

Cystopteris protusa **in most respects is quite similar to** *C. fragilis*, **but its rhizomes are covered with golden hairs and protrude well beyond the cluster of fronds. The protruding rhizome can usually be detected by probing the soil in front of the fern cluster with fingers. (Neither of these characters is of much use when examining only a few picked fronds or dried specimens consisting only of fronds.) The pinnules closest to rachises on the lowest pinnae are cuneate on short (ca. 0.5–1 mm) stalks.**

Cystopteris tenuis (Michaux) Desvaux

Nephrodium tenue Michaux; *Cystopteris fragilis* (Linnaeus) Bernhardi var. *mackayi* G. Lawson

Mackay's Brittle Fern, Brittle Bladder Fern

Etymology. Latin *tenuis*, thin, fine, slender, referring to the narrow lanceolate blades. Vernacular name, as with the Latin name, refers to the narrow blades of this species. Mackay from the

Cystopteris tenuis: Mackay's brittle fern: A. Cluster of fronds on rhizome. B. Single frond. Informative arrows: 1. Basiscopic pinnules of lowest pinnae wedge-shaped. 2. Blade widest just below middle of blades. 3. Pinnae tips often curving toward tips of blade.

variety *mackayi*—this fern was once thought to be a variety of *Cystopteris fragilis*.

Plants. Deciduous.

Fronds. Clustered at rhizome tips, 15–40 × 2.5–7.5 cm, erect, arching, or prostrate.

Rhizomes. Short-creeping, covered with old stipe bases, hairs lacking, scales lanceolate, tan to light brown.

Stipes. Shorter than or same length as blade, dark brown or black at base, green to straw-colored distally, scales sparse at base.

Blades. 1-pinnate-pinnatifid to 2-pinnate-pinnatifid, lanceolate to narrowly elliptic, **widest at or just below middle,** tapering to narrow tips (short-attenuate); bulblets lacking; rachises and costae lacking glandular hairs; almost all blades fertile.

Pinnae. 12 ±, **typically at less than 90° angle to rachis curving toward blade tips**, ± opposite, nearly all with sori.

Pinnules. Lobed with rounded margins, crenulate or with rounded teeth, basiscopic pinnules of lowest pinnae nearest rachis narrowly triangular, sessile or nearly so (resembling very small white oak leaves).

Veins. Free, simple or forked, ending in teeth and some in sinuses.

Sori. Round, medial; indusia ovate to cup-shaped, glandular hairs absent, shriveling and disappearing with maturity.

Habitat. Shaded rock and cliff faces and forest floors.

Distribution. Canada: From Ontario to Labrador. USA: Minnesota to the Atlantic Ocean, south to Oklahoma, Arkansas, Louisiana, Mississippi, Tennessee, and South Carolina, also disjunct in Nevada, Utah, and Arizona.

Chromosome #: $2n = 168$

Cystopteris tenuis is similar in appearance to *C. fragilis* and was once considered to be a variety of that species. It is probably an allotetraploid with *C. protusa* and a hypothetical extinct diploid related to *C. fragilis* as parents. In the middle part of its range *C. tenuis* is easily distinguished from *C. fragilis* by its narrow elliptical pinnae angled toward the blade tip and its rounded teeth. In Michigan they may be difficult to distinguish.

Cystopteris tenuis **is distinguished from** *C. bulbifera* **and** *C. laurentiana* **by the absence of glandular hairs and bulblets and by fronds that are widest at or just below the middle of the blades. It is most similar to** *C. fragilis* **and** *C. protrusa* **but may be distinguished from them by pinnae angled toward the blade tips at acute angles (less than 90°) to rachises (vs. perpendicular to rachis on** *C. fragilis* **and** *C. protusa*)**, by pinnae that often curve toward the blade tips; by its somewhat narrower fronds; and by its narrow, elliptic pinnae with rounded teeth. It differs from** *C. protrusa* **in having the fronds clustered at rhizome tips that lack hairs (vs. rhizomes with profuse golden hairs protruding beyond fronds in** *C. protrusa*)**.**

DENNSTAEDTIA Bernhardi Dennstaedtiaceae

Hay-Scented Fern

See Dennstaedtiaceae in appendix I for a description of the family and a key to the Michigan genera.

Etymology. Name honors August Wilhelm Dennstaedt (1776–1826), a German botanist.

Plants. Monomorphic, medium-sized to large, often forming colonies.

Rhizomes. Subterranean, short- to long-creeping, bearing dark reddish-brown hairs, scales absent.

Fronds. Clustered or scattered, erect to arching, ovate to lanceolate, often large, 0.4–1.5 m tall.

Stipes. Glabrous to pubescent; vascular bundles U-shaped in cross section.

Blades. 2-pinnate-pinnatifid; rachises hairy with glandular hairs; scales lacking.

Pinnules. Ovate to lanceolate, margins dentate or lobed.

Veins. Free, pinnately branched, ending in bulbous expansions short of margins.

Sori. Terminal on vein tips or marginal connecting veins, marginal or submarginal, round or elongate.

Indusia. Cup-shaped and submarginal; indusia formed by fusion of true indusium and minute blade tooth to form circular or slightly 2-valvate cups.

Chromosome #: $x = 34, 46, 47$

Dennstaedtia is a mostly tropical genus of about 70 species. Three species are treated in *Flora of North America North of Mexico*, vol. 2 (1993). It is represented in Michigan by a single species that is rare and very localized in the state.

***Dennstaedtia punctilobula* (Michaux) T. Moore**
Nephrodium punctilobulum Michaux; *Aspidium punctilobum* (Michaux) Swartz
Hay-Scented Fern

Etymology. Latin *punctum*, small spot, + *lobulus*, alluding to the very small sori on the pinnules (lobula). The vernacular name, hay-scented, refers to the fronds, which are aromatic when bruised and emit a strong hay

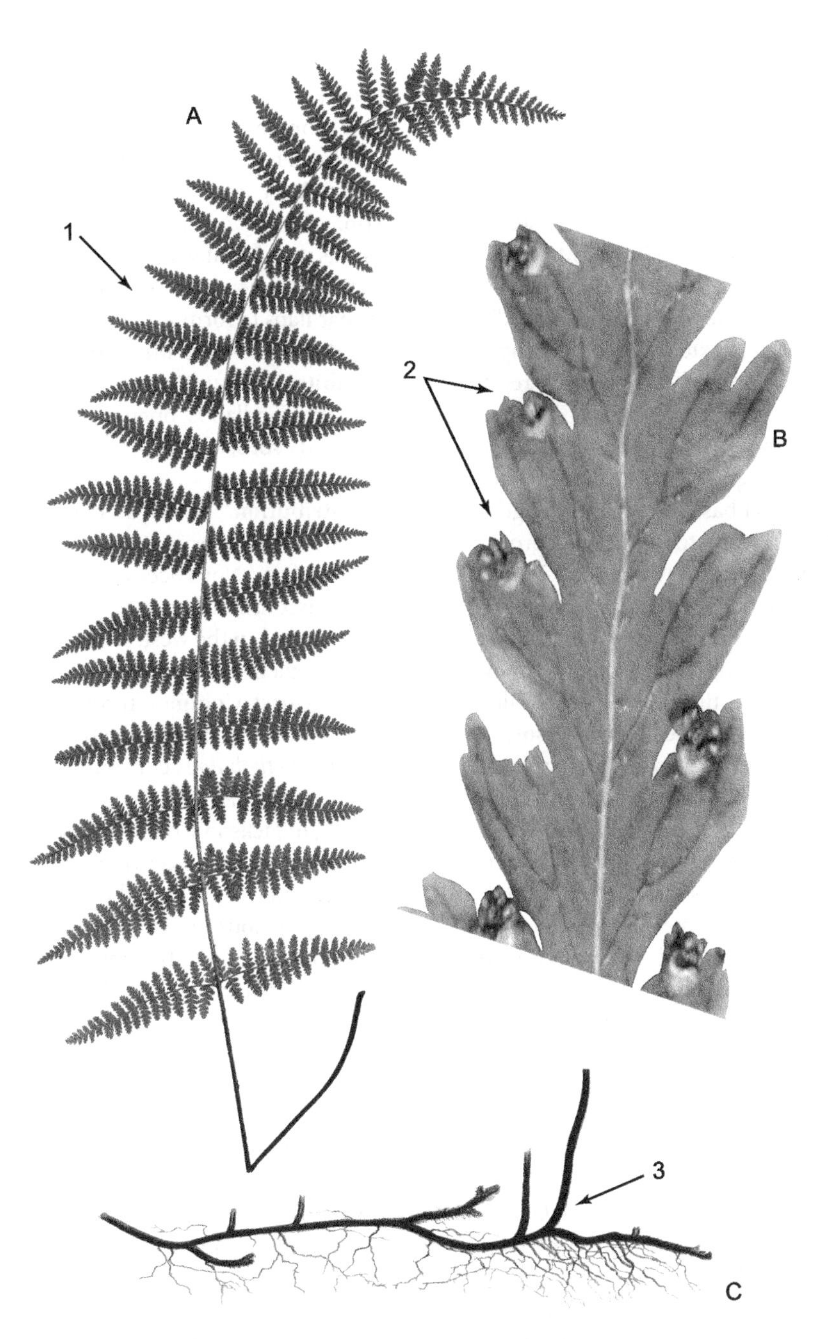

Dennstaedtia punctilobula: Hay-scented fern: A. Frond. B. Pinna section. C. Rhizome. Informative arrows: 1. Look for long narrow frond with silvery hairs on both surfaces. 2. Cup-shaped indusia. 3. Long-creeping slightly subterranean rhizomes.

or alfalfalike odor that is particularly noticeable in drying plants.

Plants. Deciduous, monomorphic, **forming colonies.**

Fronds. 30–90 × 9–25 cm, slightly narrowed at the base, not clustered, **spread out along horizontal rhizomes.**

Rhizomes. Long-creeping, 1.5–3 mm thick, extensively branching just below soil surface, hairs dark reddish-brown, jointed.

Stipes. Scattered along rhizome 1–4 cm apart, straw-colored to brown, darker at base, 1/3–1/2 the length of the blade, up to 35 cm long, grooved, pubescent with white, soft, silvery jointed hairs.

Blades. 2-pinnate-pinnatifid, 3× as long as wide, long-triangular to ovate-lanceolate, tips pointed, pale yellowish-green to pale green, papery, **soft silver-gray hairs on both surfaces;** rachises light brown to straw-colored with short spreading hairs.

Pinnae. 10–25+ pairs, subopposite, broadest at the base, tapering to tips, lowest pair somewhat twisted out of plane of blade, silvery soft hairs on upper and lower surfaces.

Pinnules. 6–20 pairs, ovate to lanceolate, nearly sessile, 2–3 oblique teeth on each side, margins deeply lobed, sinuses extending almost to the costae, serrate-crenate, upper and lower pinnules nearly same size.

Veins. Free, pinnately branched.

Sori. Very small, <0.5 mm in diam., globose to almost cylindrical, in sinus margins at vein tips; **indusia** formed by fusion of inner and outer laminar flaps to form a **circular, small, delicate, whitish cup under the sorus.**

Habitat. Full sun to partial shade, open dry woodlands, open meadows with sandy soils, and rocky slopes with acidic soils.

Distribution. Eastern and north-eastern North America. Canada: Newfoundland and Nova Scotia to Ontario. USA: Wisconsin, Illinois, Missiouri, and Arkansas in the west, Alabama and Georgia in the south, east to the Atlantic Coast. Common throughout much of its range. In Michigan *Dennstaedtia punctilobula* is currently known from a single colony that was found growing in Jackson County in 2006. It is the only known occurrence in the state and is far removed from any other populations. Previously noted collections are of questionable venue. It is considered a Michigan threatened species.

Chromosome #: $2n = 68$

Dennstaedtia punctilobula is very rare in Michigan, but it is certainly globally secure.

Dennstaedtia punctilobula **may be distinguished by fronds that are yellowish-green or pale green and covered with plentiful silver-gray hairs. Its sori are tiny, marginal, and round with cup-shaped indusia under the sori, and its fronds are spread out, not clustered.**

D. punctilobula **may be confused with ferns that have fronds somewhat similar in both shape and size, such as** *Athyrium filix-femina*, *Dryopteris intermedia*, *D. carthusiana*, **and** *D. marginalis*.

A. filix-femina **has dark brown scales instead of hairs at stipe bases. It grows**

in clumps and has linear to J-shaped sori with a flaplike indusium. *D. intermedia*, *D. carthusiana*, **and** *D. marginalis* **grow in rosettes, forming clusters, and have round sori covered with round indusia. In contrast** *D. punctilobula* **forms spreading colonies and has cup-shaped sori.**

DEPARIA Hooker & Greville Athyriaceae

See Athyriaceae in appendix I for a discussion of the family and a key to Michigan genera.

Etymology. Greek *depas*, cup or bowl, alluding to the marginal cup-shaped sori of the first described species in the genus from Hawaii (*Deparia prolifera*). The sorus of that species is actually very atypical in the genus.

Plants. Deciduous; monomorphic, small to medium-sized, occasionally large.

Fronds. 1-pinnate-pinnatifid.

Rhizomes. Erect to moderately long-creeping.

Stipes. 1/3–2/3 length of blade, grooved, **with multicellular hairs, scales and hairlike scales; cross sections at bases reveal 2 ribbon-shaped vascular bundles appearing as slightly linear, curved parallel lines.**

Blades. 1-pinnate-deeply-pinntifid, triangular to lanceolate, papery to herbaceous; *rachises* **adaxially grooved, the grooves not connecting to those of pinna costae**, hairy or scaly, as are stipes and costae.

Pinnae and pinnules. Multicellular septate hairs and hairlike scales at bases.

Veins. Free, forked or pinnate, not reaching margins.

Sori. Linear along veins, sometimes paired along both sides of a vein.

Indusia. Linear (cup-shaped only in 1 Hawaiian species if sori extend beyond margins).

Habitat. Rich, moist, well-drained, semishaded woods, often on slopes.

Chromosome #: $x = 40$

Deparia is a genus of about 50 species widespread in tropical and warm-temperate parts of the Old World. Two species are treated in *Flora of North America North of Mexico*, vol. 2 (1993). One of these (*Deparia petersenii*) is an alien escape in a few southern states. One species is found in Michigan.

Deparia **is distinguished from** *Homalosorus* **and** *Athyrium* **(all previously included in the genus** *Athyrium*) **by the presence of multicellular hairs along the costae and by grooves on the costae that do not connect to grooves on the rachises.**

Deparia acrostichoides (Swartz) M. Kato

Asplenium acrostichoides Swartz; *Athyrium acrostichoides* (Swartz) Diels; *Diplazium acrostichoides* (Swartz) Butters

Silvery Spleenwort, Silvery Glade Fern

Etymology. The name *acrostichoides* derives from the tropical genus *Acrostichum* + *-oides*, a Greek suffix that indicates resemblance. The mature sori of *Diplazium acrostichoides* appear to cover the entire undersurface of the pinnae like those of the fertile fronds of species of *Acrostichum*. The common name, silvery, reflects either the color of the young sori or the sheen caused by this fern's light-colored hairs, a character conspicuous on the undersurface of some plants but less pronounced in others. Spleenwort reflects the fact that this fern was originally recognized as an *Asplenium*—the spleenwort genus.

Plants. Deciduous, somewhat dimorphic.

Fronds. 30–100 × 13–25 cm, arising in short rows from rhizomes; fertile fronds appearing later in summer are more erect, taller, and narrower than the sterile ones.

Rhizomes. 5–7 mm diam., short-creeping, decumbent to erect.

Stipes. 20–40 cm long, usually much shorter than blade, dark red-brown at swollen bases, straw-colored distally; scales at base light brown, linear-lanceolate to lanceolate, white hairs on younger stipes.

Blades. 30–100 × 10–25 cm, **1-pinnate-pinnatifid,** oblong-lanceolate, narrowed to base, broadest near middle, tapering to narrow tips, hairs and narrow scales present on younger fronds; fertile fronds taller and narrower than sterile fronds; rachises pale green.

Pinnae. 0.7–2 × circa 0.4 cm; narrow, 10–20 pairs, opposite and alternate, linear-lanceolate, bases cut sharply, narrowing to long-pointed tips; **midribs with multicellular hairs.**

Veins. Pinnate, lateral veins unforked or occasionally 1-forked, reaching margins.

Ultimate segments. Oblong, tips round to slightly pointed, margins entire to slightly lobed.

Sori. 3–7 per segment, **linear, straight or slightly curved, in a herringbone pattern,** silvery initially, blue-gray when older; indusia linear, silvery when young, light brown with age.

Habitat. Rich, moist, well-drained, semishaded woods, often on slopes.

Distribution. Canada: Ontario to the Atlantic Ocean. USA: Minnesota south to Arkansas, east to Georgia and the Atlantic Ocean.

Chromosome #: $2n = 80$

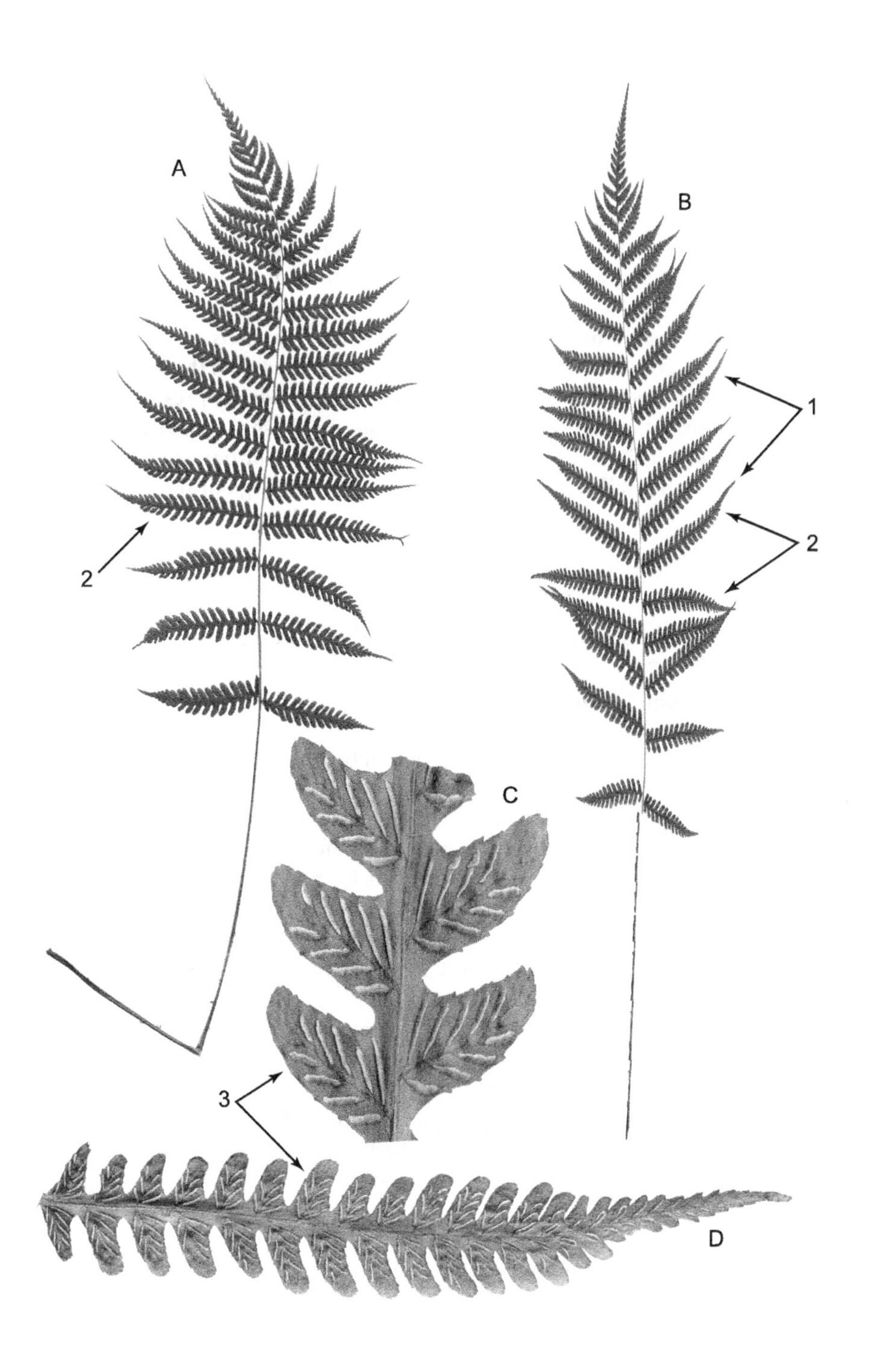

Deparia acrostichoides: Silvery glade fern: A. Sterile frond. B. Fertile frond. C. Pinnule enlargement. D. Fertile pinna.

Informative arrows: 1. Narrower fertile pinnae. 2. Look for often shiny hairs on midribs. 3. Linear sori in "herringbone" pattern.

Some plants of *D. acrostichoides* are notably finely hairy and stand out clearly, especially early in the summer.

Deparia acrostichoides **may be distinguished from** *Athyrium filix-femina* **by the presence of multicellular hairs along the costae, and by grooves on the costae that do not connect with grooves on the rachises. The sori of** *D. acrostichoides* **are usually longer and straighter than those of** *A. filix-femina* **that tend to be J-shaped.**

D. acrostichoides **may be distinguished from** *Dryopteris carthusiana* **by its linear sori (vs. round in** *Dryopteris***). It may be confused with the dimorphic** *Osmunda cinnamomea* **or** *O. claytoniana* **if no fertile fronds are present, but it can be distinguished by hairs that give the fronds a pale sheen and by prominent linear, silvery sori.** *O. cinnamomea* **has small clumps of light tan hairs at the base of its pinna stalks, and** *O. claytoniana* **is essentially glabrous.**

DRYOPTERIS Adanson Dryopteridaceae

Woodferns, Shield Ferns, Buckler Ferns

See Dryopteridaceae in appendix for a description of the family and a key to the Michigan genera.

Etymology. Greek *dryos*, oak or tree, + *pteris*, fern; the ferns of this genus often grow in woodlands. The vernacular name woodfern derives from the scientific name and its preferred habitat. The names shield and buckler (a small shield) imply a resemblance to those objects.

Plants. Monomorphic or dimorphic, evergreen or deciduous, rarely on rock.

Rhizomes. Short-creeping to erect, stout, scaly.

Fronds. Usually close and forming rosettes.

Stipes. Usually 1/4–1/3 frond length, grooved, **partially or heavily clothed with brown scales at base; cross section revealing 3 or more round vascular bundles arranged in a semicircle.**

Blades. 1-pinnate-pinnatifid to 3-pinnate-pinnatifid, deltate-ovate to lanceolate, gradually tapering to pinnatifid tips, herbaceous to somewhat leathery, glandular or not, bearing nonclathrate scales and hairlike scales but not needlelike (acidular) hairs; rachises and midribs of pinnae and pinnules grooved, grooves usually continuous from rachis to costae.

Pinnae. Segment margins entire, scalloped, serrate, or spinulose, sessile to stalked; costae adaxially grooved, grooves continuous from rachises to costae to costules.

Pinnules. (Basiscopic) pinnules often longer than upward-facing ones (acroscopic), sessile or stalked.

Veins. Free, forked.

Sori. Kidney-shaped to round, medial, submedial to very near margins, in

1 row on each side of pinnule midrib, sometimes on upper parts of fronds only; indusia kidney-shaped to round, attached at narrow sinuses, white to transparent when immature, persistent or deciduous.

Habitat. Found in diverse habitats, including woodlands, rocky slopes, and wetlands.

Distribution. Worldwide, species most numerous in temperate Asia.

Chromosome #: $x = 41$

Dryopteris is genus of about 240 species worldwide with 14 included in the *Flora of North America North of Mexico*, vol. 2 (1993). Ten species and around 18 sterile hybrids (2 of which are common enough to be treated here) are found in Michigan.

Dryopteris is a very well studied genus of North American ferns. The experimental study of its hybrids has clarified the interrelationships and origin of its species. See the material under *Dryopteris* hybrids following this generic treatment.

The Michigan *Dryopteris* species can be clearly divided into three groups.

1. A 2-pinnate (sometimes 1-pinnate-pinnatisect) group (*D. celsa*, *D. clintoniana*, *D. cristata*, *D. filix-mas* subsp. *brittonii*, *D. goldiana*, and *D. marginalis*).
2. A 2-pinnate-pinnatifid to 3-pinnate group (*D. carthusiana*, *D. intermedia*, and *D. expansa*).
3. *Dryopteris fragrans*, a very distinctive small, very aromatic-glandular species of cliff faces and talus slopes in the far north.

KEY TO THE SPECIES OF *DRYOPTERIS* IN MICHIGAN

Species identifications in the field is complicated by the frequent presence of sterile hybrids, a few of which are quite common. These hybrids may be differentiated from fertile species by their intermediate morphology and distorted spores. They are not included in the key, but their parentage may be inferred by observing combinations of characters in the key (e.g., glands from *Dryopteris intermedia*, nearly marginal sori from *D. marginalis*, narrow blades from *D. cristata*, and large scales with a dark central stripe from *D. goldiana*).

1. Blade undersurfaces densely scaly; aromatic glands plentiful; old fronds forming conspicuous gray or brown clumps; fronds mostly less than 25 cm long; found on granitic cliffs or talus slopes (western half of the Upper Peninsula only) . . . . *D. fragrans*

1. Blade undersurfaces glabrous to sparsely scaly; glandular or not, glands if present not aromatic; old fronds not persisting in conspicuous gray or brown clumps; fronds mostly more than 25 cm long; found in diverse habitats, not normally on granitic cliffs or talus slopes (not limited to western half of the Upper Peninsula) (2).

2(1). Blades 2-pinnate-pinnatisect to 3-pinnate-pinnatifid at base (3).

2. Blades 2-pinnate to 2-pinnate-pinnatifid (sometimes 1-pinnate-pinnatisect) at lowermost pinnae (5).

3(2). Downward-pointing basal pinnules on lowest pinnae closest to the rachis usually shorter or same size as adjacent downward-pointing pinnules; rachises and midribs of pinnae and indusia with tiny glandular hairs; blades evergreen . . . . *D. intermedia*

3. Downward-pointing pinnules of basal pinnae on lowest pinnae closest to the rachises longer than adjacent pinnules; glandular hairs lacking or sparse; blades die back in winter or evergreen (4).

4(3). Blades ovate-lanceolate; downward-pointing pinnules nearest rachis on lowest pinnae up to twice as long and not much wider than the upward-pointing pinnule nearest rachis; upward- and downward-pointing pinnules on the lowest pinnae closest to the rachis are nearly opposite; stipe scales tan (Upper and Lower Peninsulas) . . . . *D. carthusiana*

4. Blades broadly triangular; downward-pointing pinnules nearest rachis on lowest pinnae 3–5× the length of the upward-pointing pinnules closest to rachises and as wide as the space occupied by the first 2 upward-pointing pinnules closest to rachis; downward-pointing pinnules on the lowest pinnae closest to the rachis are noticeably farther from the rachis than the upward-pointing ones; stipe scales tan, with dark central stripe (northwestern Upper Peninsula and Isle Royale only) . . . . *D. expansa*

5(2). Sori at or near margins of pinnules; fronds with dull bluish-green upper surface and lighter grayish-green undersurface; stipe scales at base densely clustered, pale yellowish-brown . . . . *D. marginalis*

5. Sori medial or closer to midribs of segments; fronds green to dull green; stipe scales sparse or with scattered tan to dark brown scales at base (6).

6(5). Fronds widest at the middle, clearly narrowed at base; stipe scales of 2 types: broad scales mixed with hairlike scales and stipes mostly less than 1/4 length of fronds. Found mostly in rocky forests (Upper Peninsula and Alpena County) . . . . *D. filix-mas* subsp. *brittonii*

6. Fronds not or only somewhat narrowed at base; stipes with 1 scale type, broad to narrow, hairlike scales lacking; stipes 1/4–1/3 length of fronds; mostly in rich, moist forests and swamps (Upper and Lower Peninsulas) (7).

7(6). Blades monomorphic; larger blades 11–22 (–25) cm wide; basal pinnae ovate; blades ovate to ovate-lanceolate; scales at base of stipes dark brown or tan with dark brown stripe (8).

7. Blades dimorphic or somewhat dimorphic; larger blades 6–12.5 cm wide; basal pinnae triangular; fertile blades lanceolate with parallel sides; scales at base of stipes tan, mostly without dark brown stripe (9).

8(7). Blades ovate, tapering abruptly at tips; sori nearer midvein than margins of segments; found in damp beech-maple-basswood woods, on wet springy slopes, and on margins of or hummocks in swamps (uncommon, scattered throughout Michigan, rare in Upper Peninsula) . . . . *D. goldiana*

8. Blades ovate-lanceolate, gradually narrowed to tips; sori halfway between midveins and margins; found on logs and hummocks in acidic swamps (southwestern Lower Peninsula only, rare in Michigan) .... *D. celsa*

9(7). Basal pinnae narrowly elongate-deltate; pinnae of fertile fronds nearly in plane of blade; fertile fronds not or only slightly different from sterile ones .... *D. clintoniana*

9. Basal pinnae deltate; pinnae of fertile fronds twisted nearly at right angles to plane of blade (like a venetian blind); fertile fronds distinctly different from sterile ones .... *D. cristata*

Dryopteris carthusiana (Villars) H. P. Fuchs

Polypodium carthusianum Villars

Spinulose Woodfern

Etymology. *Dryopteris carthusiana* was named (as *Polypodium carthusianum*) for the village of Grand Chartreuse near Grenoble in France, where Dominique Villars supposedly collected it. An alternative theory is that it is named after Johan Friedrich Cartheuser (1704–77), a professor of medicine, botany, and pharmacology in Frankfurt, Germany. The specific epithet *Dryopteris spinulosa* was widely used, and is still used occasionally, but in 1980 Christopher Fraser-Jenkins established that *D. carthusiana* was the older name and therefore the correct one. The common name alludes to the species' spiny-toothed pinnules.

Plants. Deciduous, monomorphic, common.

Fronds. 20–80 × 12–30 cm, lacy, growing in clusters.

Rhizomes. Short-creeping to ascending; roots black, wiry.

Stipes. 1/4–1/2 length of frond; sparsely scaly at base, scales light brown, concolorous or with dark center.

Blades. 2-pinnate-pinnatifid to 3-pinnate, ovate-lanceolate, **not glandular**; light green to yellowish-green; rachises straw-colored, glands absent.

Pinnae. Narrowly triangular, lowest pinnae more broadly triangular and slightly reduced in size.

Pinnules. 13–18, downward-pointing pinnules of the basal pinnae closest to the rachis are noticeably longer than the adjacent downward-pointing pinnules and up to twice as long as the upward-pointing pinnules closest to the rachis; leaf margins serrate with spiny teeth.

Veins. Free, ending in teeth.

Sori. Round, medial; indusia kidney-shaped, **lacking glands.**

Habitat. Swamps, marshes, wet meadows, and bog margins, occasionally in rich moist forests,

Dryopteris carthusiana: **Spinulose woodfern: A. Pinnule. B. Frond. C. Basal pinna.** Informative arrows: 1. Spiny marginal teeth. 2. Small glandular hairs on rachis and costae absent (compare to *D. intermedia*). 3. Basiscopic pinnule on basal pinna closest to the rachis longer than adjacent pinnule.

commonly associated with *Athyrium filix-femina*.

Distribution. Circumboreal. Eurasia. Canada and USA: Alaska to Newfoundland, south to Washington, Idaho, and Montana, Upper Midwest south to Arkansas and east to Maryland.

Chromosome #: $2n = 164$ allotetraploid

Dryopteris carthusiana is a fertile tetraploid hybrid with *D. intermedia* ($2n = 82$) as 1 parent and a hypothetical, and presumably extinct, "*D. semicristata*" (presumably $2n = 82$) as the other.

Dryopteris ×*triploidea*, a sterile triploid hybrid between *D. carthusiana* and *D. intermedia*, frequently occurs when both parental species are present. This hybrid is often difficult to identify with certainty by means of examination in the field. It has the glandular hairs found on *D. intermedia*. Microscopic examination of its spores shows deformed, infertile spores. (See the description under *Dryopteris* hybrids below.)

Dryopteris carthusiana **is similar in appearance to** *D. intermedia* **but may be distinguished by examining pinnules on the lower pinnae. In** *D. carthusiana* **the basiscopic pinnules on the lowest pinna closest to the rachis are longer than the next set farther out. In contrast, in** *D. intermedia* **on the lower pinnae the basiscopic second pinnule out from the rachis is usually longer than the one closest to the rachis. Some fronds are difficult to place using this character, and it is best to look at several fronds if the identification is not clear.**

D. carthusiana **lacks glandular hairs, while** *D. intermedia* **has them on the indusia, rachises, and costae. These hairs are easy to see on fresh fronds, but they tend to wear off as the fronds age and often require use of a hand lens. (A little experience makes the glands relatively easy to find.)**

Fronds of *D. intermedia* **are evergreen, and those of** *D. carthusiana* **are not, a character helpful in the spring and winter.**

Dryopteris celsa (W. Palmer) Knowlton

Dryopteris goldiana (Hooker) A. Gray subsp. *celsa* W. Palmer

Log Fern

Etymology. Latin *celsus*, high, lofty, referring to the plant's habit of growing perched on logs or hummocks above the water level in swamps. The vernacular name derives from the plant's tendency to grow on logs in swamps.

Plants. Monomorphic; deciduous.

Fronds. 65–120 × 15–30 cm.

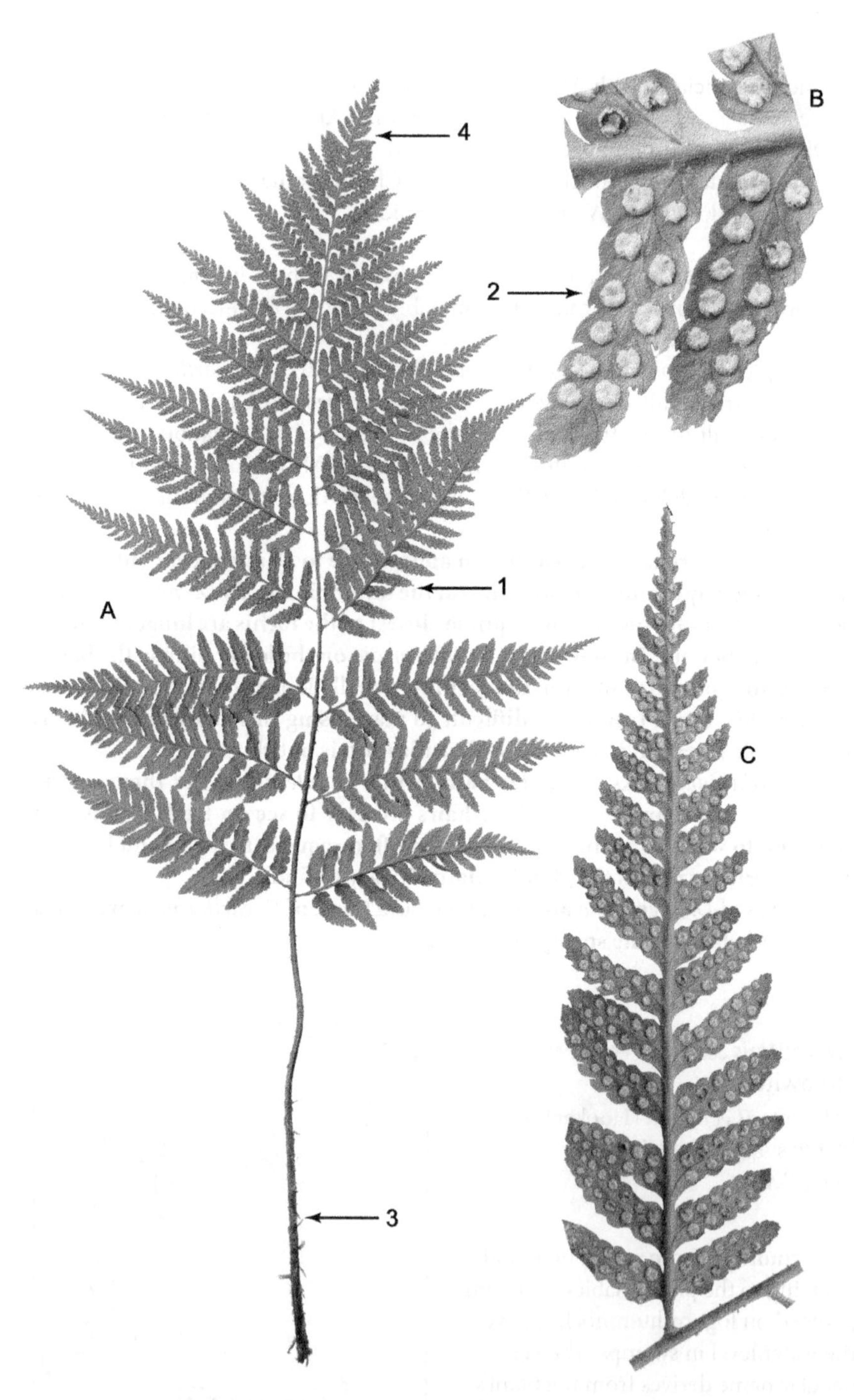

Dryopteris celsa: Log fern: A. Frond. B. Fertile pinnules. C. Fertile pinna.
Informative arrows: 1. Blade is 2-pinnate. 2. Sori medial with peltate indusia. 3. Dark stipe scales. 4. Gradually tapering tip (compare with *D. goldiana*).

Rhizomes. Short-creeping.

Stipes. 1/3 to 1/2 length of blades, grooved, scaly at least at base, scales scattered, dark brown or tan with dark center.

Blades. 2-pinnate to 2-pinnate-pinnatifid, dark green, ovate-lanceolate, **tapering gradually to tips,** glands absent, linear to ovate scales on undersurface, absent on upper surface.

Pinnae. 15–20 pairs, ovate-lanceolate, gradually tapering to long, slender pinna tips, basal pinnae often much smaller, ovate with the first segments shorter than the adjacent ones; midribs grooved, lobe margins crenate.

Pinnules. Basiscopic pinnules and acroscopic pinnules of equal length; margins crenately toothed.

Sori. Round, **medial;** indusia reniform.

Habitat. Logs and hummocks in acidic swamps.

Distribution. USA: Mostly in the Piedmont and Coastal Plain, extending west to Arkansas, Missouri, and Louisiana. Rare and local in Michigan—small disjunct populations are found only in 4 southwestern counties (Berrien, Cass, Kalamazoo, and Van Buren) of the Lower Peninsula. These disjunct populations are around 400 miles from the nearest known site of this species.

Chromosome #: $2n = 164$ allotetraploid

Dryopteris celsa is a Michigan threatened species. It is considered globally secure.

Dryopteris celsa is a fertile allotetraploid hybrid species resulting from hybridization between *D. goldiana* ($2n = 82$, diploid) and *D. ludoviciana* ($2n = 82$, a diploid not found in Michigan). It hybridizes with 5 other *Dryopteris* species that are found in Michigan.

Dryopteris celsa **appears somewhat similar to** *D. goldiana*, **but its blades are narrower, gradually tapering to the tips (vs. abruptly tapered on** *D. goldiana*), **and its sori are medial (vs. nearer the midvein on** *D. goldiana*). **Both** *D. celsa* **and** *D. goldiana* **have stipe scales with dark centers.**

Dryopteris celsa **differs from** *D. clintoniana* **in its more ovate blades (vs. lanceolate in** *D. clintoniana*) **and its dark stipe scales (vs. tan on** *D. clintoniana*).

Its basal pinnae are more ovate-lanceolate (rather than triangular), and its rachises are slightly scaly (rather than glabrous or nearly so).

Dryopteris clintoniana (D. C. Eaton) Dowell

Aspidium cristatum (Linnaeus) Swartz var. *clintonianum* D. C. Eaton; *Dryopteris cristata* (Linnaeus) A. Gray var. *clintoniana* (D. C. Eaton) Underwood

Clinton's Woodfern

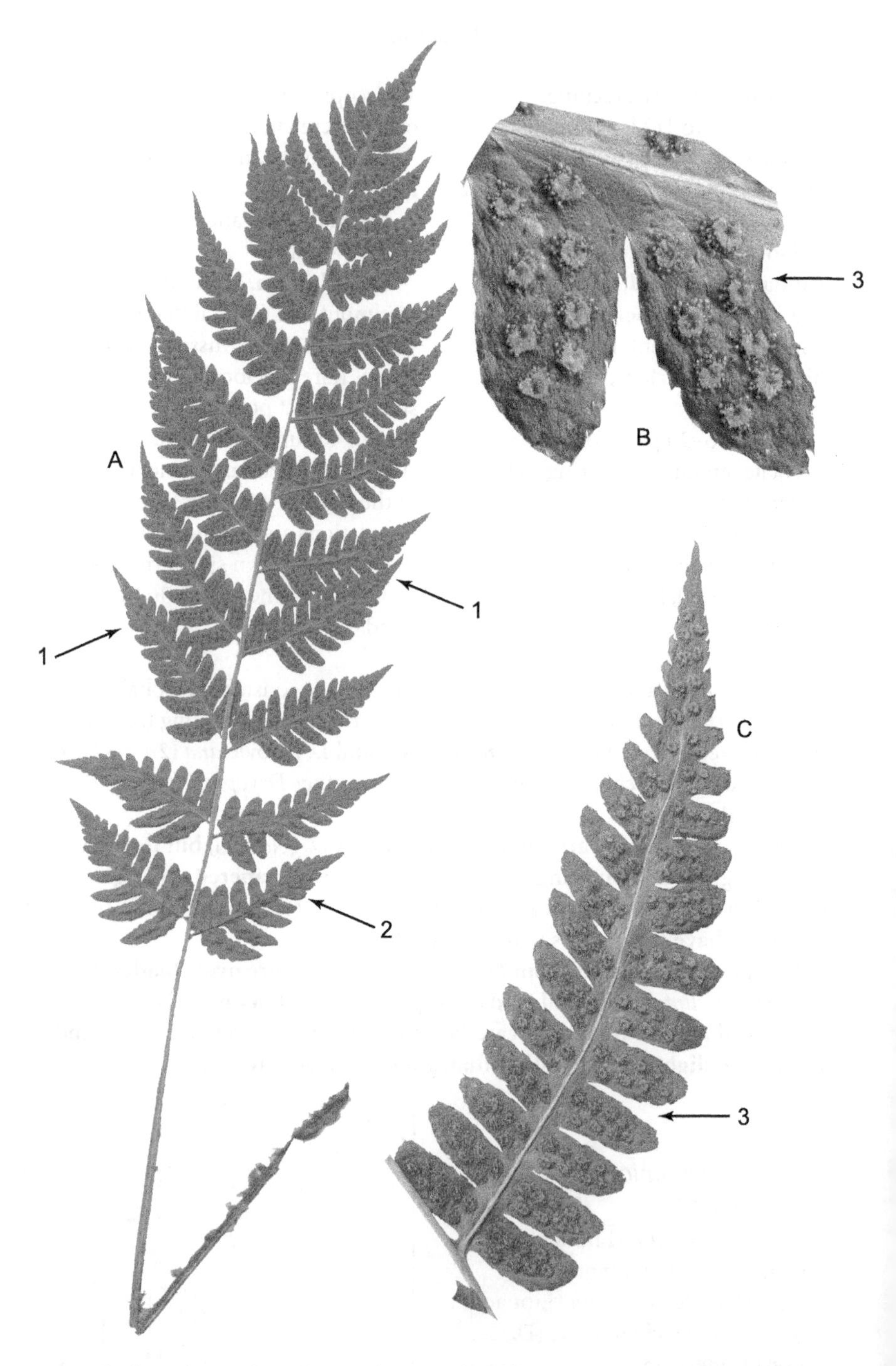

Dryopteris clintoniana: Clinton's woodfern: A. Sterile frond. B. Fertile pinnules. C. Fertile pinna.

Informative arrows: 1. Narrow parallel-sided frond. 2. Narrow-triangular basal pinnae. 3. Sori near midrib with peltate indusia.

Etymology. Name honors Judge G. W. Clinton (1807–85), who botanized near Buffalo, New York. He served as mayor of Buffalo, was first president of the Buffalo Society of Natural history, and prepared a catalog of the plants of the Buffalo area.

Plants. Somewhat dimorphic; fertile fronds deciduous, sterile fronds evergreen; uncommon.

Fronds. 45–100 × 12–20 cm, clustered; **sterile fronds smaller, spreading; fertile fronds taller, somewhat narrower, erect, but not conspicuously different.**

Rhizomes. Stout, short-creeping to ascending, covered with light brown scales, sometimes with dark brown centers.

Stipes. 1/4–1/3 length of leaf, scales scattered, mostly at base, tan, sometimes darker brown, usually lacking a central dark stipe.

Blades. 1-pinnate-pinnatifid, lanceolate, **with nearly parallel sides,** green, herbaceous, glands absent; rachises green, sparsely scaly.

Pinnae. 10–16 pairs, long-triangular, lowermost pinnae triangular; **fertile pinnae minimally twisted out of plane of blade (a venetian blind effect not prominent).**

Pinnules. Inner pinnules longer than or equal to adjacent pinnules, lower and upper pinnules of equal length, margins serrate with spiny teeth.

Sori. Medial; indusia round, glands lacking.

Habitat. Swamp margins or hummocks, occasionally in wet woods.

Distribution. Canada: Ontario, Quebec, New Brunswick. USA: Northeastern United States south to New Jersey, west to Michigan and Indiana.

Chromosome #: $2n = 246$ allohexaploid

Dryopteris clintoniana is a fertile allohexaploid hybrid ($2n = 246$) derived from a cross between *D. cristata* ($2n = 164$) and *D. goldiana* ($2n = 82$). It hybridizes with 6 other *Dryopteris* species. Hybrids arising from *D. clintoniana* may be recognized by their somewhat narrow blades and long-triangular lower pinnae; hybrids of *D. clintoniana* are also larger than corresponding *D. cristata* hybrids.

D. clintoniana is often associated with *D. cristata*, *D. carthusiana*, and *D. intermedia*.

Dryopteris clintoniana **somewhat resembles a large** *D. cristata;* **however, its sterile and fertile blades are not as clearly different as those of** *D. cristata*. **The venetian blind effect is not as evident as in** *D. cristata,* **and its lowest pinnae are more narrowly triangular.**

Dryopteris cristata (Linnaeus) A. Gray

Polypodium cristatum Linnaeus, *Aspidium cristatum* (Linnaeus) Swartz; *Filix cristata* Farwell; *Lastrea cristata* Presl; *Nephrodium cristatum* (Linnaeus) Michaux, *Polystichum cristatum* (Linnaeus) Roth; *Thelypteris cristata* (Linnaeus) Nieuwland

Crested Woodfern, Crested Shield Fern, Crested Buckler Fern

Etymology. Latin *crista*, tuft, comb, plume, ridge, + *-ata*, a suffix implying likeness—possibly alluding to the appearance of the tufted fronds and the tall, venetian-blind-like fertile fronds of this plant. Linnaeus never explained his choice of this specific name. Crested derives from the specific name; shield and buckler (a small shield) are names applied to some *Dryopteris* species. (This large synonymy is retained here to demonstrate the various genera the *Dryopteris* species have been placed in. It shows the changing opinions taxonomists have had dealing with this genus. Most of these generic combinations have been eliminated from the synonymies of the other *Dryopteris* species treated here.)

Plants. Dimorphic; *fertile fronds* deciduous; *sterile fronds* partially evergreen.

Rhizomes. Creeping to erect, occasionally branching, thick, dark brown, scales abundant, brown; *roots* black, wiry.

Stipes. 1/4–1/2 length of fronds; bases swollen, tan, grooved; scales sparse persistent, light brown at base.

Sterile fronds. 35–70 × 6–12 cm, **broader, more spreading and 1/2–3/4 length of fertile fronds,** forming a rosette of outer fronds that persists partially evergreen over the winter.

Fertile fronds. 25–70 × 8–15 cm; **narrow, parallel-sided, tall, rigidly erect,** deciduous, found in central area of frond rosettes.

Blades. 10–20 pinna pairs; 1-pinnate-pinnatifid to 2-pinnate, narrowly lanceolate or with parallel sides, thick, somewhat leathery, green to bluish-green, glands lacking; rachises green, slightly scaly at base, glabrous above, scales linear to ovate.

Sterile pinnae. Similar to fertile pinnae but longer and narrower.

Fertile pinnae. 10–20 pairs, alternate, lowest pair somewhat smaller, well separated, **short equilateral-triangular**, acroscopic pinnae closer, long-triangular; **all pinnae twisted 90° out of plane of blade and perpendicular to it (a venetian blind effect).**

Pinnules. Oblong, rounded with spiny teeth near tips, downward-facing pinnules and upper-facing pinnules equal in length.

Veins. Free, forked, embossed.

Sori. Medial, round, indusia kidney-shaped, glands lacking.

Habitat. Swamps, marshes, wet

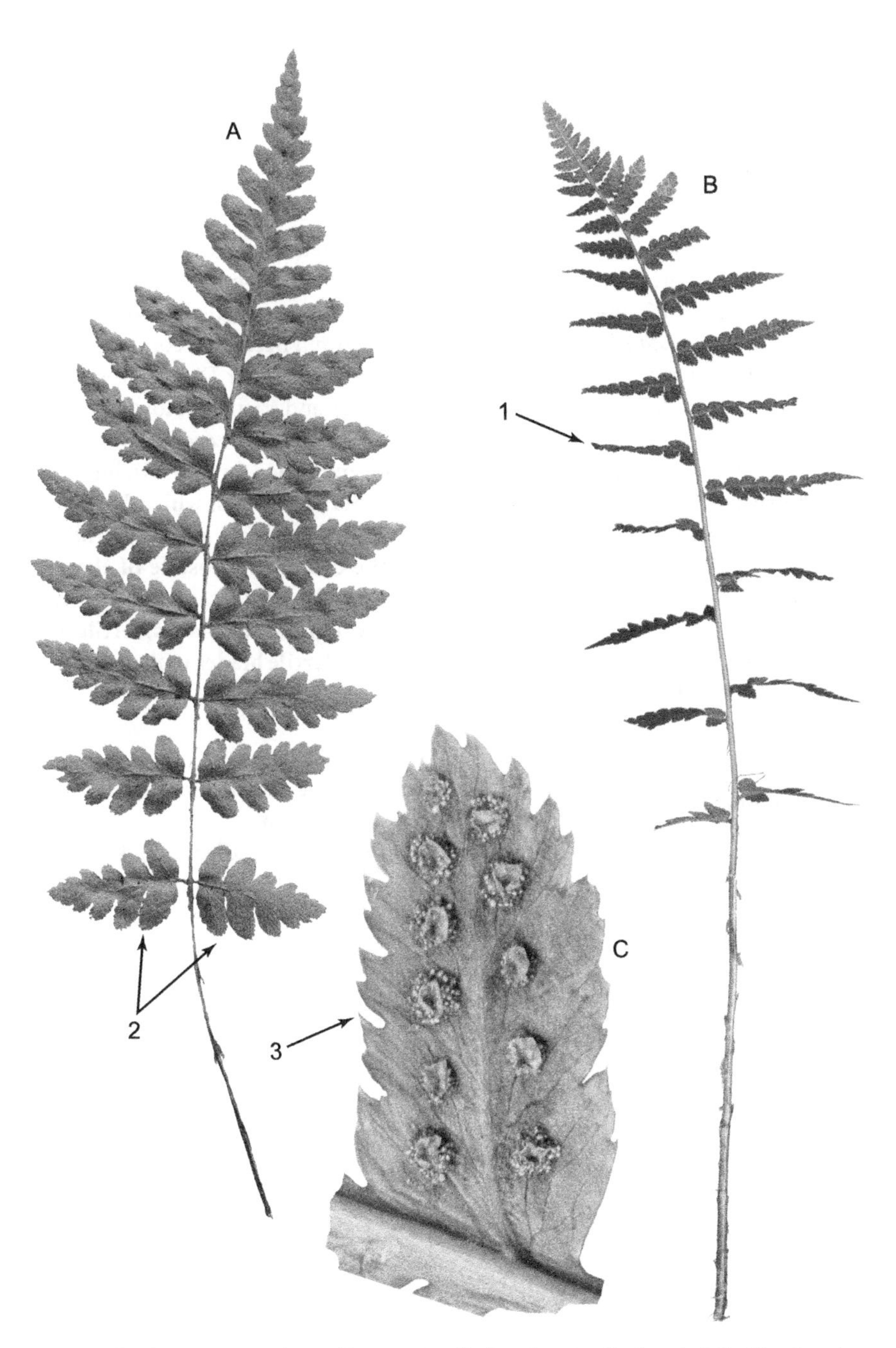

Dryopteris cristata: Crested woodfern: A. Sterile frond. B. Fertile frond. C. Fertile pinnule showing sori.

Informative arrows: 1. Longer, narrower fertile frond with "venetian blind" arrangement of pinnae. 2. Short-triangular lowest pinnae. 3. Sori near midrib.

meadows, bog margins, and moist woods.

Distribution. Circumboreal. Eurasia. Canada: British Columbia to Newfoundland. USA: Entire northern tier of states south to Nebraska, Missouri, Alabama, and Georgia.

Chromosome #: $2n = 164$ allotetraploid

Dryopteris cristata, a fertile allotetraploid hybrid (a tetraploid formed with 2 separate species as parents), is thought to have originated from a cross between *D. ludoviciana* ($2n = 82$, not present in Michigan) and a theoretical, and probably extinct, *D.* "*semicristata*" ($2n = 82$) (named *semicristata* because it constitutes half of the *D. cristata* genome). *D. cristata* hybridizes with 5 other *Dryopteris* species, producing plants identifiable by their narrow blades and triangular lower pinnae.

Dryopteris cristata often grows in association with *D. carthusiana.*

Dryopteris cristata **is clearly distinct from any other fern in Michigan. It may be recognized by its narrow, stiffly erect fertile fronds, which have pinnae tilted nearly 90° from the plane of the frond, suggesting a venetian-blind-like" arrangement. The fertile fronds sit in the center of a rosette of shorter, spreading sterile fronds. It is most easily confused with** *D. clintoniana,* **which has narrower and longer basal pinnae, wider straight-sided blades, fertile and sterile fronds that are not markedly different, and fertile blades with a minimal venetian blind effect.**

Dryopteris expansa (C. Presl) Fraser-Jenk. & Jermy

Nephrodium expansum C. Presl; *Dryopteris assimilis* S. Walker; *D. dilatata* auct. non (Hoffm.) A. Gray

Northern Woodfern, Spreading Woodfern, Broad Buckler Fern

Etymology. Latin *expansus*, expanded, spread out, in reference to its broad fronds. The common names derive from its relatively wide fronds and boreal distribution.

Plants. Monomorphic, subevergreen, dying in fall or not.

Fronds. 25–100 × (15–) 20–40 cm, **broad,** flat, clustered in rosettes, erect to slightly arching.

Rhizomes. Erect or ascending, short, stout, branching, covered with old stipe bases storing starches, densely scaly, scales broad, tan with dark central stripe, soft.

Stipes. 0.4–1.1 × length of blade, grooved, dark brown at the base and pale or green above, scaly mostly at the base, scales broad, ovate-lanceolate to lanceolate, brown with medial dark stripe (sometimes uniformly brown).

Dryopteris expansa: **Northern woodfern: A. Frond. B. Basal pinna. C. Fertile pinnule showing sori.**

Informative arrows: 1. Frond 2-pinnate-pinnatisect to 3-pinnate-pinnatifid. 2. Downward-pointing (basiscopic) pinnule of lowest pinna closest to the rachis longer than the next ones out, 3–5× the length of the upward-pointing (acroscopic) pinnules closest to rachises, and so wide as to match the width of the two short acroscopic ones immediately above it.

Blades. 2-pinnate-pinnatisect to 3-pinnate-pinnatifid, broadly triangular to ovate, herbaceous to somewhat leathery, glands absent or occasionally glandular, green; rachises scaly below, scales absent upward, scales tan, linear to ovate, glands absent; fertile blades usually have sori on all pinnae.

Pinnae. 12–25 pairs, longest 6–20 cm long, opposite, lanceolate-oblong; lowest pinna pair larger or slightly smaller than those above, triangular; scales on lower surface narrow, margins serrate, spiny.

Pinnules. Innermost pinnules of lowest pinnae distinctly alternate, the downward-pointing (basiscopic) pinnules of lowest pinnae closest to the rachises longer than the adjacent ones, 3–5× the length of the upward-pointing (acroscopic) pinnules closest to rachises, and so wide as to match the width of the 2 short acroscopic ones immediately above it; lanceolate; margins toothed, spiny.

Veins. Free, ending in marginal teeth.

Sori. Medial or submarginal, round; in 2 rows along pinnules; *indusium* kidney-shaped; lacking glands or sparsely glandular; evanescent.

Habitat. Cool moist woodlands, rocky slopes, and ravine slopes.

Distribution. Circumboreal. Eurasia. Greenland. Canada and USA: Western North America from Alaska and Yukon to California, east to Idaho; eastern North America from Newfoundland through Quebec and Ontario to Michigan, Wisconsin, and Minnesota. A large gap exists between the western and eastern North American populations, but they are also scattered in higher altitudes farther south. *Dryopteris expansa* is uncommon to rare in its range.

Chromosome #: $2n = 82$ diploid

This species was included in *Dryopteris dilatata* (a European fern), prior to 1951 when Irene Manton showed that *D. dilatata* could be divided into diploid (*D. expansa*) and hexaploid (*D. dilatata*) species separated by chromosome counts, morphology, and ecology.

A rare species in Michigan *Dryopteris expansa* **may be distinguished from** *D. carthusiana* **by examining the basiscopic (downward-pointing) pinnules closest to rachis on the lowest pinnae. In** *D. expansa* **the basiscopic pinnules in this location are 3–5× the length and 2× the width of the acroscopic (upward-pointing) pinnules and are so wide as to match the width of the 2 short acroscopic pinnules immediately above them. Additionally, the basiscopic pinnules here are attached noticeably farther from the rachis than the acroscopic pinnules. In contrast, in** *D. carthusiana* **the basiscopic pinnules in this location are usually less than twice as long as the acroscopic, and the basiscopic and acroscopic pinnules are nearly opposite each other (equidistant from the rachis).** *D. expansa* **often remains green longer in the winter than** *D. carthusiana.*

D. expansa **may be distinguished from** *D. intermedia* **by comparing the length of the basiscopic pinnules. In** *D. expansa* **the basiscopic pinnules closest to the rachis on the lowest pinnae are much longer and wider than the basiscopic pinnules lateral to them. In contrast, in** *D. intermedia* **the pinnules in this location are shorter or the same size as the pinnules lateral to them. Addi-**

tionally the blades of *D. intermedia* **are much more glandular than those of** *D. expansa*.

Dryopteris filix-mas (Linnaeus) Schott subsp. *brittonii* Fraser-Jenkins & Widen

Polypodium filix-mas Linnaeus

Male Fern

Etymology. Latin *mas*, male, + *filix*, fern—the male fern. The coarsely cut, scaly, somewhat leathery fronds contrast this fern with the more delicately dissected, thinner fronds of the lady fern (*Athyrium filix-femina*).

Plants. Deciduous, monomorphic.

Fronds. 30–120 × 10–30 cm; erect to somewhat arching, forming a rosette.

Rhizomes. Erect, stout, the apex very scaly with long, light brown scales covered with old stipe bases.

Stipes. Less than 1/4 length of blade; grooved, straw-colored, **scaly mostly at base with both brown and distinct hairlike scales.**

Blades. 1-pinnate-pinnatisect to 2-pinnate at base, ovate-lanceolate, widest at the middle, **clearly narrowed at base,** the lower pinnae are less than 1/2 the length of the longest pinnae; somewhat leathery, not glandular; midgreen; rachises green, scales thin out toward tips, glands absent.

Pinnae. 16–30 alternate pairs; narrow, lanceolate, straight to slightly curved upward, basal pinnae much smaller; midribs grooved.

Pinnules. Acroscopic pinnules nearly same size as basiscopic ones, margins slightly toothed.

Sori. Medial; round, prominent; mostly found on the upper half of the blade; *indusium* reniform, lacking glands.

Habitat. Limestone or calcareous cliffs, sinkholes, ravines, cracks and crevices in limestone pavements, mesic northern woodlands.

Distribution. Canada and USA: Newfoundland to Ontario, south to Wisconsin, Michigan, Maine, Pennsylvania, and Vermont. Further investigations may widen this distribution. Most abundant in the western half of the Upper Peninsula where it is uncommon.

Chromosome #: $2n = 164$ tetraploid

Dryopteris filix-mas subsp. *brittonii* is a Michigan plant of special concern; its status is uncertain.

The *Dryopteris filix-mas* subsp. *filix-mas* has a very wide distribution in Eurasia, Africa, and North America. The subspecies *brittonii*, found in the northeastern United States and eastern Canada, was long recognized as distinct and was

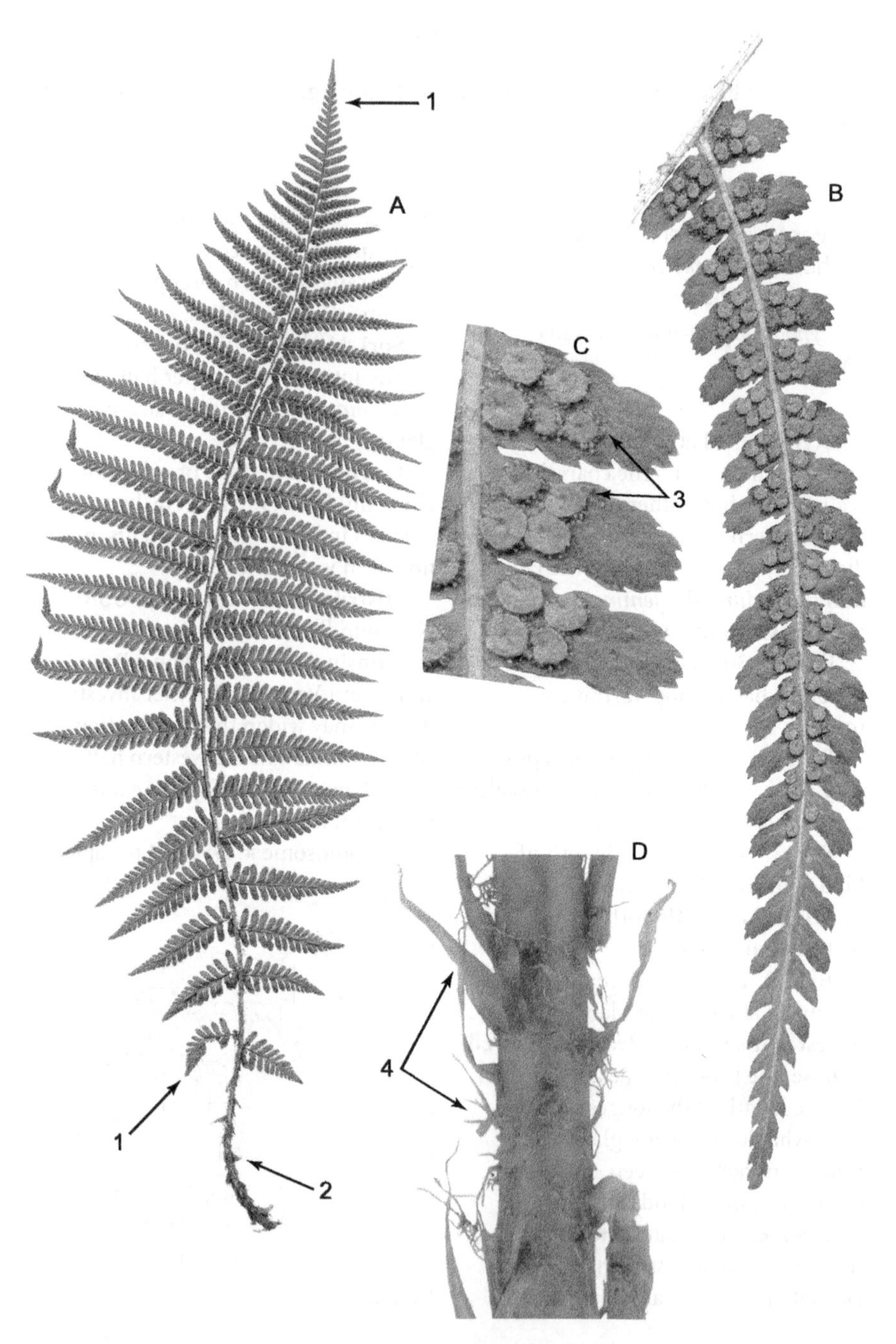

Dryopteris filix-mas: **Male fern: A. Frond. B. Fertile pinna. C. Fertile pinnules with sori and indusia. D. Portion of stipe.**

Informative arrows: 1. Frond tapering at both ends. 2. Short stipe. 3. Peltate indusia. 4. Two types of scales, narrow and wide.

formally recognized by Fraser-Jenkins as a subspecies, in a difficult to access publication, as distinct from the European and western US *D. filix-mas* subspecies.

This subspecies differs from var. *filix-mas* in that the pinnules in the lower middle part of the middle pinnae are more sloping to their bases and are more obviously narrowed on their acroscopic side. Its pinnules are rather long, and their tips are more acute; the pinnule teeth are rather short; the bases of the lowest opposite pair of pinnules on each of the lower and lower middle pinnae are widened into prominent, rectangular lobes; and the sori are arranged in 2 slightly more separated lines in each pinnule.

Aspidium oleoresin, produced by extracting powdered *Aspidium filix-mas* (now *Dryopteris filix-mas*) with ether, was used as an anthelminic to treat miscellaneous other diseases. Its laxitive effect is powerful, but it is very toxic. For decades it had an official status in the US pharmicopea.

Dryopteris filix-mas **subsp.** *brittonii* **may be recognized by its thick-textured, 1-pinnate-pinnatifid to 2-pinnate deciduous fronds. The fronds have short stipes densely covered with brown scales and distinct hairlike scales (this is the only Michigan** *Dryopteris* **with two distinct forms of scales). It somewhat resembles** *D. marginalis* **but is easily distinguished by its medial (rather than marginal) sori and narrower pinnae. The base of the blades is clearly narrowed, and the lower pinnae are less than 1/2 the length of the longest pinnae. The stipe is less than 1/4 the length of the frond.**

Dryopteris fragrans (Linnaeus) Schott

Polypodium fragrans Linnaeus

Fragrant Woodfern, Fragrant Shield Fern, Fragrant Cliff Fern

Etymology. Latin *fragrans*, fragrant, alluding to a sweet, fruity odor produced when the fronds are bruised or crushed.

Plants. Evergreen; monomorphic, aromatic, small, tussock-forming.

Fronds. 8–25 cm × 2–5 cm, strongly and long-tapered at the base, old fronds persistent as gray or brown masses at rhizome tips.

Rhizomes. Short, thick, erect, **covered with shiny brown scales, withered old fronds, and stipe bases.**

Stipes. Up to 1/3 length of fronds, scales dense throughout, large, pale tan to reddish-brown, sparsely glandular.

Blades. 6–20 × 2–5 cm, 1-pinnate-pinnatifid to 2-pinnate, linear-oblanceolate, narrowed at base, tips acute, leathery, adaxial surface smooth, **undersurface densely scaly, glandular**; green, **blades aromatic when bruised.**

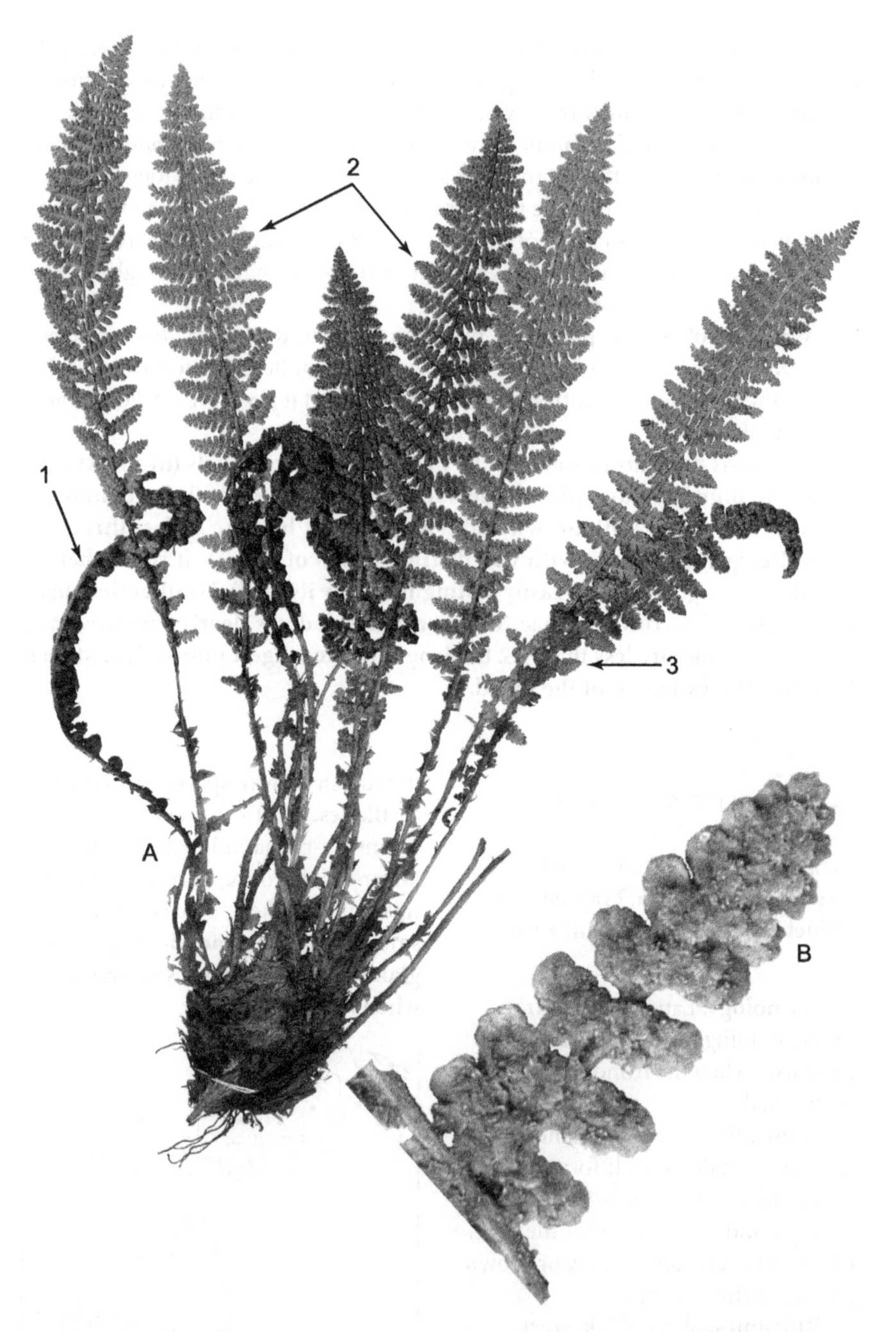

Dryopteris fragrans: Fragrant woodfern: A. Cluster of fronds on rhizome. B. Pinna. Informative arrows: 1. Old withered fronds. 2. Look for aromatic fragrance when bruised. 3. Fronds narrow at base.

Pinnae. 3–1.2 × 8–4 cm; 15–30 pairs, alternate to subopposite, linear-oblong, crowded, lowest pairs much smaller, tips subacute to rounded, often overlapping and inrolled.

Pinnules. 6–10 pairs, pinnules closest to rachis longer than adjacent pinnules, acroscopic pinnules same size as basiscopic, pinnule margins scalloped with rounded teeth.

Sori. Medial, large; crowded, often overlapping, covering most of pinnae undersurfaces; indusia large, round-reniform, margins glandular.

Habitat. Usually on shady, well-drained, north-facing, vertical cliffs and talus slopes and in rock crevices. (In Michigan it probably never grows on limestone or dolomite, although much descriptive material from other sources states that calcareous rock is a preferred habitat.)

Distribution. Interrupted circumboreal. Europe and Asia. Canada and USA: Alaska to Newfoundland and Greenland, south to upper tiers of US states and east to New England. Uncommon in the western half of the Upper Peninsula. About half of the recorded collections are from Isle Royale.

Chromosome #: $2n = 82$

Dryopteris fragrans is a Michigan fern of special concern. It is globally secure.

Dryopteris fragrans **is a northern species of cliff faces and talus slopes not closely related to or resembling other North American** *Dryopteris* **species. It is the smallest North American woodfern (usually less than 25 cm long and 5 cm wide), with blades tapering gradually toward both ends. The undersides of its thick fronds have many fragrant, aromatic glandular hairs. The densely scaly rhizome usually retains a mat of gray dead fronds resembling a mop.**

Dryopteris fragrans **is distinguished from** *Woodsia ilvensis* **by its densely crowded fronds; its fronds tapering at the base: its short, unjointed stipes; its lack of broken-off persistent stipe bases; its larger densely crowded sori; its round-reniform indusia (not hairlike as in** *W. Ilvensis*)**; and its fruity fragrance when bruised.**

Dryopteris goldiana (Hooker) A. Gray

Aspidium goldianum Hooker; *Thelypteris goldiana* (Hooker) Nieuwland

Goldie's Woodfern

Etymology. Name honors John Goldie (1793–1886), a Scottish botanist who on a journey to North America discovered this fern near Montreal in 1819.

Plants. Deciduous, monomorphic.

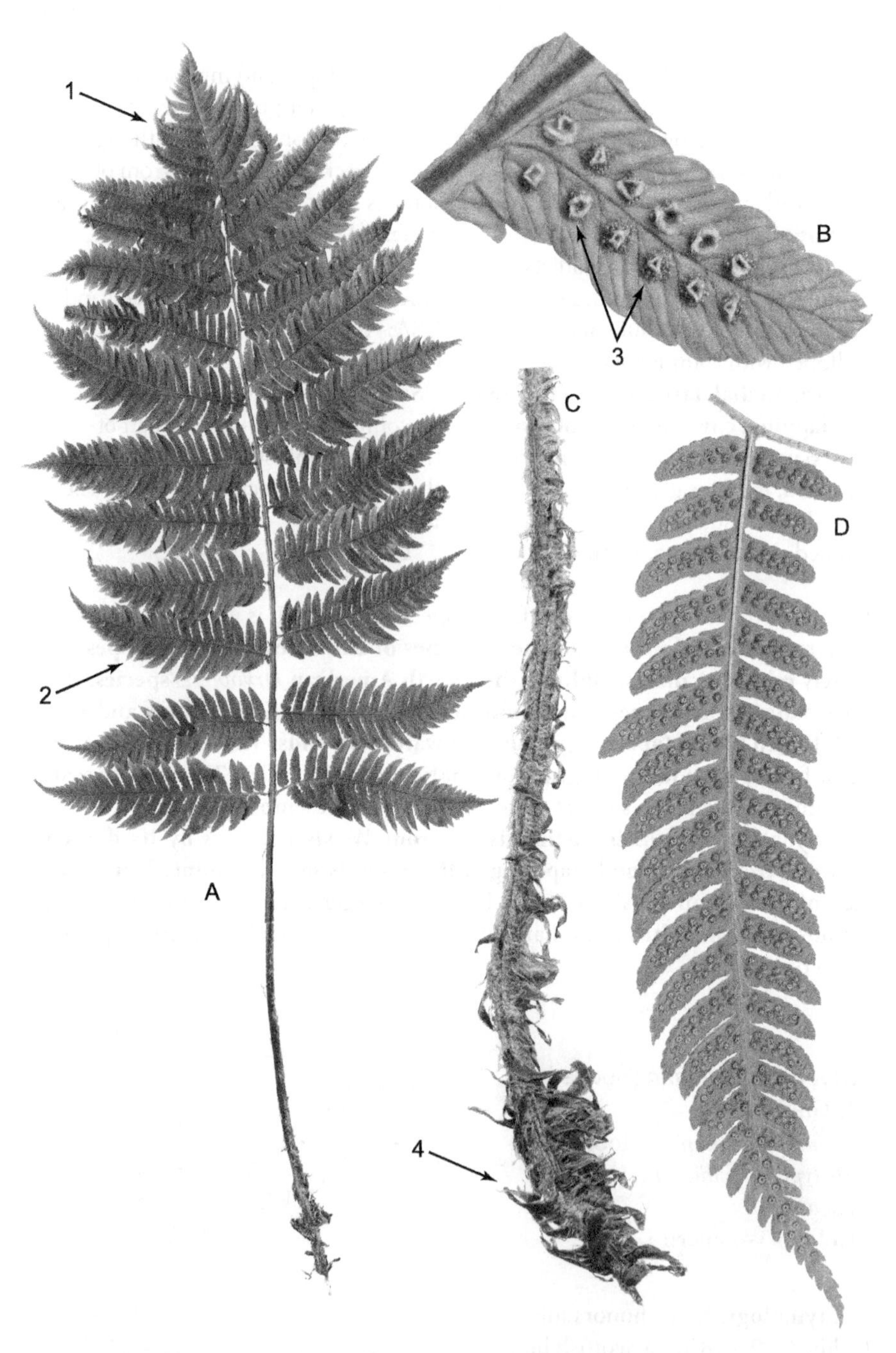

Dryopteris goldiana: Goldie's woodfern: A. Frond. B. Fertile pinnule. C. Stipe. D. Fertile pinna.

Informative arrows: 1. Abruptly tapering tip. 2. Frond 2-pinnate. 3. Sori with peltate indusia close to midrib. 4. Dense scales, many with dark centers.

Fronds. 35–120 × 15–40 cm, arching in vase-shaped clusters.

Rhizomes. Creeping to erect, stout, very scaly.

Stipes. 1/3 length of leaf, grooved, conspicuously scaly, especially at base; **scales scattered, long, pointed, dark glossy brown to nearly black centers with pale margins.**

Blades. 2-pinnate to 2-pinnate pinnatifid at base; ovate-lanceolate, broad, flat; **tips abruptly tapered,** not glandular, lustrous green, often mottled white near tips; scales linear to ovate below, absent above; rachises green, with small pale scales.

Pinnae. 15–20 pairs, ovate-lanceolate, narrow at base, widest in the middle, basal pinnae broadly oblong-lanceolate, slightly smaller than those above, midribs grooved.

Pinnules. Acroscopic pinnules same size as basiscopic pinnules, pinnule closest to rachis same size as adjacent pinnule, margins crenulate to broadly toothed.

Sori. 6–10 pairs, **nearer midvein than margin.**

Habitat. Damp beech-maple-basswood woods, wet springy slopes, margins of or hummocks in swamps.

Distribution. Northeastern United States and eastern Canada south to Alabama and west to Missouri and Minnesota. In Michigan it is most common in the southern counties, where it is local and uncommon. It is rare in the northern three tiers of counties in the Lower Peninsula and recorded only a few times in the Upper Peninsula.

Chromosome #: $2n = 82$ diploid

Dryopteris goldiana, a diploid species, hybridizes with 5 other *Dryopteris* species. These hybrids can usually be identified by their glossy dark scales and large blade size. It is one of the parents of the fertile hybrid species *D. celsa* and of *D. clintoniana.*

The largest of Michigan's woodferns, *Dryopteris goldiana* **is easily identified by its large, 2-pinnate to 2-pinnate-pinnatifid, broad, ovate blades with abruptly tapering tips. The lower pinnae are usually noticeably narrowed at the base. The sori are closer to the midveins than the margins. The stipes are about 1/3 the length of the frond and have large, glossy, dark brown scales with pale borders.**

Dryopteris intermedia (Muhlenberg) A. Gray

Aspidium intermedium Muhlenberg; *D. spinulosa* (O. F. Müller) Watt var. *intermedia* (Muhlenberg) Underwood

Evergreen Woodfern, Fancy Woodfern, Glandular Woodfern

Etymology. Latin *intermedia*, intermediate, referring to the leaf appearance intermediate between that

Dryopteris intermedia: **Evergreen woodfern: A. Pinnule. B. Frond. C. Pinna. D. Line drawing of glandular hair.**

Informative arrows: 1. Basiscopic pinnule on basal pinna closest to the rachis shorter than adjacent pinnule. 2. Look for small glandular hairs on rachis and costae.

of *Dryopteris carthusiana* and other species of *Dryopteris.*

Plants. Evergreen; monomorphic.

Fronds. 32–90 × 10–20 cm, somewhat arching; arranged in circular clusters.

Rhizomes. Erect, thick, covered with persistent leaf bases, scaly.

Stipes. 1/4–1/3 length of leaf, scales tan, mostly at base, scattered above.

Blades. 2-pinnate-pinnatisect to 3-pinnate-pinnatifid, ovate to narrowly triangular, lacy, herbaceous, green, **glands present on rachises and midribs of pinnae (often disappearing as fronds age or dry).**

Pinnae. 10–15 pairs, lanceolate-oblong, basal pinnae linear-lanceolate, somewhat asymmetrical, not reduced; subopposite, midribs with glandular hairs.

Pinnules. Basiscopic pinnules of lowest pinnae closest to rachises shorter or same size as adjacent pinnules. (On average this character holds, but there is great variability, and nonconforming examples are frequently found.) Basiscopic pinnules of lowest pinnae longer than opposite acroscopic pinnules, pinnule margins serrate, teeth spiny.

Sori. Medial, small, *indusia* **with minute glandular hairs when young,** glands less visible when older.

Habitat. Moist woods, where it is often the most frequent fern, wet meadows, dry woods, hummocks in swamps, hardwood-hemlock forests, usually avoids standing water.

Distribution. Canada and USA: Ontario to Newfoundland, south to Minnesota, Iowa, Missouri, Tennessee, Alabama, and Georgia. It is the most common *Dryopteris* in Michigan.

Chromosome #: $2n = 82$ diploid

Dryopteris intermedia, a diploid species, is one of the parents of the allotetraploid species (a fertile hybrid resulting from mating of 2 separate species) *D. carthusiana. D. intermedia* also hybridizes with 7 other *Dryopteris* species. The sterile hybrids may be recognized by their intermediate morphology, by the presence of glandular hairs on the indusia and usually on the rachis and midribs, and by abortive spores.

Dryopteris intermedia is the most abundant woodfern in Michigan, and it has the widest range of habitats. Its evergreen marcescent fronds are found flattened on the forest floor in spring and gradually become dry and shriveled as they appear to give up their nutrients to the new fronds being produced. The glandular hairs of this species are very small and resemble a minute hatpin (a pin with a spherical end). A hand lens and good lighting are usually required to see these hairs, and you must think small—then think even smaller. They are always present on young fronds but may be lost over the summer due to stress, dryness, or wind. However, they really are there and once recognized are easy to find.

Dryopteris xtriploidea is a common sterile hybrid ($3n = 123$, 42 single chromosomes and 42 pairs) between *D. intermedia* ($2n = 82$ diploid) and *D. carthusiana* ($2n = 164$ tetraploid). It may be difficult to recognize without examining for abortive spores.

Dryopteris intermedia **may be distinguished from the closely related and quite similar** *D. carthusiana* **and** *D. expansa* **by the basiscopic (downward-**

pointing) pinnules of the lowermost pinnae closest to the rachises, which are usually shorter or the same length as the adjacent ones. Some fronds are difficult to place on this basis, and it is wise to look at a large sample (and check for the presence of glands) if there is any question.

In contrast to *D. intermedia*, *D. carthusiana* **lacks glandular hairs on the indusia, rachises, and pinna midribs.** *D. expansa* **usually lacks glands but is sometimes somewhat glandular. It takes a little practice to see these glands.**

The fronds of *D. intermedia* **persist through the winter, while those of** *D. carthusiana* **do not.**

The sterile hybrid *D.* ×*triploidea* (*D. carthusiana* × *D. intermedia*) **is often present when both parental species are nearby. See discussion of this hybrid under "Sterile Hybrids" below.**

Dryopteris marginalis (Linnaeus) A. Gray

Polypodium marginale Linnaeus

Marginal Woodfern

Etymology. Latin *margo*, border, + *-alis*, pertaining to, alluding to the placement of the sori near the margins of the pinnae. The epithet is not quite accurate as the sori are actually submarginal (near but not touching the margins).

Plants. Evergreen, monomorphic.

Fronds. 25–100 × 12–25 cm, clustered, borne in rosettes.

Rhizomes. Thick, short, erect or ascending, covered with long yellow-brown scales.

Stipes. 1/4–1/3 length of fronds, covered with golden brown scales sometimes 2 cm long, dense at bases, less scaly upward.

Blades. 30–50 × 9–25 cm; **2-pinnate to 2-pinnate-pinnatifid at bases of large fronds**, ovate-lanceolate, **thick, somewhat leathery, dull bluish-green upper surface with lighter grayish-green undersurface;** rachises straw-colored, somewhat scaly.

Pinnae. 15–20 pairs, lanceolate, tapering to pointed tips, lowest pinna pairs slightly smaller, alternate.

Pinnules. 15–20+ pairs and lobes; blunt-tipped, margins shallowly crenate to entire, basiscopic pinnules longer than acroscopic ones.

Veins. Free, obscure.

Sori. Round-reniform, circa 1.5 mm diam., prominent, **close to margins**, single or in rows, indusia prominent and conspicuous, kidney-shaped, light-colored when young, dark with age, opening outward, margins smooth.

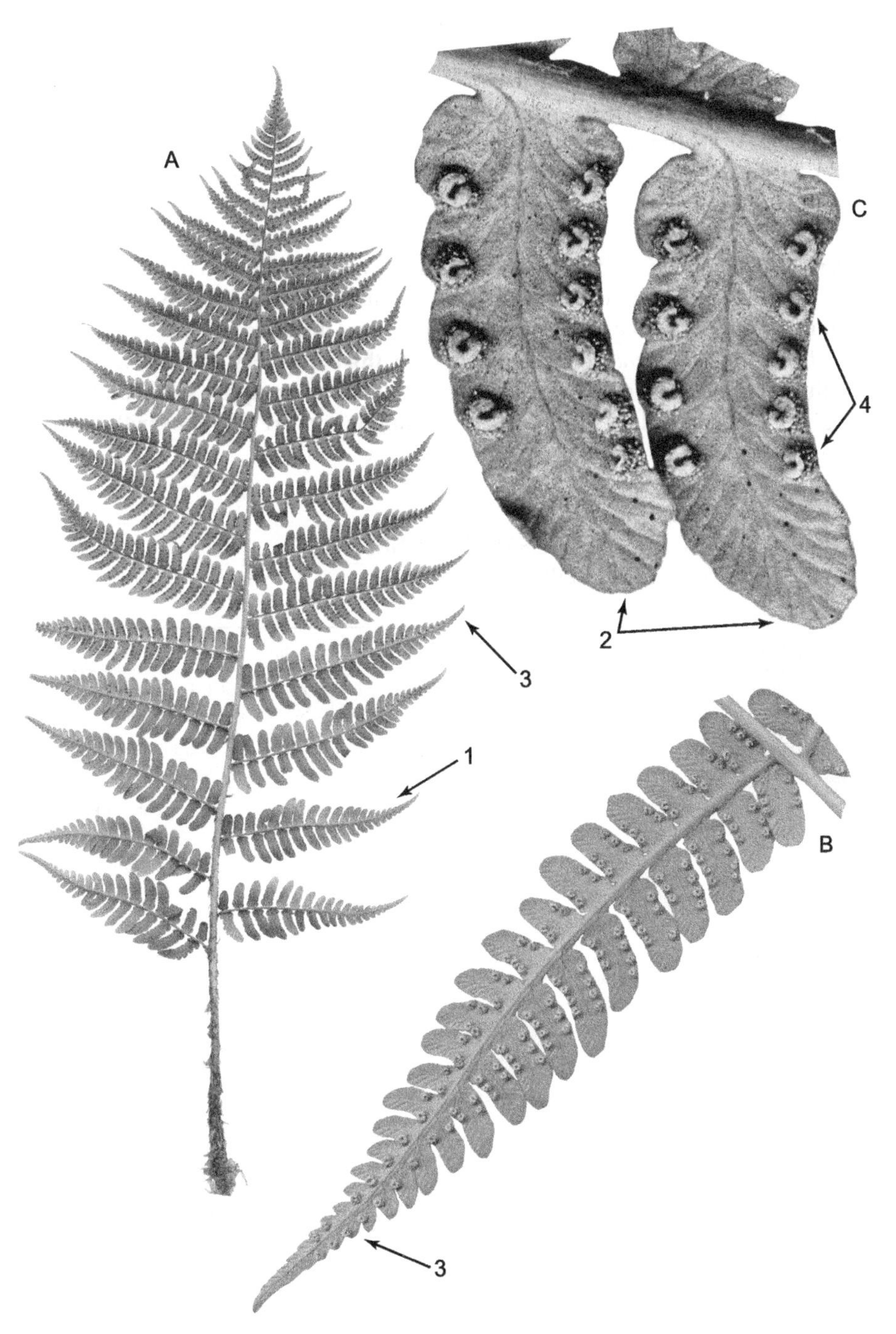

Dryopteris marginalis: **Marginal woodfern: A. Frond. B. Fertile pinna. C. Pinnules.**
Informative arrows: 1. Blade 2-pinnate. 2. Pinnules blunt. 3. Pinnae tapering to pointed tip. 4. Sori near margin.

Habitat. Mostly on relatively dry, north- or east-facing slopes in deciduous woods; sometimes on dry, exposed rock ledges, swamp margins, or hummocks in swamps.

Distribution. Greenland. North America: Ontario to Newfoundland, south to Georgia, west to Mississippi, and north to Wisconsin.

Chromosome #: $2n = 82$ diploid

Dryopteris marginalis hybridizes with 10 other *Dryopteris* species, producing sterile hybrids, some of these hybrids have been found in Michigan. They can be recognized by the presence of nearly marginal sori and distorted spores.

Dryopteris marginalis **is recognized by its thick, leathery, evergreen fronds with dull grayish-green undersurfaces and bluish-green upper surfaces. It has 1-pinnate-pinnatifid to 2-pinnate fronds, and its sori are found near the margins of the pinna lobes.**

CHART RELATING MICHIGAN *DRYOPTERIS* SPECIES TO IDENTIFYING CHARACTERISTICS

Species characters	*D. carthusiana*	*D. celsa*	*D. clintoniana*	*D. cristata*
Frond dimensions in cm[a]	L 20–80 W 10–30	L 90–120 W 20–30	L 45–100 W 12–20	L Sterile 25–75 W Sterile 8–15
Frond proportions[b]	B 3/4–1/3 S 1/4–1/3	B 1/2–2/3 S 1/2–1/3	B 2/3–3/4 S 1/4–1/3	B 2/3–3/4 S 1/3–1/4
Monomorphic, dimorphic -,+,++,+++	Monomorphic	Monomorphic	Dimorphic ++	Dimorphic +++
Blade shape	Ovate-lanceolate	Ovate-lanceolate, tapering toward tip	Sides nearly parallel	Fertile fronds narrowly lanceolate with parallel sides
Blade cutting	2-pinnate-pinnatisect to 3-pinnate	2-pinnate	1-pinnate-pinnatifid	1-pinnate-pinnatifid to 2-pinnate
Pinna pairs	13–18	15–20	10–16	10–20
Basal pinna shape	Broadly triangular	Ovate-lanceolate	Long-triangular, same as pinnae above	Triangular
Length of pinnules of basal pinnae, acroscopic vs. basiscopic	Basiscopic up to 2× longer than acroscopic	More or less same length	More or less same length	More or less same length
Length of basiscopic pinnules closest to rachis vs. adjacent basiscopic pinnules	Pinnules closest to rachis longer	More or less same length	Slightly longer	Slightly longer

D. expansa	D. filix-mas	D. fragrans	D. goldiana	D. intermedia	D. marginalis
L 25–100 W 20–40	L 30–120 W 10–30	L 6–20 W 2–6	L 75–120 W 15–40	L 35–90 W 10–25	L 25–100 W 12–25
B 1/2 to 2/3 S 1/2 to 1/3	B 3/4 + S 1/4-	B 2/3 + S 1/3-	B 2/3 S 1/3	B 2/3 S 1/3	B 3/4–2/3 S 1/4–1/3
Monomorphic	Monomorphic	Monomorphic	Monomorphic	Monomorphic	Monomorphic
Broadly triangular to ovate	Ovate-lanceolate, widest in middle, narrow at base	Linear-lanceolate, narrow at base, tips acute	Ovate-lanceolate, broad, tips abruptly tapered	Ovate to narrowly triangular	Ovate-lanceolate
3-pinnate to 4-pinnate	2-pinnate	2-pinnate	2-pinnate	2-pinnate-pinnatisect to 3-pinnate-pinnatifid	2-pinnate
12–25	15–30	6–10	15–20	10–15	15–20
Broadly triangular	Same as pinnae above	Same as pinnae above	Ovate-lanceolate, same as pinnae above	Same as pinnae above	Same as pinnae above
Basiscopic longer	Equal	Mostly equal	Equal	Basiscopic longer	Basiscopic somewhat longer
2–5× longer than acroscopic pinnules closest to rachis	More or less same length	More or less same length	More or less same length	Significantly shorter to same length	More or less same length

(*continues*)

Chart Relating Michigan *Dryopteris* Species to Identifying Characteristics—*Continued*

Species characters	*D. carthusiana*	*D. celsa*	*D. clintoniana*	*D. cristata*
Sorus location	Medial	Medial	Medial	Medial
Glands +, ++, +++	Absent	Absent	Absent	Absent
Scales on rachises	Light brown, concolorous or with dark centers	Dark brown to tan, with dark centers	Light brown, sometimes with dark centers	Light brown
Habitat	Swamps, wet soils, occasionally in rich moist forests	Logs and hummocks in acidic swamps	Swamp margins, hummocks, sometimes in wet woods	Swamps, marshes, bog margins, wet meadows and woods
Distribution	Circumboreal, upper United States	Found in 4 counties in the southeastern Lower Peninsula	More common in southern half of the state	Throughout Michigan
Rarity in Michigan	Common	Rare	Uncommon to rare	Uncommon but common in preferred habitat
Genome ploidy[c]	Tetraploid IISS $2n = 164$	Tetraploid GGLL $2n = 164$	Hexaploid GGLLSS $2n = 246$	Tetraploid LLSS $2n = 164$
Special characters	Basal pinnae distinctive (check description), no glands	Large, gradually tapering frond tips, dark scales	Sterile fronds evergreen, basal pinnae triangular	Sterile frond short-spreading, fertile fronds erect, "venetian blind" pinnae

[a]W = width, L = length.

[b]B = blade S = stipe.

[c]Genomes: E = *D. expansa*, F_1F_2 = *D. filix-mas*, Fr = *D. fragrans*, G = *D. goldiana*, I = *D. intermediate*, L = *D. lucoviciana* (not found in Michigan): M = *D. marginalis*, S = "*D. semicristata*" (hypothetical fern extinct or not yet found but well established scientifically)

D. expansa	*D. filix-mas*	*D. fragrans*	*D. goldiana*	*D. intermedia*	*D. marginalis*
Medial to submarginal	Medial	Medial	Nearer midrib than margin	Medial	Near margins
Absent to +	Absent	+++	absent	++	absent
Light brown, with dark centers	Light brown, 2 distinct types, broad and narrow	Light brown	Glossy dark brown to black	Light brown, with dark centers	Light brown, tan
Cool moist woodlands and rocky slopes	Calcareous cliffs, cracks, and sinkholes	Noncalcareous, usually north-facing cliffs and talus slopes	Damp woods, wet springy slopes, hummocks in swamps	Moist woods, dry woods, hummocks in swamps	Usually dry north- or east-facing slopes, also on hummocks in swamps
Four counties in the Upper Peninsula	Upper Peninsula, scattered, more common in the west; Lower Peninsula, Alpena County	Western Upper Peninsula, Isle Royale	Mostly in southern counties, rare in the Upper Peninsula	Throughout Michigan	Throughout Michigan
Rare	Rare	Rare	Uncommon	Common	Common
Diploid EE $2n = 82$	Tetraploid $F_1F_1{}_1F_1$? $2n = 164$	Diploid FrFr $2n = 82$	Diploid GG $2n = 82$	Diploid II $2n = 82$	Diploid MM $2n = 82$
Similar to *D. carthusiana* and *D. intermedia*, basal pinnae distinctive	Narrow pinnae, blades widest in middle, tapered at both ends	Abundant glandular hairs, cliff habitat, small size	Fronds large with abruptly tapering tips, scales large, dark	Lacy fronds, minute glandular hairs on indusia and rachises, basal pinnae distinctive	Sori near margin, leathery, glaucous

FERTILE HYBRIDS

A new sporophyte, the spore-producing plant that we recognize as the fern, is formed when a sperm fertilizes an egg on the gametophyte. Even if the sperm and egg are from different species, they may rarely produce a sporophyte. Ordinarily hybrids between 2 species are sterile because the chromosomes are unmatched and unable to pair, which is required in meiosis.

At some time in the past a sterile triploid hybrid was produced that had 2 chromosome sets from 1 species and 1 set from the other (sterile because the single chromosome of 1 species has no chromosome to pair with in meiosis, for which chromosome pairing is required). For mitosis to occur and a cell to divide, it is only necessary that the chromosomes duplicate themselves. This they can do whether the chromosomes are matched or not. This allows production of sterile, mature spore-producing plants. Virtually all the spores produced by these plants are sterile and misshapen and are incapable of producing another plant. Rarely, within the millions of spores produced by these sterile plants, an error in cell division occurs, producing a spore containing all the chromosomes from both parents (one set from each parent of this sterile hybrid), doubling the chromosome number. Pairing of chromosomes produces a plant with chromosomes from both parents that, when mature, has the ability to produce spores by normal meiosis (the chromosome pairing needed for meiosis to occur). This fertile fern plant will share the characteristics of both parent species and is recognized as a regular (allopolyploid) species (Greek *allo-*, other, + *poly*, many, + *-ploidy*, a suffix implying number of chromosome sets). The term *allopolyploid* is used when the parents are of different species.

Some entirely fertile *Dryopteris* species are the result of hybridization between 2 species with subsequent doubling of their chromosome numbers. *Dryopteris clintoniana* ($2n = 246$) is a fertile hexaploid resulting from a cross between *D. cristata* ($2n = 164$) and *D. goldiana* ($2n = 82$).

When a species is formed by the doubling of its own chromosome number it is called a homopolypoid.

The species produced in this manner (called nothospecies) are good species in every sense of the word, but they have arisen in a non-Darwinian, nonevolutionary fashion.

DRYOPTERIS HYBRIDS IN MICHIGAN

Dryopteris species are well known for hybridizing. There are 10 species of *Dryopteris* in Michigan. Theoretically 45 hybrids could be produced if each species is capable of hybridizing with the other 9. Some hybrid combinations found in other regions have not yet been found in Michigan.

While almost any combination of Michigan's 10 *Dryopteris* species may produce sterile hybrids, this book treats only the 2 hybrids that are most likely to be found. These are *D. cristata* × *D. intermedia* (*D.* ×*boottii* [Tuckerman] Underwood) and *D. carthusiana* × *D. intermedia* (*D.* ×*triploidea* Wherry).

How to recognize hybrids:

Within populations of species, hybrids usually look odd and atypical. They are sporadic within colonies of typical plants and appear singly or as a cluster derived from an older hybridization event.

Spores. Examining for abortive spores is the ultimate test but will often require at least a 40× magnification. In the *Dryopteris* hybrids virtually 100% of the spores are misshapen.

Blade cutting. *Dryopteris filix-mas* and *D. marginalis* are 2-pinnate, and *D. carthusiana* and *D. intermedia* are 2-pinnate-pinnatifid to 3-pinnate. These characters will be represented in hybrids of these species. A triploid hybrid such as *D. carthusiana* × *D. intermedia* will have a tendency to more closely resemble the parent contributing 2/3 of the genes—in this case *D. carthusiana*.

Glandular hairs. The infamous glandular hairs of *D. intermedia* can help identify hybrids of that species. While present on the rachises and costae of the fronds, these hairs are most likely to be found on the indusia. They may be absent from older fronds. Don't be discouraged; they really do exist and once recognized are easy to find.

Scales at the stipe bases. See if the scales are uniform in color, are large with a dark central area at the base, or have a central dark streak from *D. goldiana*. Note their size.

Caveats:

Don't try to identify young or sterile plants. Without sori you will usually get nowhere.

Entire fronds, including the base of the stipes, must be examined.

Hybrids are quite variable and often express the full range of variation between the parental species.

During the past 60 years there has been an enormous increase in our understanding of *Dryopteris* species and their hybrids. Warren "Herb" Wagner and Florence Wagner were responsible for much of the advancement of knowledge of the reticulate speciation in this group. The sexual behavior of *Dryopteris* species is extreme by human standards. Miscegenation is rampant, interspecific hybrids are common, and even an intergeneric hybrid (with *Polystichum*) has been found.

FERTILE *DRYOPTERIS* POLYPOID HYBRIDS FOUND IN MICHIGAN

Hybrid origin of D. carthusiana, a fertile tetraploid species

D. "semicristata"[a]		*D. intermedia*		*Sterile hybrid*		*D. carthusiana fertile*
2*n* = 82	×	2*n* = 82	→	2*n* = 82	Chromosome doubling →	2*n* = 164
SS		II		SI		SSII

Hybrid origin of D. clintoniana, a fertile hexaploid species

D. goldiana		*D. cristata*		*Sterile hybrid*		*D. clintoniana fertile*
2*n* = 82	×	2*n* = 164	→	2*n* = 123 unpaired	Chromosome	2*n* = 246 paired
GG		LLSS		GLS	doubling →	GGLLSS

Hybrid origin of D. cristata, a fertile tetraploid species

D. ludoviciana[b]		*D. "semicristata"*		*Sterile hybrid*		*D. cristata fertile*
2*n* = 82	×	2*n* = 82	→	2*n* = 82 unpaired	Chromosome	4*n* = 164
LL		SS		LS	doubling →	LLSS

Note: These hybrids are treated as fertile species in the treatments above. I = *D. intermedia*; S = *D. semicristata*; L = *D. ludoviciana*; G = *D. goldiana*.

[a]*Dryopteris "semicristata"* is a hypothetical diploid species, probably now extinct, that is proposed as a diploid progenitor of the allotetraploids *D. cristata* and *D. carthusiana*. It has never been collected, but studies have validated its prior existence. Using morphologic studies and molecular methods including isozyme, chloroplast DNA, and phylogenetic studies; and by extrapolating likely characteristics from related living species a plausible morphology for *D. "semicristata"* has been described. *Dryopteris filix-mas* subsp. *brittonii* 2*n* =164, recently identified as separate from its European counterpart and found in North America, may be an allopolyploid of two closely related taxa or an homopolyploid. *D. filix-mas* subsp. *filix-mas*, found in Europe, is known to be a fertile tetraploid hybrid between *D. caucasica* and *D. oreodes*.

[b]*D. ludoviciana* does not occur in Michigan.

See the material in the Introduction regarding polyploidy (both allopolyploidy, and homopolyploidy) in the production of species resulting from interspecific hybridization, and from doubling of the chromosome number of a species.

STERILE HYBRIDS

Dryopteris intermedia and *D. marginalis* are the most likely species to produce sterile hybrids. Identification of these hybrids is aided by the presence of glands derived from the 2-pinnate-pinnatisect to 3-pinnate-pinnatifid *D. intermedia* and the dull bluish-green color and 2-pinnate dissection of *D. marginalis*.

Sterile hybrids most likely to be found in Michigan:

D. cristata ×*intermedia* = *D.* ×*boottii* (Tuck.) Underwood
D. intermedia ×*carthusiana* = *D.* ×*triploidea* Wherry (most common Michigan hybrid)

Uncommon to rare hybrids that have been found in Michigan but will not be treated in this book:

D. carthusiana ×*clintoniana* = *D.* ×*benedictii* (Farwell) Wherry
D. carthusiana ×*cristata* = *D.* ×*uliginosa* (A. Braun) Druce
D. carthusiana ×*marginalis* = *D.* ×*pittsfordensis* Slosson
D. celsa ×*goldiana*
D. celsa ×*marginalis* = *D.* ×*leedsii*
D. clintoniana ×*cristata*
D. clintoniana ×*goldiana* = *Dryopteris* ×*mickelii* J. H. Peck
D. clintonia ×*intermedia* = *D.* ×*dowellii* (Farwell) Wherry
D. clintoniana ×*marginalis* = *D.* ×*burgessii* Boivin
D. cristata ×*marginalis* = *D.* ×*slossoniae* (Hahne) Wherry
D. expansa ×*intermedia*
D. expansa ×*marginalis*
D. filix-mas ×*marginalis* = *D.* ×*montgomeryi* Fraser-Jenkins
D. goldiana ×*intermedia*
D. goldiana ×*marginalis* = *D.* ×*neowherryi* W. H. Wagner
D. intermedia ×*marginalis*

Dryopteris ×triploidea **Wherry**
Dryopteris carthusiana ×intermedia
Dryopteris spinulosa (O. F. Müller) Watt var. *fructuosa* (Gilbert) Trudell
Triploid Woodfern

Etymology. Latin *triplex*, triple, + *-ploideus*, a suffix used in botany to state the number of chromosome sets in a hybrid.

Plants. Partially evergreen, remaining green well into winter before turning brown.

Fronds. 3-pinnate, rachis and costae somewhat glandular.

Pinnule costae. Glandular.

Indusia. Glandular.

Habitat. Swamps and damp woods—overlapping habitats of both parents.

Abundance. Very common.

Chromosome #: $3n = 123$ 42 pairs and 42 single chromosomes. *D. intermedia* $2n = 82$; *D. carthusiana* $2n = 164$

This sterile hybrid has 3 sets of chromosomes, 2 from *D. carthusiana* and 1 from *D. intermedia*.

Dryopteris ×triploidea **is the most common** *Dryopteris* **hybrid but is difficult to identify. It is almost always found where both parents are present. It is often the largest plant in the population, appearing somewhat more like** *D. carthusiana*, **but with the glandular hairs of** *D. intermedia*. **The lower innermost pinnules on the basal pinnae are usually about equal in length to the adjacent pinnules. Finding distorted spores confirms the identification.**

Dryopteris ×boottii **(Tuckerman) Underwood** ***Dryopteris cristata ×intermedia***
Aspidium boottii Tuckerman
Boott's Hybrid Woodfern

Etymology. Name honors Francis Boott (1792–1863), an American-born English botanist and physician who worked extensively in America between 1820 and 1840.

Plants. Evergreen.

Fronds. Less than 20 cm wide.

Blades. 2-pinnate-pinnatifid, dark green, rachises glandular.

Pinnae. Basal triangular.

Pinnules. With minute glandular hairs.

Indusia. With minute glandular hairs.

Habitat. Moist woods, swamps, and wet thickets.

Abundance. Common where the 2 species occur together. One of the most common hybrids.

Chromosome #: $3n = 123$ 123 single chromosomes; *D. cristata* $2n = 164$; *D. intermedia* $2n = 82$

Dryopteris ×boottii **is commonly found where both parents occur together and sometimes in the absence of one. It nicely combines the characters of both parents, producing a frond intermediate in character.**

GYMNOCARPIUM Newman Cystopteridaceae

Oak Ferns

See Cystopteridaceae in appendix for a discussion of family and a key to Michigan genera.

Etymology. Greek *gymnos*, naked, + *karpos*, fruit, alluding to the lack of indusia.

Plants. Monomorphic; deciduous; forming small to large clones of close fronds.

Fronds. Blades tilting almost horizontally to the ground from erect stipe tip.

Rhizome. Slender, long-creeping, subterranean.

Stipes. About 1.5–3 times length of blades, bases not swollen; cross section of stipe reveals 2 ± oblong vascular bundles.

Blades. 2-pinnate-pinnatifid to 3-pinnate-pinnatifid, 3–14 cm, broadly triangular; divided into 3 parts, or ovate, tapering to pinnatifid apices; rachises grooved.

Pinnae. Lowest pair longest and stalked, costae grooved, grooves not continuous with those of rachis, minute (0.1 mm) glandular hairs lacking, or glands present on undersurface and sometimes along costae adaxially.

Pinnules. Margins entire to crenate; pinnules on basiscopic side of pinnae longer than those on acroscopic side.

Veins. Free, simple or forked, pinnately branching.

Sori. Medial, round; indusia absent.

Habitat. Forest floors and rocky places.

Chromosome #: $x = 40$.

Gymnocarpium is a genus of 8 species found in temperate North American and Eurasia. Five species are treated in the *Flora of North America North of Mexico*, vol. 2 (1993). In the past this genus was variously placed in the genera *Polypodium*, *Dryopteris*, *Phegopteris*, *Thelypteris*, and *Currania* and in the families represented by those genera. It is now clearly recognized as a distinct genus, and a recent phylogenetic classification has placed it in the family Cystopteridaceae.

Gymnocarpium is represented by 3 somewhat similar species in Michigan. If a *Gymnocarpium* is found in Michigan it will probably be the common *Gymnocarpium dryopteris*. The other 2 species and hybrids are rare, or very rare, and scattered.

KEY TO THE SPECIES OF *GYMNOCARPIUM* FOUND IN MICHIGAN

1. Abaxial and adaxial blade surfaces and rachis essentially glabrous; second pinna pair from base sessile. . . . . *G. dryopteris*

1. Abaxial blade surface and rachis moderately or densely glandular, adaxial blade surface glabrous or moderately glandular; second pinna pair from the base sessile or stalked (2).

2(1). Blade moderately glandular on adaxial surface; proximal pinnae and basiscopic pinnules of proximal pinnae more or less perpendicular to rachis and costa, respectively; second pinnae pair usually stalked or if sessile with basal pinnules shorter than adjacent pinnules . . . . *G. robertianum*

2. Blade glabrous on adaxial surface; proximal pinnae and basiscopic pinnules of proximal pinnae curving toward leaf tip and tip of pinnae respectively; second pinnae pair almost always sessile with basal pinnules more or less equal in length to adjacent pinnules. . . . . *G. jessoense* subsp. *Parvulum*

Gymnocarpium dryopteris (Linnaeus) Newman

Polypodium dryopteris Linnaeus; *Aspidium dryopteris* Baumgarten; *Dryopteris dryopteris* (Linnaeus) Britton; *Lastrea dryopteris* (Linnaeus) Bory; *Nephrodium dryopteris* (Linnaeus) Michaux; *Phegopteris dryopteris* (Linnaeus) Fée; *Polystichum dryopteris* (Linnaeus) Roth; *Thelypteris dryopteris* (Linnaeus)

Oak Fern

Etymology. Greek *dryos*, oak or tree, + *pteron*, wing, which describes the shape of the pinnae. The word *pteris* was used by the ancient Greeks for all ferns. The vernacular name is a translation of the Greek specific epithet, although this species is not particularly associated with oak trees. (The large synonymy is retained here to demonstrate the various genera in which the *Gymnocarpium* species have been placed, and it shows the changing opinions taxonomists have had dealing with this genus. Most of these genera have been eliminated from the synonymies of the other *Gymnocarpium* species treated here.)

Plants. Monomorphic; deciduous; fiddleheads appear throughout the summer.

Fronds. 12–40 × 10–22 cm., small, delicate, in three parts; spreading in colonies of horizontally held blades forming a layer 10–20 cm above the ground, fronds not densely aggregated.

Rhizomes. Dark brown to black, slender, 0.5–1.5 mm diam., long-creeping and branching; scales brown, stiff, 2–5 mm diam.

Stipes. 9–30 cm long, **usually longer than blades, very slender, perpendicular to the ground, shiny,** straw-colored, dark at base, glands absent or sparse distally; scales light

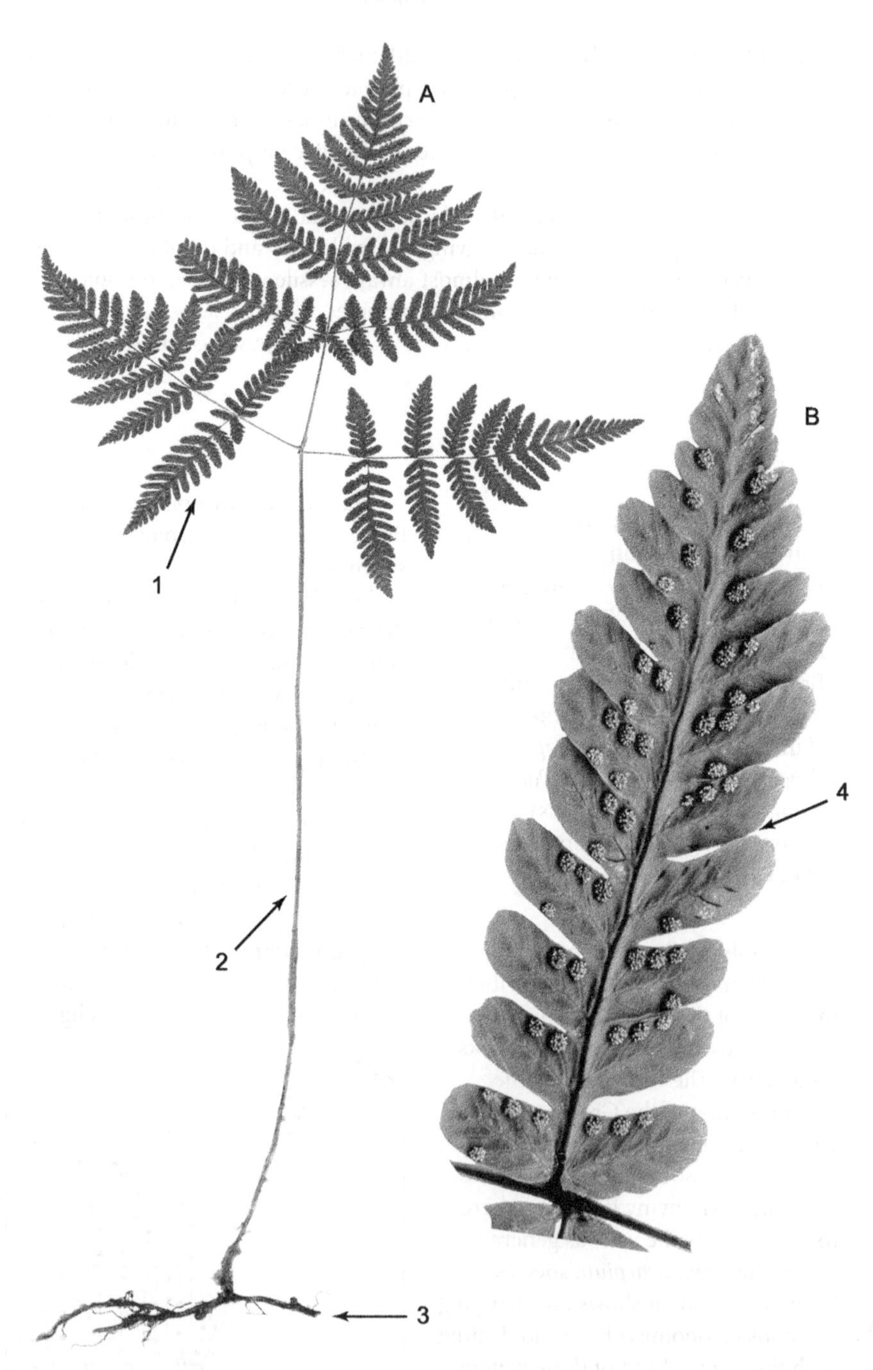

Gymnocarpium dryopteris: Oak fern: A. Frond. B. Fertile pinna.
Informative arrows: 1. Three equal-part triangular blade (horizontal to ground). 2. Long stipe. 3. Long-creeping rhizome. 4. Sori lacking indusium.

brown, scattered, mostly at base, 2–6 mm long.

Blades. 2-pinnate-pinnatifid to 3-pinnate, **broadly triangular, usually longer than wide,** 5–18 × 5–13 cm, **divided into three triangular, approximately equal divisions,** lax, **delicate, yellow-green, tilting almost horizontally to ground from stipe tip,** undersurface and rachis glabrous or sometimes with sparse glandular hairs; rachises green, delicate, with a narrow, thin, translucent wing in spring, slight swelling present at the junction of the lower pinna stalks and the rachis at the point where the blade tilts toward horizontal.

Pinnae. Pinna pairs opposite, perpendicular or only slightly angled to rachises, each lower pinna more or less equal in size to the blade distal to it, distal 6–8 pinna pairs sessile.

Pinnules and lobes. 6–10 pairs, opposite, sessile, acroscopic pinnules smaller than basiscopic pinnules, pinnules closest to rachis more divided than those farther out, margins entire to crenate.

Ultimate segments. Margins entire to crenate, tips entire, rounded.

Sori. Small, round, submarginal; indusia absent.

Habitat. Moist to wet sites in cool, mixed conifer and northern hardwood forests, stream banks, and acidic woodlands.

Distribution. Circumboreal. Europe. Asia. Greenland. Canada and USA: Alaska to Newfoundland, south to Oregon, northern Idaho, Montana, Minnesota, Iowa, Wisconsin, Ohio, West Virginia, and Maryland. Most commonly found in the Upper Peninsula and upper Lower Peninsula.

Chromosome #: $n = 80$

Gymnocarpium dryopteris is a tetraploid with *G. appalachianum* (2×) and *G. disjunctum* (2×)—neither of which is found in Michigan—as parents. *Gymnocarpium dryopteris* hybridizes with both *G. jessoense* subsp. *parvulum* and *G. robertianum*. It spreads freely in open areas, forming colonies that are several m in diam. and have rounded margins.

Gymnocarpium dryopteris **(common oak fern) may be identified by its small size and fronds that are broadly triangular, 3-lobed (ternate), delicate, and yellow-green. Its sori are round and lack indusia. It may be distinguished from** *G. robertianum* **by the absence, or sparseness, of glands on the fronds (***G. robertianum* **is quite glandular) and because its 3 frond segments are more or less equal in size. The central segment, the blade beyond the first pair of pinnae, is significantly larger than the 2 side lobes in** *G. robertianum***.**

Gymnocarpium dryopteris **is distinguished from the rare** *G. jessoense* **by its pinnae, which that are mostly perpendicular to the rachis (those of** *G. jessoense* **are curved toward the frond tips) and by the absence or sparseness of glandular hairs on the fronds (***G. jessoense* **is very glandular to sparsely glandular on its undersurface).**

Gymnocarpium jessoense (Koidzumi) Koidzumi subsp. *parvulum* Sarvela
Dryopteris jessoensis Koidzumi
Gymnocarpium continentale (Petrov) Pojarkova
Nahanni Oak Fern, Asian Oak Fern

Etymology. *Jesso*, + Latin possessive suffix *-ense*, "of Jesso." Jesso is an older spelling of Yezo Island, the Japanese island now known as Hokkaido. Koidzumi described this species from a specimen from Hokkaido. The vernacular name derives from the Nahanni National Park Reserve in the Northwest Territories of Canada, a boreal site where this fern grows.

Plants. Deciduous, monomorphic.

Fronds. 8–39 cm long, usually spread out, not aggregated.

Rhizomes. Slender, long-creeping, 0.5–1.5 mm diam., scales 1–4 mm long.

Stipes. 5–25 cm long, straw-colored, usually longer than blade, **with moderately abundant glandular hairs distally,** scales 2–6 mm long.

Blades. 2-pinnate-pinnatifid, 3–14 cm, narrowly triangular to narrowly ovate, divided into 3 roughly equal segments at the base, lowest pinnae curved, arching toward tip of blade; **abaxial surface of blade moderately to densely covered with short, glandular hairs; adaxial surface glabrous;** rachises moderately to densely glandular.

Pinnae. 2–12 cm long, **lowest pinna pair stalked, strongly curved toward frond tips,** second pinna pair sessile.

Pinnules. Oblong, entire to slightly crenate, tips round, **basiscopic pinnules of lowest pinnae strongly curved toward pinna tips**, basiscopic pinnule of the lowermost pair of pinnae closest to the rachis only slightly longer than the opposite acroscopic pinnule.

Sori. Round; indusia absent.

Habitat. Cool, noncalcareous rocky outcrops, shale talus slopes, and granite cliffs.

Distribution. Circumboreal. Northern Eurasia. Canada: throughout. USA: Alaska, Minnesota, Iowa, Wisconsin, northern Maine, Vermont, and Michigan.

Chromosome #: $n = 80$

Gymnocarpium jessoense subsp. *parvulum* is a Michigan endangered species. It is considered globally secure. *Gymnocarpium jessoense* subsp. *jessoense* is widespread in Eurasia.

Hybrids between *G. jessoense* subsp. *parvulum* and *G. dryopteris* (*G. ×intermedium* Sarvela) are often found where these taxa occur together.

Gymnocarpium jessoense **is a very rare fern in Michigan, having been found only a few times in the Upper Peninsula. It is smaller and more slender than** G.

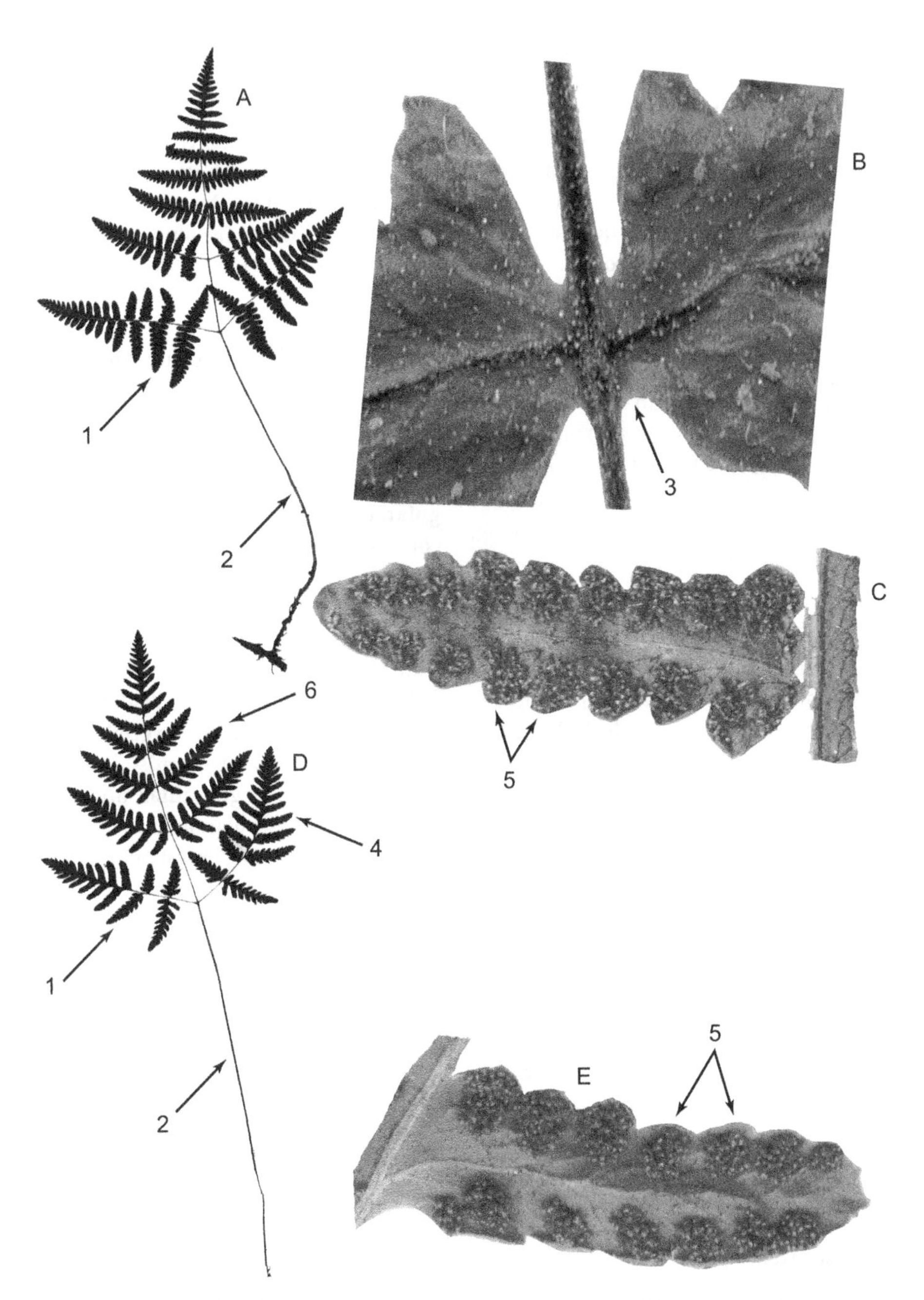

Gymnocarpium robertianum: Northern oak fern: A. Frond. B. Rachis and pinna base showing glandular hairs. C. Pinna.

Gymnocarpium jessoense: Nahanni oak fern: D. Frond. E. Pinnae.

Informative arrows: 1. Blades horizontal to ground. 2. Long stipes. 3. Look for undersurface of blades densely covered with short glandular hairs, upper surface moderately glandular. 4. Lowest pinna pair strongly curved toward frond tips. 5. Indusia lacking. 6. Look for blades glandular on the undersurface and glabrous on the upper.

***robertianum*, its pinnae curve toward the blade tips, and its blades are glandular on the undersurface and glabrous on the upper. In contrast, *G. dryopteris* is glabrous on both surfaces, and *G. robertianum* is glandular on both surfaces. In *G. robertianum*, the lower pair of pinnae is usually perpendicular to the rachises, and the basiscopic pinnules are perpendicular to the pinnae—not pointing toward the tips of the fronds.**

Gymnocarpium robertianum (Hoffman) Newman

Polypodium robertianum Hoffman

Northern Oak Fern, Limestone Oak Fern, Robert's Oak Fern, Scented Oak Fern

Etymology. Hoffman applied the specific name *robertianum* because the faint odor emitted from glands on its fronds is similar to that of *Geranium robertianum* (Herb Robert) and there is a certain likeness in the leaves. The original Herb Robert was probably named for Abbot Robert of Molesme (Holy Robert) who founded the Cistercian order in France in 1098 and whose festival date in April occurs at about the time the flowers bloom in Europe. Some say it was named for Robert Goodfellow, who is known as Robin Hood. The vernacular names derive from the scientific name, geologic preference, fragrant glands, and oak association.

Plants. Deciduous, monomorphic.

Fronds. 10–35 (–50) cm long, fronds well separated.

Rhizomes. Long-creeping, branching, 1–2 mm diam., dark brown or black, scales 2–4 mm.

Stipes. 5–30 cm long, **usually longer than blade,** 0.5–1 mm diam.; **glandular hairs numerous,** grooved; dull green, scaly at the base, scales 2–6 mm long, broadly lanceolate, thin, pale reddish-tan, soon falling.

Blades. 5–20 × 10–20 cm, 2-pinnate-pinnatifid to 3-pinnate-pinnatifid; **broadly triangular,** usually firm and robust, clearly composed of 3 parts, 2 large basal pinnae and a larger central frond tip significantly larger than the 2 lower pinnae, bends toward the horizontal from the rachis (but not as much as in *G. dryopteris*); surface dull; **abaxial surface of blade densely covered with short, glandular hairs; rachises' undersurface densely glandular,** upper surface moderately glandular.

Pinnae. 9–12 pairs, lowest pinnae 3–13 cm; more or less perpendicular to rachises, second pinna pair usually stalked or if sessile with basal pinnules shorter than adjacent pinnules, similar in form but smaller than lowest pair, tips acute, margins entire to slightly crenate.

Pinnules. Oblong, tips rounded, margins entire to slightly crenate, basiscopic pinnule of the lowermost pair of pinnae closest to the rachis usually much longer than the opposite acroscopic pinnule and divided to the same extent as the adjacent pinnule.

Sori. Round on both sides of pinnules, closer to the margin than midrib; indusia absent.

Habitat. Mainly in northern white cedar swamps, also calcareous substrates, woodlands, cool rocky swamps, limestone pavement, and cliffs.

Distribution. Circumboreal. Europe. Asia. Canada: from Ontario east and a few locations in Manitoba. USA: Iowa, Minnesota, Wisconsin, and Michigan. Uncommon in Michigan. A single collection from St. Clair County was made in 1888.

Chromosome #: $n = 80$

Gymnocarpium robertianum is a Michigan threatened species. It is considered locally imperiled but globally secure.

Gymnocarpium robertianum **may be recognized by blades that are densely covered with short, glandular hairs, which, when rubbed, emit a distinctive odor. It also has a larger and longer central lobe. (The 3 lobes of the** *G. dryopteris* **are roughly equal in size, and its fronds are smooth.)**

G. dryopteris **may be distinguished from** *G. jessoense* **by its lower pair of pinnae, which are more or less perpendicular to the frond axis, and by its basiscopic pinnules, which are more or less perpendicular to the midrib of the pinnae. On** *G. jessoense* **the lower pair of pinnae and their basiscopic pinnules curve strongly toward the tips of the fronds and pinnae respectively.**

GYMNOCARPIUM HYBRIDS

Hybrids with *Gymnocarpium robertianum* are extremely rare.

Gymnocarpium ×brittonianum (Sarvela) K. M. Pryor & Haufler (*G. dryopteris* × *G. disjunctum*) has been found in Houghton, Mackinac, and Ontonagon counties.

Gymnocarpium ×intermedium Sarvela (*G. jessoense* × *G. dryopteris*) has been found in Marquette County.

HOMALOSORUS Pichi Sermolli Diplaziopsidaceae

Glade Ferns

See Diplaziopsidaceae in appendix for discussion of this family.

Etymology. The scientific name derives from the Greek *homalo*, of the same kind or like, plus *sorus*. John K. Small, who original proposed the name, stated, "The name is from the Greek alluding to the regular lines of sori."

Plants. Weakly dimorphic (fertile fronds have narrower pinnae); deciduous.

Rhizomes. Short-creeping or erect.

Blades. 1-pinnate, tips pinnatifid, glabrous, lamina soft-herbaceous.

Rachises. Grooves V-shaped, not continuous with costae, the sulcus wall of the rachis continuing as a prominent ridge onto the sulcus wall of the costae.

Stipes. V-shaped grooves adaxially, straw-colored throughout; cross sections at bases reveal ends of 2 curved, linear vascular bundles appearing as linear and parallel lines that unite forming a single V-shaped bundle in upper stipes.

Veins. Free, forked.

Sori. Medial, **linear, single, usually on only 1 side of vein.**

Indusia. Linear, attached along lengths of sori.

Chromosome #: $x = 40$

Homalosorus is a genus consisting of 1 species limited to North America. Until recently it was treated as *Diplazium pycnocarpon* in most if not all floras and *Flora of North America North of Mexico*, vol. 2 (1993). It is treated here as belonging to the genus *Homalosorus*, but it may well belong in the genus *Diplaziopsis*. Further DNA and morphological studies are needed to clarify this. See the discussion under Diplaziopsidaceae in appendix I for a discussion of the problem.

Homalosorus pycnocarpos (Sprengel) Pichi-Sermoli

Asplenium pycnocarpon Sprengel; *Athyrium pycnocarpon* (Sprengel) Tidestrom; *Asplenium angustifolium* Michaux; *Dipalziopsis pycnocarpa* (Sprengel) Price; *Diplazium pycnocarpon* (Sprengel) M. Broun

Glade Fern, Narrow-Leaved Spleenwort

Etymology. Greek *pyknos*, dense, thick, crowded, compact, + *karpos*, fruit, alluding to the denseness of the sori on the narrow fertile pinnae. The common name narrow-leaved spleenwort refers to the narrow pinnae of this fern. Glade fern is a somewhat inappropriate name as this fern prefers the shade. (It was originally described as *Asplenium angustifolium* by Michaux [1803], but because that name had previously been taken Sprengel used *Asplenium pycnocarpon* [1804].)

Plants. Deciduous, somewhat dimorphic; fertile fronds appearing later in summer.

Fronds. 50–100 cm long, clustered with 5–6 fronds in linear to nearly circular groups near rhizome tips; *fertile fronds* appearing in late summer

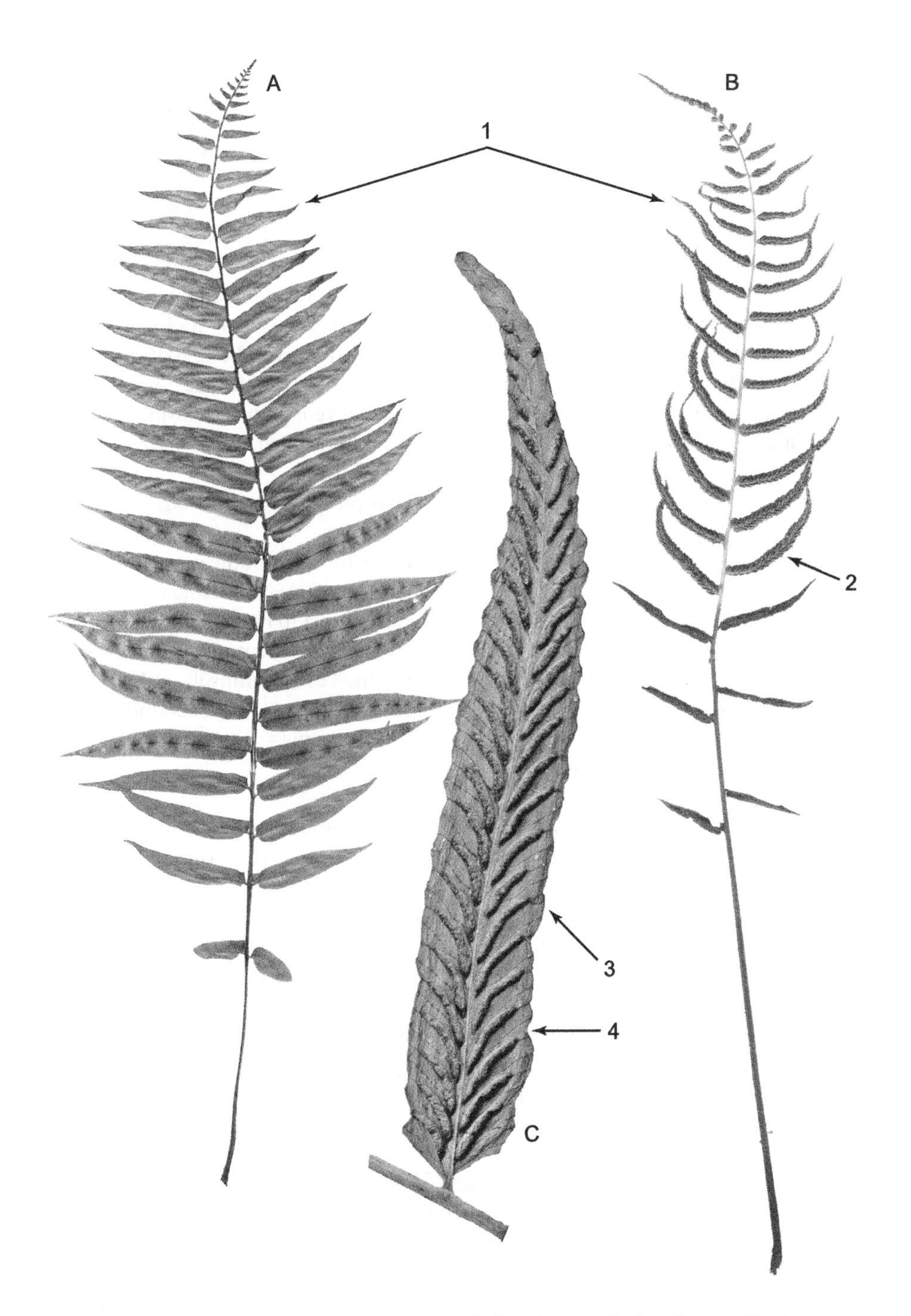

Homalosorus pycnocarpos: Glade fern: A. Sterile frond. B. Fertile frond. C. Fertile pinna. Informative arrows: 1. 1-pinnate frond. 2. Fertile frond narrower and with narrower pinnae than sterile frond. 3. Long narrow pinnae with wavy margins. 4. Linear sori.

are narrower, more erect, with longer stipes.

Rhizomes. 4–6 mm diam., short-creeping, older parts covered with stipe bases.

Stipes. 15–40 cm long, about 1/2 as long as blade, reddish-brown at base, greenish to straw-colored above, slightly scaly mostly at base; stipes of fertile fronds longer, more erect, and stiffer than sterile ones.

Blades. 25–75 × 8–25 cm, **1-pinnate**, narrowly lanceolate with long narrow tips, delicate, soft, herbaceous, wilting rapidly when cut; rachises glabrous, adaxially grooved, grooves not continuous with pinna costae; multicellular hairs on costae lacking.

Pinnae. 7–12 × 1.5 cm long, 20–35 alternate pairs, flaccid, soft; nearly linear, gradually tapering to pointed tips, **margins entire to somewhat wavy,** lower pinnae shorter, glabrous; *fertile pinnae* **much narrower and shorter** (appearing later in season).

Veins. Free, forked near midrib and sometimes beyond, expanding near margins, reaching margins at shallow sinuses between crenulations.

Sori. Long, linear or slightly curved with herringbone pattern, some sori at base of pinnae borne back to back on vein; indusia conspicuous, thin, brownish, opening toward midrib.

Habitat. Rich moist woods and shady slopes.

Distribution. Southeastern Canada. Eastern USA: Minnesota to Louisiana and east to the Atlantic Ocean. In Michigan found mostly in the lower half of the Lower Peninsula and scattered in its upper 3 tiers of counties.

Chromosome #: $2n = 80$

In most, possibly all, fern floras *Homalosorus pycnocarpos* (now in the family Diplaziopsidaceae) has been treated as *Diplazium pycnocarpon* (usually in the family Athyriaceae).

It was recently placed in the family *Diplaziopidaceae* as *Homalosorus pycnocarpos* (Sprengel) Pichi-Sermoli. The name *Dipalziopsis pycnocarpa* (Sprengel) Price has been proposed. Recent DNA and morphologic studies have suggested this change, and future studies of the uncommon ferns in the Diplaziosidaceae may well support the change. Certain morphological and chromosome differences have caused hesitation in making this change now. For an explanation of this change see Diplaziopsidaceae in appendix I.

Homalosorus is a genus with a single species native to eastern North America. There is a large geographic separation between this genus and the other members of its family, which are found in China and Southeast Asia.

Homalosorus pycnocarpos **is one of the few Michigan ferns with 1-pinnate fronds and pinnae with entire to crenulate margins and the only one of these with sori linear along the veins. It may resemble** *Polystichum acrostichoides* **but can be distinguished by sori that are linear with a herringbone pattern and pinnae with mostly smooth margins. In contrast,** *P. acrostichoides* **has round sori pinnae with "ears" at the bases, toothed margins, and small fertile pinnae limited to the frond tips.**

LYGODIUM Swartz Lygodiaceae

Climbing Ferns

This is a family with a single genus, *Lygodium*. The treatment of this family is identical to the treatment of the genus below.

Etymology. Greek, *lygodes*, pliant or flexible, in reference to the winding, flexuous, climbing rachis.

Plants. Growth indeterminate; climbing; terrestrial.

Rhizomes. Subterranean; slender, with dark, dense hairs.

Fronds. Vinelike, twining and climbing; usually 3–10 m long, separated by 1–4 cm.

Stipes. Twining, branching, shiny black, shading to brown and green above, hairy at base, 1 cylindrical vascular bundle.

Blades. Long, rachises often very long, 2-pinnate or more divided, straw-colored, glabrous.

Pinnae. Alternate, short, 1–3 cm long, represented by pinna rachises that are short stubs, often with dormant buds at apices, buds immediately branching at 180° angles into 2 opposite frondlike pinnules, fertile pinnae borne toward apex; fertile and sterile pinnae similar or fertile pinnae greatly contracted.

Pinnules. More or less entire to palmately or pinnately lobed.

Ultimate segments. Palmate, pinnate, or entire, alternate, usually either entirely vegetative or entirely fertile; *fertile segments* narrower, with contracted, fingerlike lobes.

Sporangia. Borne in 2 rows on either side of marginal lobes; *false* indusia ovate, hoodlike flaps, facing outward.

Chromosome #: $x = 29, 30$

Lygodium is a mostly pantropical and subtropical genus of about 35–40 species. Three species are treated in *Flora of North America North of Mexico*, vol. 2 (1993). It is represented in Michigan by a single rare (in Michigan) species with a limited distribution.

Lygodium palmatum (Bernhardi) Swartz

Gisopteris palmata Bernhardi

Climbing Fern, Hartford Fern

Etymology. Latin, *palmatus*, hand-shaped or lobed, alluding to the handlike shape of the pinnules. The vernacular name describes its vinelike climbing growth habit. Hartford fern references the capital of Connecticut, where the state legislature passed a conservation law in 1869 that made this species the first legally protected plant in the United States.

Plants. Partially evergreen (fertile fronds dying back in winter), dimorphic in that fertile pinnae are very different from sterile, more like a climbing vine than a fern.

Fronds. Trailing, vining; up to 3

m long, ca. 13 cm wide, fertile toward tips, sterile toward base.

Rhizomes. Slender, long-creeping, branching, black, with short blackish or red hairs.

Stipes. 1/30 the length of blades, separated by 1–4 cm.

Blades. 2-pinnate, rachises twining, 120+ cm long, growing from rhizome and branches above ground (not fern-like), round, shiny black at base, brown to green upward.

Sterile pinnae. 2–6 × 1–4 cm, on 1–2 cm stalks, alternate, with a single pair of lobed pinnules.

Sterile pinnules. Palmately divided with 4–10 lobes; handlike, held horizontally; triangular-elongate to oblong, margins entire, finely hairy on undersurface.

Fertile pinnae. 3–8 × 1–3 cm, narrower and shorter than fertile pinnae, on 1–2 cm stalks, always toward tips of the fronds.

Sterile pinnules. Smaller than fertile pinnules, with 3–6 pinnules, 1–2 × 0.8–1.2 cm, irregularly palmately divided; lobes triangular-elongate to oblong, lobe tips round to pointed.

Veins. Free, forking.

Sori. In 2 rows of 6–10 on each fertile pinna lobe; *false indusium* scale-like, an outgrowth of the leaf lamina, opening along a longitudinal slit, covering the surface of the ultimate segment.

Habitat. Open woods, low-shade habitats, moist acidic soils, bogs, and marshes.

Distribution. USA: eastern United States, Michigan to New Hampshire, south to Florida, west to Mississippi, north to Michigan; generally local and rare except on the Cumberland Plateau of Kentucky and Tennessee. In Michigan *Lygodium palmatum* has been found in only two disjunct colonies; both in wet and shaded sites in the southwestern Lower Peninsula.

Lygodium palmatum is listed a Michigan endangered species. It is globally secure. In 1985 Burlington County, New Jersey, moved a planned road 200 feet to avoid 2 patches of this fern.

***Lygodium palmatum* resembles no other fern in Michigan and is very unfernlike. It appears to be a slender climbing, twining vine bearing alternate leaves with short petioles. The long twining stem is composed of the twining stipe and rachis, and the apparent leaves are pinnae with palmately lobed pinnules.**

***Lygodium palmatum* is recognized by its long twining rachises, alternate pinnae well separated on short stalks, subterranean rhizomes, and stipes only 1/30 as long as the blade. The pinnae stalks end abruptly after producing 2**

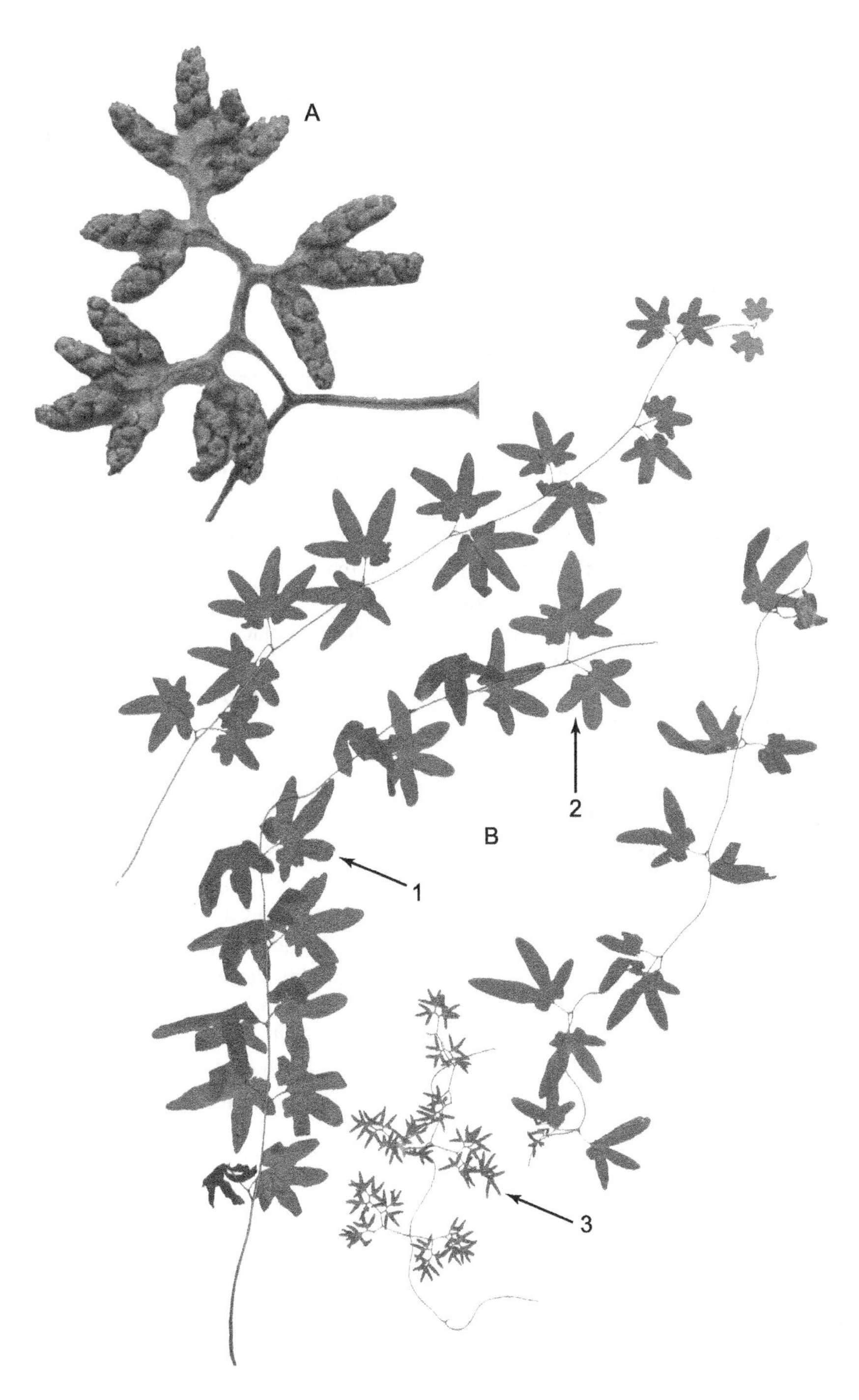

Lygodium palmatum: Climbing fern: A. Enlargement of fertile pinnae. B. Vinelike frond (frond is 200 cm long) with fertile and sterile pinnae.

Informative arrows: 1. Frond vinelike. 2. 3–5 fingered sterile pinnae. 3. Fertile pinnae at frond tip.

hand-shaped, alternate pinnules. The fertile pinnae are found only near the tips of the blade and bear 1 to 3 pinnate pinnules with linear ultimate segments that are smaller and narrower than the fertile ones.

MARSILEA Linnaeus Marsileaceae

Water Clovers

Marsileaceae is a family with 3 genera. The genus *Marsilea* is the only one found in Michigan and the only genus in the family treated here.

Etymology. Generic and family names honor Count Luigi Ferdinando Marsigli (1656–1730), an Italian botanist from Bologna.

Plants. Small, deciduous, monomorphic, aquatic or amphibious, fronds floating, forming clones in transient pools or moist flooded places.

Rhizomes. Slender, mostly long-creeping, growing on soil surface or subterranean in mud, often hairy; roots arising at nodes or along internodes.

Fronds. Narrow.

Stipes. Round, long, green.

Blades. 7–21 × 0.5–1.9 cm, with 4 closely attached pinnae at tips of stipes giving the **appearance of a four-leaf clover**.

Veins. Dichotomously branched, often anastomosing near their tips.

Sporocarps. Elliptic, firm, hairy (glabrous with age), single or multiple with firm, thick walls, inserted on bases of stipes or on rhizomes near stipe bases. Sporocarps are probably derived from highly modified blades or pinnae fused at the margins to form the capsules. Their stalks are probably derived from stipes. The sporocarps and enclosed sorophores contain multiple sporangia of 2 kinds, *megasporangia* containing a single *megaspore* (female) and *microsporangia* containing 20–64 *microspores* (male). Sporocarps are extremely drought resistant and may remain dry and dormant for many decades, reviving when rains return.

Chromosome #: $x = 20$

Marsileaceae is a widespread family of 3 genera and 45–50 species of nearly worldwide distribution, absent only from very cold climates. Two genera and 7 species (6 *Marsilea*, 5 native and 1 introduced) are treated in *Flora of North America North of Mexico*, vol. 2 (1993). The family is represented in Michigan by a single introduced species of *Marsilea*.

The species in Marsileaceae are often difficult to distinguish because of their similar morphology and often sterile condition.

The leaves and sporocarps of some *Marsilea* species are and were used as food in various parts of the world. Sporocarps of some Australian species are edible and were eaten by Aborigines and early white settlers, who knew them as **nardoo**.

The sporocarps contain an enzyme that destroys thiamine (vitamin B_1), leading to brain damage in sheep and horses, and to beriberi in humans, if not properly prepared. During floods in Australia sheep have died after eating nardoo. Sheep that received thiamine injections recovered.

Thiamine deficiency caused by eating incorrectly prepared nardoo was blamed for the deaths of Robert O'Hara Burke and William John Wills on their exploratory expedition across central Australia in 1860.

Marsilea quadrifolia Linnaeus
European Water Clover

Etymology. Latin *quadri-*, fourfold, + *-folius*, leaved, alluding to the 4 pinnae of the frond. The vernacular name refers to the fronds' resemblance to a four-leaf clover and to the fern's watery habitat.

Plants. Deciduous, forming clones in shallow water or wet soil.

Rhizomes. Slender, long-creeping; roots both at nodes and sparsely along internodes; anchored on muddy soils or pond bottoms.

Blades. Resembling a four-**leaf clover,** with 4 closely attached pinnae at tips of stipes; floating blades usually larger than those on land.

Stipes. 5.5–16.5 cm, narrow; green; sparsely pubescent to glabrous, variable in length depending on depth of water, mostly trailing when along ground on wet substrate.

Pinnae. 7–21 × 6–19 mm, **wedge-shaped with rounded outer edges,** glabrous or sparsely pubescent, **arranged in a whorl of 4, resembling a four-leaf clover.**

Veins. Free, forked, spreading fanlike.

Sporocarps. 4–5.6 × 3–4 mm, 2.3–2.8 mm thick, 2–3 per branched or unbranched stalk, attached 1–12 mm above base of stipe, pubescent, soon glabrous, rounded, oval, or elliptic in lateral view; sometimes with a small bump or 2 on the upper surface; stalks attached to base of stipes, frequently branched, common trunk of branched stalks at 1–4 mm, appearing from June to October.

Spores. Macrospores (female) and microspores (male) are contained in a gelatinous sorophore, which when released from the sporocarp shoots out, dissolves in the water, and releases the spores.

Distribution. Native to Europe, first noted in Connecticut in 1860, since spread throughout New England to Ontario, in the Midwest to Ohio and Missouri, and south to Georgia. Humans probably extend its range, planting it as a curiosity. In Michigan it has been found in Grand Traverse and several southern counties thus far.

Chromosome #: $2n = 40$

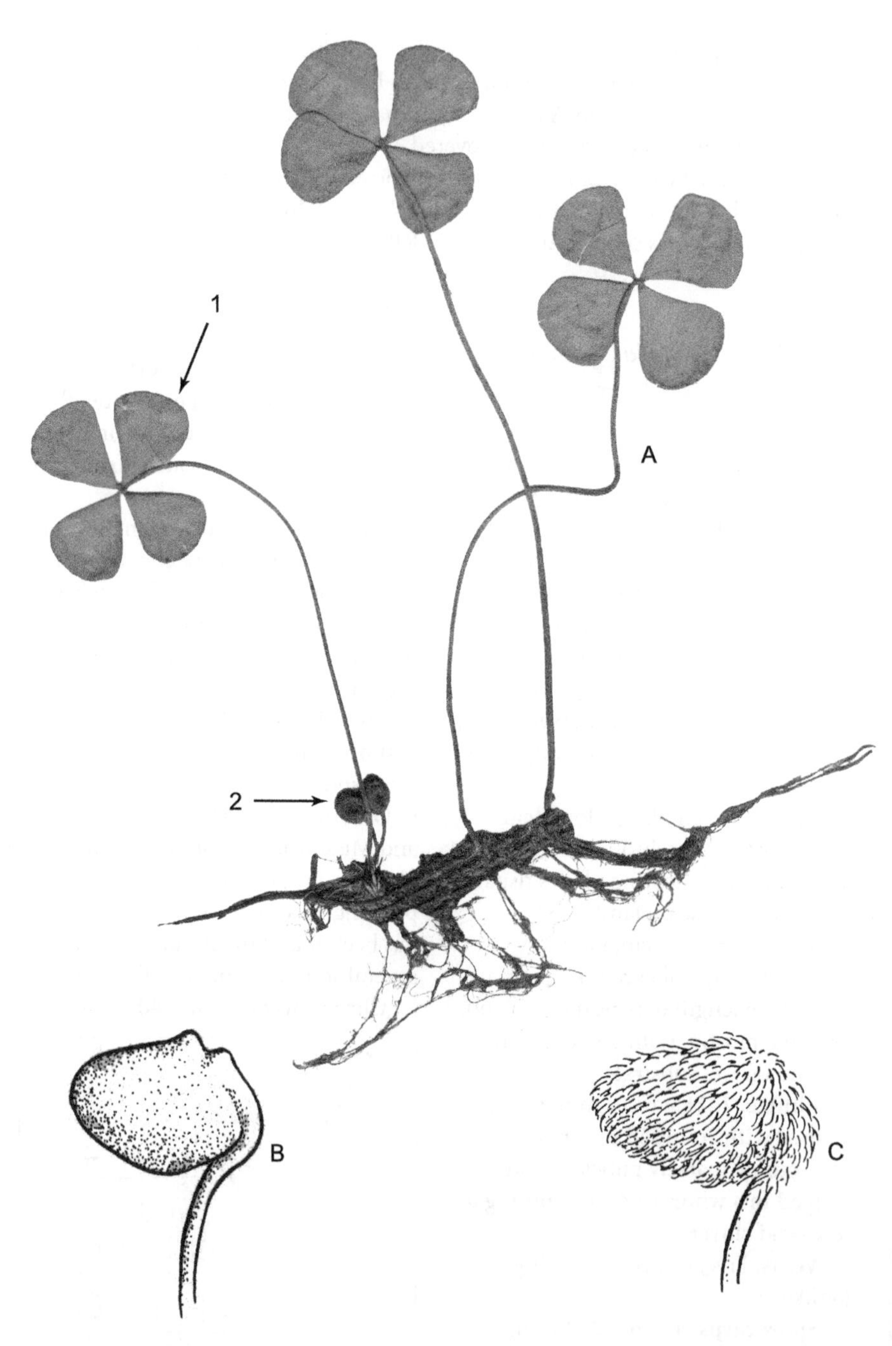

Marsilea quadrifolia: European water clover: A. Plant with rhizome. B. Old sporocarp. C. Young hairy sporocarp.

Informative arrows: 1. Blades resembling a four-leaf clover. 2. Sporocarps.

Native to Eurasia, *Marsilea quadrifolia* may be the only alien species of fern found escaped in Michigan.

Fertile leaves are produced mainly on dry land, seldom in water.

A part of the hard capsule of a dry sporocarp may be clipped off. If it is then soaked in water in a short time it will produce a visible, long gelatinous sorophore—an interesting simple demonstration.

A patch of *Marsilea quadrifolia* **resembles a group of four-leaf clovers growing in shallow water or on mud. The sporocarps are tiny and nutlike objects attached to the base of the stipes.**

MATTEUCCIA Todaro Onocleaceae

Ostrich Ferns

See Onocleaceae in appendix I for a description of the family and a key to the Michigan genera. Pteretis Rafinesque is an older legitimate generic name, but it is not currently used.

Etymology. The name honors Carlo Matteucci (1811–63), an Italian nationalist, politician, electrophysiologist, and naturalist.

Plants. Dimorphic; sterile fronds deciduous, fertile fronds persistent through winter; clonal and spreading, with many shuttlecock-shaped ramets.

Rhizomes. Ascending to erect, narrow stolons present, giving rise to a new plant and forming a colony, persistent thick stipe bases (trophopods that store starch) present.

Sterile fronds. Much longer and with wider pinnae than fertile ones.

Stipes of sterile fronds. Short, circa 1/10–1/5 blade length; cross section reveals 2 curved vascular bundles.

Sterile blades. 1-pinnate-pinnatisect, oblanceolate, base tapering gradually to small basal pinnae, gradually to somewhat abruptly reduced to pinnatifid tips (with an ostrich feather shape); thick, leathery.

Sterile pinnae. Several pairs of lower pinnae gradually and greatly reduced, sessile, margins of lobes entire, glabrous or sometimes with deciduous hairs on undersurface, midribs with shallow grooves, grooves not continuous from rachis.

Veins. Free, pinnately branched in lobes.

Fertile blades. 40–60 cm long, 1-pinnate, oblanceolate (like sterile fronds, but much smaller and narrower), becoming stiff, erect, brown and persisting over winter.

Fertile pinnae. Lined with 2 alternate rows of hard, pealike pinnules.

Fertile pinnules. Green, with horizontal slits, becoming dark brown and hard at maturity.

Stipes of fertile fronds. About the same length as fertile blades, bases thick.

Sori. Obscure, 1 row, medial, covered with hardened downward-rolled pinnule; indusia persistent, difficult to see in mature pinnules.

Spores. Green.

Distribution. East Asia and northern temperate regions of the eastern United States and Canada.

Matteuccia is a circumboreal genus of 3 species. One species is treated in *Flora of North America North of Mexico*, vol. 2 (1993). The genus is represented in Michigan by this single species.

Chromosome #: $x = 40$

Matteuccia struthiopteris (Linnaeus) Todaro var. *pensylvanica* (Willdenow) C. V. Morton

Osmunda struthiopteris Linnaeus; *Matteuccia pensylvanica* (Willdenow) Raymond; *Pteretis struthiopteris* (Linnaeus) Nieuwland

Ostrich Fern

Etymology. Latin *struthio*, ostrich, + *pteris*, botanical Latin for fern (related to Latin *pteron*, feather or wing)—ostrich fern, alluding to the sterile fronds' shape and size, which resemble those of ostrich feathers. The vernacular name derives from the scientific name and the shape of the sterile fronds.

Plants. Large, **dimorphic, sterile fronds dying in winter, fertile fronds appearing in mid- to late summer, persisting through winter,** shedding spores the following spring.

Rhizomes. Thick, erect, covered with persistent, thick stipe bases (trophopods that store starch), with many narrow subterranean stolons.

Sterile fronds. 50–145 × 12–35 cm, 1-pinnate-pinnatifid, arranged in **vaselike clusters.**

Sterile blades. Oblanceolate to oblong, much narrower at the base (like ostrich feathers), gradually to nearly abruptly reduced distally to pinnatifid tips; rachises grooved, sometimes hairy.

Stipes of sterile blades. 4.5–45 cm long, 1/4 or less as long as blade, dark brown, thick, rigid, flattened at base, deeply grooved distally; scales pale orange-brown.

Sterile pinnae. Longest 6.5–13.5 cm, 20–60 pairs; alternate, linear, narrow, pointed, gradually becoming shorter toward base of blade, lower pinnae much smaller; sinuses deeply cut.

Lobes of sterile pinnae. 20–40 pairs, tips rounded.

Fertile fronds. 15–46 × 2.5–6.5 cm, 1-pinnate, oblong to oblong-lanceolate (same general outline as sterile frond), tough and durable; appearing mid- to late summer, persisting over winter and shedding spores in the spring.

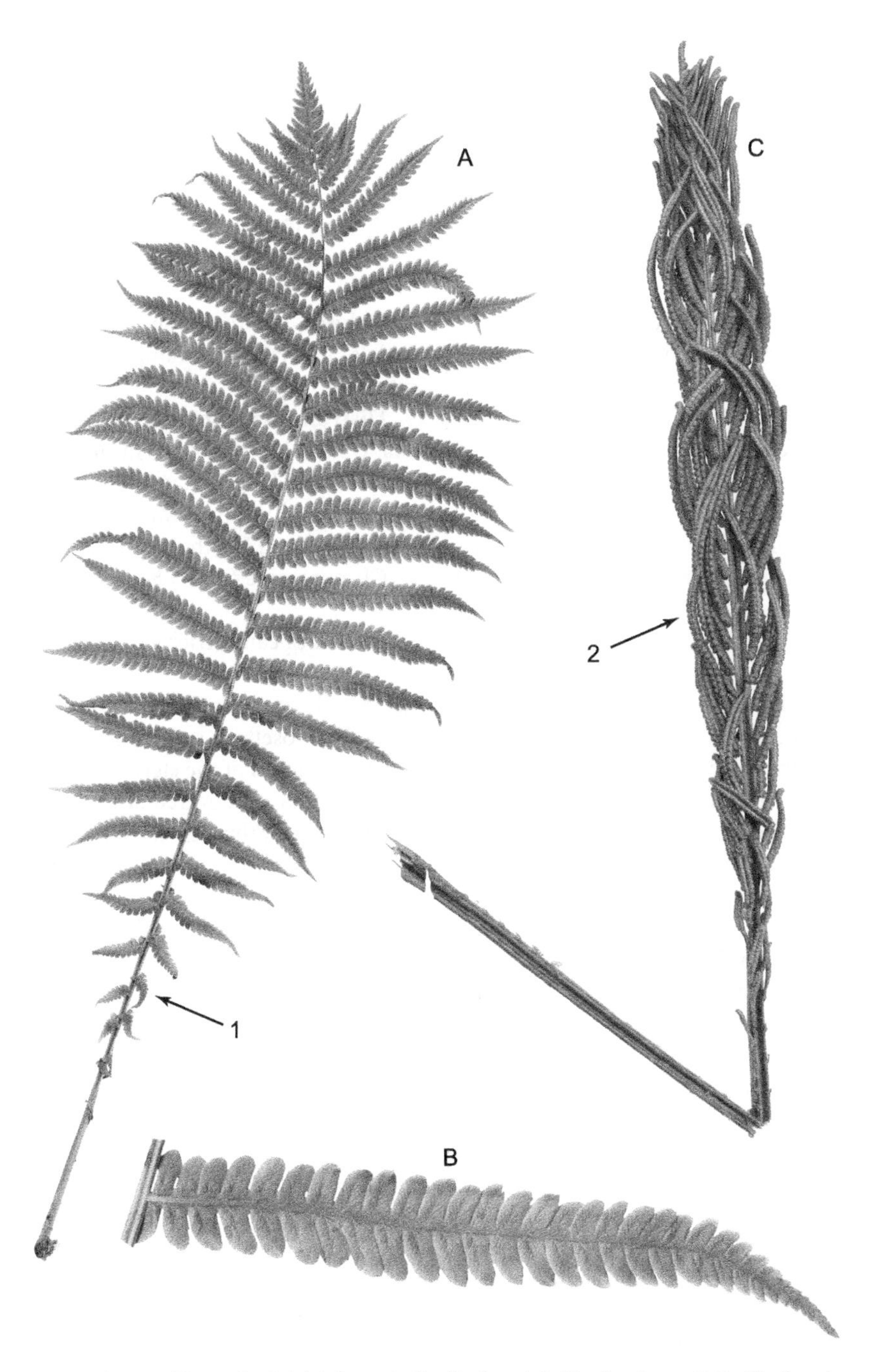

Matteuccia struthiopteris: Ostrich fern: A. Sterile frond. B. Sterile pinna. C. Fertile frond. Informative arrows: 1. Tapering to very small lower pinnae (with an ostrich feather shape). 2. Fertile frond becomes dark and woody in fall, persistent over winter.

Stipes of fertile blades. 11–24 cm, erect, approximately equal to blade length, dark brown to black, stiff, bases scaly.

Fertile pinnae. 3–5.6 × 0.2–0.5 cm, 30–45 pairs, linear, strongly ascending, becoming shorter toward base of blade, greenish, becoming dark brown at maturity, lined with 2 alternate rows of pinnules.

Fertile pinnules. 10–40+, consisting of contracted hardened, durable, pealike spheres with horizontal slits.

Sori. Borne within tightly inrolled edges of hardened pinnules clustered on the upright fertile fronds, separated by narrow slits; indusia small, cuplike, not easily seen on mature fronds. (Sori protected more by pinnules than by indusia.)

Spores. Green.

Veins. Free; pinnately branched in lobes.

Habitat. Deciduous and mixed woods, alluvial flats, swamps, and calcareous soils.

Distribution. Throughout Canada. USA: Alaska, north central and northeastern United States south to Missouri and Virginia. Most common in northeastern North America.

Chromosome #: $2n = 80$

Matteuccia struthiopteris var. *struthiopteris* is the European ostrich fern. It differs in scale color and pinna lobe shape.

Matteuccia struthiopteris var. *pensylvanica* has short fertile fronds with tough dry pinnules appearing in the center of the sterile leaf rosette in summer. The shallow lobes of the fertile pinnules correspond with the lobes of the sterile pinnules, and each bears a few sori with many sporangia, which are covered at their base with a long cuplike indusium. The fertile pinnules persist through the winter and split open in early spring, exposing the sporangia and releasing the spores before the new sterile fronds have expanded.

The spores are released into an unimpeded airstream, allowing the spores to disperse easily when the wind blows. The green spores, lacking oil droplet energy reserves, must germinate in 2 to 5 days. Their germination and development require a continuously moist early spring just after the spring thaw. The persistent spore-producing leaves are designed to retain the spores until the optimal time for their germination and maturation.

In addition to the sexual spread by spores, this fern spreads asexually by underground stolons that often form large clones.

Matteuccia struthiopteris is the most common native species sold for landscaping in the United States and Canada. Vigorous and spreading, it is often used as a border plant.

The fiddleheads of this plant are edible and gathered in early spring. They are considered a delicacy (try them, they are good), and they appear at the same time as the ramps and morels. Canning the fiddleheads is a local industry in New England and adjacent Canada. In some areas the fiddleheads are sold in markets. Each year a few hundred tons are harvested exclusively from wild habitats, mostly by Native Americans. The Japanese call the fiddleheads *kogami* and consider them delicious.

The fronds of *Matteuccia struthiopteris* **var.** *pensylvanica* **are sometimes con-**

fused with *Osmunda cinnamomea* **and** *O. claytoniana* **but can easily be distinguished by the very small pinnae at the base of the frond and their ostrich-feather shape. The fertile fronds of** *Matteuccia* **are very different; they are shorter than the sterile fronds, become dark brown and firm when mature, and persist through the winter. In contrast, those of the** *Osmunda* **species either have a separate fertile frond that is soft and deciduous or have the sterile pinnae divided from the fertile pinnae on the same deciduous frond.**

See comments under *Onoclea sensibilis* **for differences between its persistent fertile fronds and those of** *M. struthiopteris.*

Thelypteris noveboracensis **is the only other Michigan fern with pinnae that gradually taper to very small at the base that might be confused with** *Matteuccia struthiopteris*. **However, the fronds are very different, being much smaller, thinner, tapered at both ends, and not shaped like ostrich feathers. It has round sori and lacks separate and very different evergreen fertile fronds.** *T. noveboracensis* **has small, white, acicular hairs scattered on its rachises and upper surfaces of its pinnae that are lacking in** *M. struthiopteris*.

ONOCLEA Linnaeus Onocleaceae

Sensitive Fern

See Onocleaceae in the appendix for a description of the family and a key to the Michigan genera. *Onoclea* is a genus with one species worldwide (there is possibly a second one in the Himalaya and East Asia). The description of the genus is the same as the species description that follows.

Etymology. Greek *onos*, vessel, + *kleiein*, to close, referring to the hard, bead-like structures enclosing the sori (or possibly referring to the areolae enclosed by the netted veins). *Onoclea* is an ancient name taken and used by Linnaeus.

Onoclea sensibilis **Linnaeus**
Angiopteris sensibilis (Linnaeus) Nieuwland
Sensitive Fern, Bead Fern

Etymology. Latin *sensibilis*, sensitive, refers to the fern's fronds, which quickly wilt after being picked, or perhaps to frost susceptibility (among the ferns it is not especially early frost sensitive). The common name is derived from the Latin. Bead Fern references the hardened bead-like structures bearing the sporangia on the fertile stems.

Plants. Dimorphic; sterile fronds deciduous; fertile fronds persisting over winter.

Rhizomes. Long-creeping, stout, 4–7 mm diam., branching extensively and spreading near the soil surface producing a fibrous mat, covered with thickened residual stipe bases (phyllopodia that store starch), brown, smooth; roots numerous, forming a fibrous mat.

Sterile fronds. 35–80 (–90) × 15–35 cm, irregularly spaced along rhizomes, often in thick clusters.

Stipes of sterile fronds. 20–50 cm long, between 1 and 1.5 times blade lengths, brittle, smooth, shallowly grooved, green to straw-colored, dark brown at bases; scales few, light brown at base; cross section reveals 2 elongate-curved vascular bundles.

Sterile blades. 12–34 × 15–35 cm, triangular, larger fronds 1-pinnate-pinnatifid at base, pinnatifid above bases, lower pinnae long and tapered at both ends, upper lobes with little or no tapering toward rachis, lobes cut deeply to winged rachises, light green to yellow-green; rachises winged, wings becoming broader toward tips, grooved.

Sterile lobes and/or pinnae. 5–12 pairs, subopposite, lanceolate, proximal pinnae largest or nearly so, 9–18 × 1–3 cm, unstalked, winged to rachis, margins entire to sinuous to shallowly lobed to pinnatifid, sometimes with very fine small teeth (hand lens needed), linear to lanceolate scales and/or multicellular hairs on rachis and costae; costae flat above (adaxially).

Veins of sterile fronds. Netted, enclosing areolae without included veins.

Fertile fronds. 2-pinnate, 20–50 cm long, erect, shorter than sterile fronds, brown, emerging August to October, persisting through the winter and releasing the green spores in spring before the sterile fronds emerge, may persist 2–3 years.

Fertile stipes. 19–40 cm long, 2–6 times length of blade, straw-colored to light brown, base sparsely scaly.

Fertile blades. 7–17 × 1–4 cm, linear to oblong, green, becoming black at maturity, borne at intervals along rhizomes, not forming clumps.

Fertile pinnae. 5–12 pairs, 2.5–5 × 0.5–0.8 cm, subalternate to alternate, linear, strongly ascending.

Fertile pinnules. 5–12, 2–4 mm diam., margins inrolled, contracted downward to form dry, hard beadlike structures surrounding sori, green becoming dark brown to black when mature, eventually splitting into circa 8 triangular segments to release spores.

Veins of fertile pinnules. Free.

Sori. Small, round, green, borne on free veins within inrolled hardened pinnules; indusia cuplike, not easily seen on mature fronds. (Protection of sori is derived more from pinnules than indusia.)

Spores. Green.

Habitat. Swamps, bogs, marshes, wet woods and meadows and stream and riverbanks in sun or shade.

Distribution. East Asia. Eastern temperate North America, Newfoundland west to Manitoba, south to Florida, and west to Texas and North Dakota. Found abundantly throughout Michigan.

Chromosome #: $2n = 74$

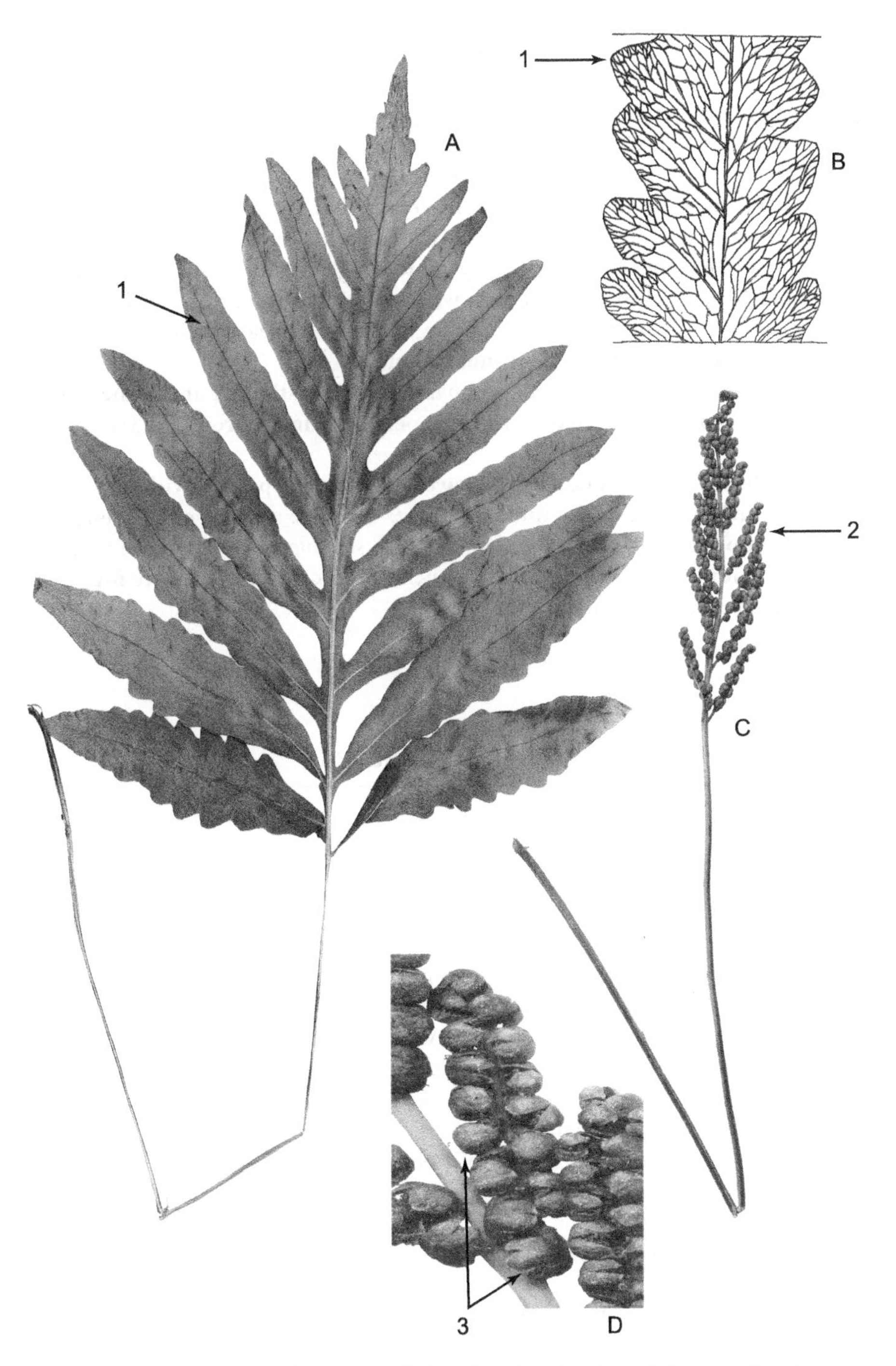

Onoclea sensibilis: Sensitive fern: A. Sterile frond. B. Line drawing of pinna section showing reticulate veins. C. Fertile frond. D. Detail of fertile pinnae and pinnules.
Informative arrows: 1. Netted veins easily seen. 2. Fertile frond becomes dark and woody in fall, persistent over winter. 3. Inrolled hardened pinnules.

As in *Matteuccia struthiopteris*, the sporophylls of *Onoclea sensibilis* persist through the winter and release the green spores in spring before the sterile leaves expand. (See the discussion under *M. struthiopteris* above.)

Onoclea sensibilis **has very distinctive, bright yellow-green, pinnatifid sterile fronds borne at intervals along the rhizomes that are not clustered. It is the only pinnate, medium-sized fern in Michigan with veins of the sterile fronds that are netted. The blades are deeply cut to winged rachises. The fertile fronds turn black or dark brown at maturity when the pinnae roll up to form hard beadlike structures that persist through the winter and release their green spores in the spring before the sterile fronds expand.**

The fertile fronds of *Onoclea sensibilis* **are quite distinctive and in the Michigan flora can be confused only with the persistent fertile fronds of** *Matteuccia struthiopteris***.**

The fertile fronds of *O. sensibilis* **have fewer than 12 pinnae, with the lowest being nearly as long as the middle pinnae. They are solitary or in groups of seldom more than 3. The fertile fronds of** *M. struthiopteris* **have more than 24 pinnae, with the lowest pinnae very much reduced in size. They are found in rosettes of 3 to 12 or more.**

OSMUNDA Linnaeus Osmundaceae

Royal Ferns

See Osmundaceae in the appendix for a description of the family.

Etymology. The real origin of this fern's name is too old to be traced, predating Linnaeus's use by half a century. It may be derived from Osmunder, the Saxon name for the god Thor. Other derivations have been proposed, including Asmund, a Nordic mythological hero who saved his family from invading Danes by hiding them among large ferns. This mythology was later incorporated into the legend of Saint Christopher—the protector of travelers. Heart of Osmund and Herb Christopher were old English names for the fern. It is also possible that it was named for King Osmund, who reigned over the southern Saxons around 758 CE If this is correct, the common name's origin is obvious. Another possibility is Latin *os*, bone, + *munda*, to clean, perhaps because the plant was used medically to clean bones. Take your pick.

Plants. Terrestrial, monomorphic or dimorphic, growing in clusters, dying back in winter.

Rhizomes. Creeping to decumbant, massive, with matted, black fibrous roots and old stipe bases, scales absent; stipe bases bearing winglike outgrowths called stipules.

Stipes. Vascular bundles U-shaped on cross section at base, with tips of arms continuing to curl.

Blades. 1-pinnate-pinnatifid to 2-pinnate, rachises grooved, indument of reddish to light brown hairs.

Pinnae. Monomorphic or dimorphic.

Veins. Free, forking.

Sori. Absent, large, spherical, thin-walled sporangia opening with a long slit at the top, attached individually to deeply contracted pinnules of fertile fronds, not clustered in sori, green becoming dark brown after spores are shed.

Spores. Green, discharged in early spring, short-lived.

Sporangia. With the annulus composed of a patch of differentiated cells (instead of the ringlike ones found in more advanced ferns).

Distribution. Nearly worldwide in temperate zones and the tropics.

Chromosome #: $x = 22$

Osmunda is a genus with 10–14 species worldwide. Three species are treated in *Flora North America North of Mexico*, vol. 2 (1993). The same 3 species are found in Michigan.

The family Osmundaceae is an isolated and primitive group not closely related to other ferns. Fossils of Osmundaceae are known from as far back as the Permian era (248 to 290 million years ago), and the family probably goes back to the Carboniferous (325 million years ago). Fossils related to the genus *Osmunda* have been found in Upper Cretaceous sediments (up to 65 million years ago).

The placement of the 3 species found in Michigan is in a state of flux. Recent DNA studies have treated these species as belonging to the separate genera *Osmundastrum* (a monospecific genus) with *Osmundastrum cinnamomea,* and *Osmunda* with 2 subgenera genera (subg. *Osmunda* with *Osmunda regalis* and subg. *Claytosmunda* with *Osmunda claytoniana*). This treatment may become widely accepted.

The sporangia of *Osmunda* species are thought to be primitive and appear to be intermediate between the large sporangia of the Ophioglossaceae, which contain thousands of spores, and the small, slender-stalked, thin-walled sporangia of "modern" ferns, which usually contain 64 or fewer spores.

Osmunda spores are green and shed from an individual leaf over the course of a few days. They must germinate and form gametophytes rapidly due to poor protection from desiccation and their high metabolic activity.

A recent publication by a group composed of a great many fern taxonomists associated with many institutions worldwide presents a phylogenetic classification derived from the best available data by those most familiar with fern taxonomy.

This study shows that the 3 species now included in the genus *Osmunda* are different enough to be separated into 3 monospecific genera. These monospecific genera and the species included in them are:

Claytosmunda Y. Yatabe, N. Murak. & K. Iwats. With *Claytosmunda claytoniana* (L.) Metzgar & Rouhan

Osmundastrum C. Presl with *Osmundastrum cinnamomea* (L.) C. Presl

Osmunda L. with *Osmunda regalis* L.

These suggestions will surely become widely accepted and will be used in future publications.

KEY TO SPECIES OF OSMUNDA IN MICHIGAN

1. All fronds bipinnate (pinnae resembling black locust leaves); fertile fronds with clusters of small fertile pinnae at the tips .... *O. regalis*
1. Dimorphic (pinnae not resembling black locust leaves); fronds 1-pinnate; fertile pinnae on wholly fertile fronds or fertile pinnae in middle of frond with sterile pinnae above and below (2).
2(1). Sterile and fertile pinnae on separate fronds; tufts of fine tan hairs persistent on undersurface of pinna stalks at junction with rachis; sterile blade tips somewhat sharply pointed .... *O. cinnamomea*
2. Sterile and fertile pinnae on same fronds when fertile, sterile fronds lacking fertile pinnae; fertile fronds with fertile pinnae in middle zone of blades with sterile pinnae above and below; tufts of tan hairs not persistent on undersurface of pinnae stalks at junction with rachis: sterile blade tips rounded .... *O. claytoniana*

***Osmunda cinnamomea* Linnaeus**
Osmundastrum cinnamomeum (L) C. Presl
Cinnamon Fern

Etymology. Latin *cinnamomeus*, cinnamon-colored, alluding to the cinnamon-colored hair on crosiers and fronds. The vernacular name has the same derivation.

Plants. Dimorphic; deciduous.

Fronds. Growing in symmetrical clusters; fiddleheads large, covered with whitish hairs that turn rusty as fronds unfurl.

Rhizomes. Large, sometimes massive, stout; short-creeping, becoming erect, covered with a dense tangled mass of roots.

Sterile fronds. 45–160 cm long; erect, arranged in a shuttlecock pattern.

Sterile stipes. About 1/3 length of blade, green to straw-colored, round, grooved adaxial surface, fuzzy with light brown soon deciduous hairs.

Sterile blades. 1-pinnate-pinnatifid to 1-pinnate-pinnatisect, 35–75 × 12–25 cm, lanceolate with **acuminate tips**, light green; rachises green with

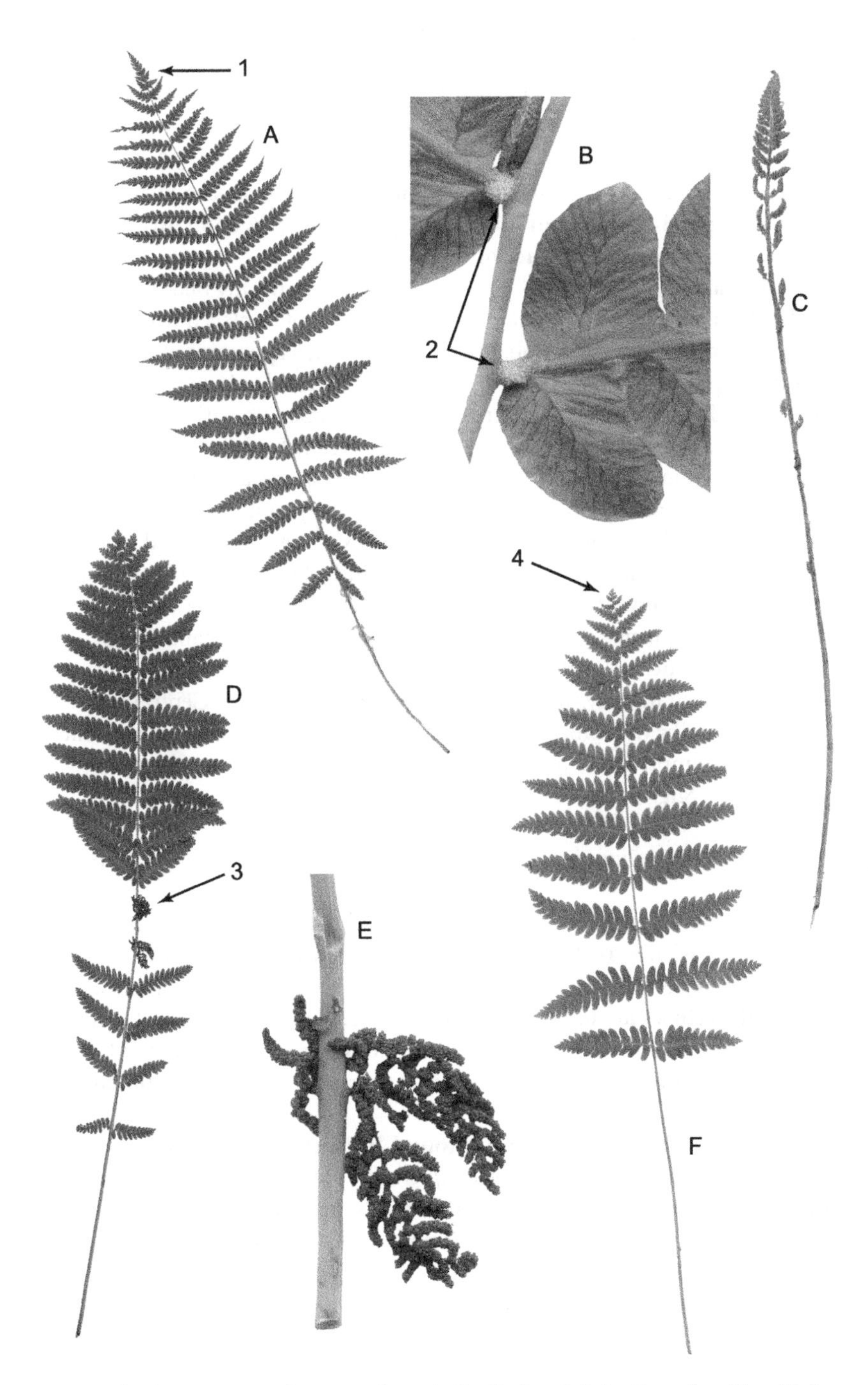

Osmunda cinnamomea: Cinnamon fern: A. Sterile frond. B. Section of rachis with bases of pinnae. C. Fertile blade.

Osmunda claytoniana: Interrupted fern: D. Fertile frond. E. Detail of fertile pinnae. F. Sterile frond.

Informative arrows: 1. Pointed tip. 2. Tufts of fine cinnamon-colored hairs. 3. Fertile pinnae between sterile pinnae. 4. Rounded tips.

scattered tan to cinnamon-colored hair early, becoming glabrous.

Sterile pinnae. Lanceolate, alternate, short-stalked, deeply cut almost to midrib, with **persistent tufts of fine rust-colored hairs on undersurface of pinnae stalks' axils.**

Sterile ultimate segments. Circa 1.3 × 0.7 cm; margins entire, tips obtuse to slightly pointed.

Fertile fronds. Appearing in center of rosettes after sterile blades are grown; soft, lax, much shorter and narrower with much smaller pinnae than sterile fronds, initially bright green, turning brown and withering.

Fertile blades. 2-pinnate, 30–60 × 2–4 cm, erect with pinnae and pinnules strongly ascending, withering after spores are shed in early summer, leaving tangled stalks with cinnamon-colored hair and empty sporangia.

Fertile stipes. Slightly shorter than blade; green; covered with light brown hairs.

Fertile pinnae. 2–5 × 0.5–1 cm; **pinnules bearing masses of sporangia, green initially, soon turning brown.**

Veins. Free, mostly once forked.

Sori. Absent, replaced with large, globose, naked sporangia attached to pinnae of fertile pinnae by a short, thick stalk, opening with a long slit at the top, greenish when young, tan or black when spores are shed in mid-spring to early summer.

Spores. Green, short-lived.

Habitat. Poorly drained soils with high organic content, wet woods, swamps, bogs, stream banks, and ditches.

Distribution. Asia. Central and South America. Mexico. Eastern United States. Canada.

Chromosome #: $2n = 44$

The rhizomes of the *Osmunda cinnamomea* may be the largest of any fern in the United States.

Osmunda cinnamomea **may be distinguished from** *Osmunda claytoniana* **by its sterile blades, which are different and entirely separate from the fertile ones. Tufts of fine colored hairs at the undersurface of the sterile pinna stalks at the junction with rachises are found on** *O. cinnamomea* **and not found on** *O. claytoniana* **and** *Matteucia struthiopteris.* **The sterile fronds may be recognized from a distance by their somewhat pointed tips in contrast to the more rounded tips of** *O. claytoniana.*

Cinnamon fern is one of 3 species found in Michigan that have separate upright fertile fronds. *Matteuccia struthiopteris* **and** *Onoclea sensibilis* **share this character, but their dark, woody, fertile fronds persist through the winter, spring, and early summer, while the soft and lax fertile fronds of** *O. cinnamomea* **usually shrivel and die before the fall.**

Matteuccia struthiopteris **is similar in size to** *O. cinnamomea* **but lacks the wooly hairs at the pinna bases and its ostrich-feather-shaped fronds are long-tapering at the base with the lowest pinna much less than half as long as the longest pinnae.** *O. cinnamomea* **is slightly tapered to the base, but the lowest pinnae are more than half as long as the longest pinnae.**

Osmunda claytoniana **Linnaeus**
Claytosmunda claytoniana (L.) Metzgar & Rouhan
Osmundastrum claytonianum (L.) Tagawa; *Stuthiopteris claytoniana* (Linnaeus) Bernhardi
Interrupted Fern

Etymology. The name honors John Clayton (1693–1773) of Gloucester County, Virginia, one of America's earliest botanists, who sent many specimens to Linnaeus. Much of Linneaus's knowledge of North American species was based on Clayton's specimens. The vernacular name refers to the fertile fronds with sterile pinnae separated (interrupted) by soon deciduous, much shorter, and darker fertile pinnae.

Plants. Dimorphic, some fronds with fertile pinnae, some without; deciduous.

Fronds. 40–155 cm long, rosette arrangement with central, longer fertile fronds surrounded by more spreading and arching sterile fronds.

Rhizomes. Large, sometimes massive, thick, with tangled roots and remnants of stipe bases.

Stipes. About 1/3 length of blades, thick; straw-colored, hairy with light brown hairs initially, becoming glabrous; fertile frond stipes longer.

Sterile blades. 1-pinnate-pinnatisect, 35–75 × 12–25 cm, lanceolate to oblong with **somewhat rounded tips.**

Sterile pinnae. 10–20 pairs, 5–12 × 2–3 cm, lanceolate widest near rachises, tapering gradually to acute tips, subalternate, lobes circa 1.2 × 0.6 cm, cut nearly to costae, margins entire.

Sterile pinnae. Lanceolate, short-stalked, alternate.

Fertile blades. Similar to sterile blades, but with 1–3 sterile pinnae at base, 2–6 pairs of very much smaller pinnae fertile in the middle, and up to 19 sterile pinnae toward tips.

Fertile pinnae. 2–3 × 1–1.5 cm, fully 2-pinnate, withering after spores released, giving the appearance of a lack of middle pinnae.

Fertile pinnules. Leafy tissue absent, bearing masses of sporangia resembling miniature grapes.

Veins. Free, pinnate in lobes, once forked near midribs.

Sori. Absent, replaced with large, globose, naked sporangia, attached to pinnae of fertile pinnae by a short, thick stalk, opening with a long slit at the top, greenish when young, tan or black when spores shed in midspring to early summer.

Spores. Green, short-lived, mature May to July,

Habitat. Wet soils, moist woods, ditches, swamp edges, sometimes in well-drained soils.

Distribution. Eastern United States and Canada from Newfoundland to Florida and west to Texas and New Brunswick.

Chromosome #: $2n = 44$

The placement of the 3 species found in Michigan has been in a state of flux. Recent studies have treated the three species found in Michigan as belonging to a separate genus *Osmundastrum* (a monospecific genus) with *Osmundastrum cinnamomea*, and *Osmunda* with *Osmunda regalis* with *Osmunda regalis*, and *Claytosmunda* with *Osmunda claytoniana*). This treatment may well become widely accepted.

The much-reduced fertile pinnae separating the sterile pinnae on the mature fertile fronds of *O. claytoniana* **are unique among the Michigan ferns. The crosiers of** *O. claytoniana* **and** *O. cinnamomea* **look very much alike when young.** *Osmunda cinnamomea* **may be distinguished by tufts of fine cinnamon-colored hairs present on the undersurface of the sterile blades at the junction of the pinna stalks with the rachises. These clusters of hairs are absent on** *O. claytoniana*. **At a distance mature fronds of** *O. claytoniana* **with its rounder blade tips may be distinguished from those of** *O. cinnamomea* **with more pointed tips.**

Osmunda regalis Linnaeus var. spectabilis (Willdenow) A. Gray
Osmunda spectabilis Willdenow
Royal Fern

Etymology. Latin *regalis*, royal, alluding to the stately appearance of this plant among the largest native ferns, or possibly to the "crown" of fertile pinnae. Asa Gray named the American variety *spectabilis*, "showy".

Plants. Deciduous; dimorphic; medium-sized, to large.

Fronds 60–160 x 30–40 cm, large, clustered, unfern-like (**resembling leaves of black locust).**

Rhizomes. Erect to semi-erect, short-creeping, massive, occasionally branching, covered with old stipe bases woven together with black, fibrous roots.

Stipes. 30–75 cm long, about same length as blades, bases flared stipule-like, pink, light brown hairs early, glabrous and smooth later.

Sterile blades. 2-pinnate; 75–100 x 30–55 cm, broadly ovate, rachises round, narrow, pale pink, straw-colored to greenish, in early spring the small crosiers are pinkish-brownish and covered with dense soon shed whitish to cobwebby hairs.

Sterile pinnae. 5–9 pairs, up to 30 cm long, oblong-lanceolate, ascending, widely spaced, sub-opposite to alternate.

Sterile pinnules. 7–13 pairs, 2.5–6. x 1–2 cm broad, widely spaced, oblong-lanceolate, tips acute to rounded, short-stalked, variable, margins entire to finely toothed.

Fertile fronds. Erect, 20–50 cm tall, shorter than sterile fronds.

Fertile blades. Usually with 2–4 pairs of sterile pinnae at the base, with

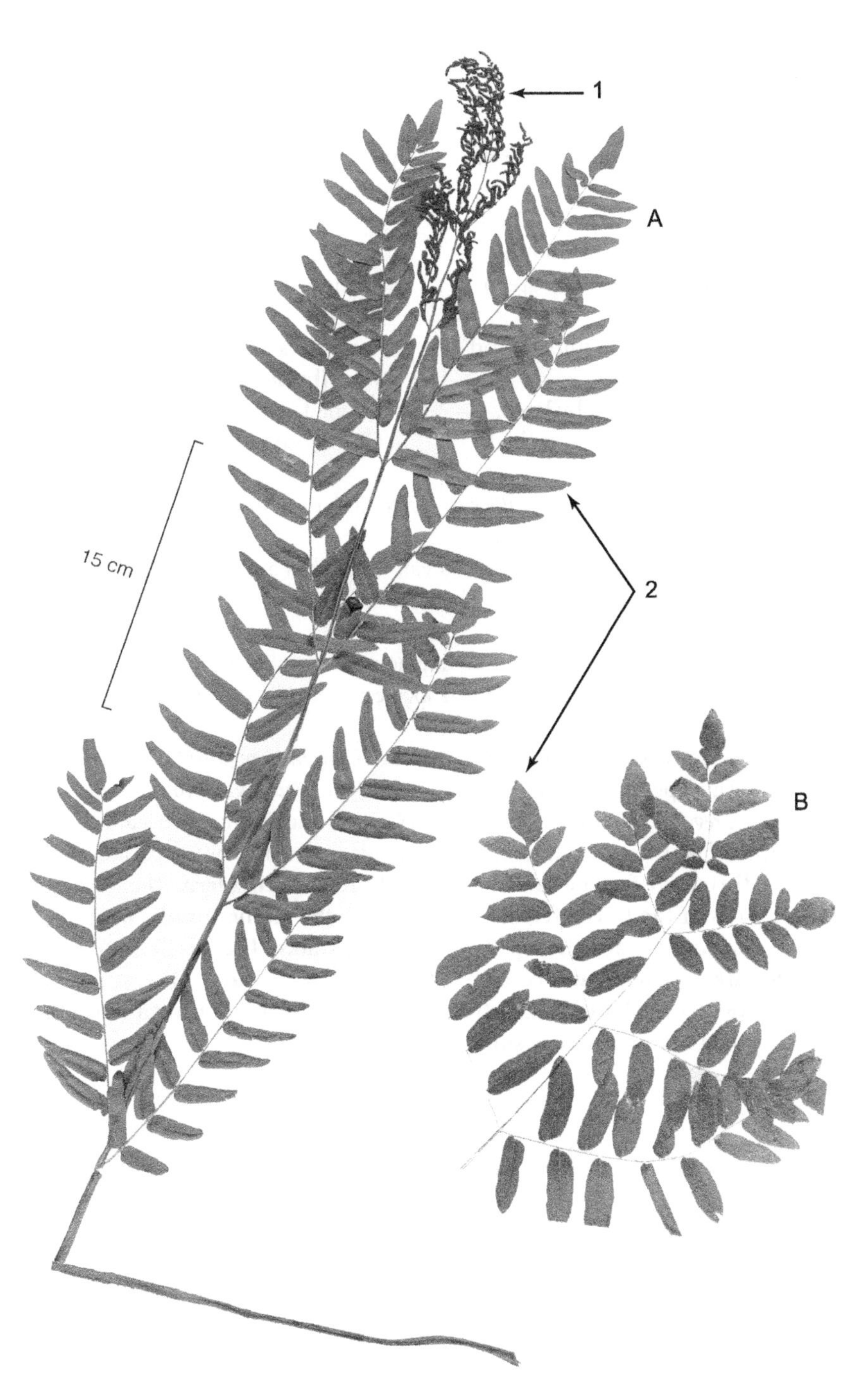

Osmunda regalis: Royal fern: A. Fertile frond. B. Tip of sterile frond.
Informative arrows: 1. Fertile pinnae at tip of frond. 2. Sterile pinnae and pinnules resemble shape and color of black locust leaves.

7–14 or more pairs of much smaller fertile pinnae at tips, fertile pinnae tassel like, bearing densely clustered sporangia.

Fertile pinnae. Narrow, 3–12 x 1–3 cm, oblong, restricted to frond tips, leafy tissue lacking, alternate, green becoming brown.

Veins. Free, forked at midrib, and forked again further out.

Sori. Absent, replaced by large, globose, naked sporangia, attached to pinnae of fertile pinnae by a short, thick stalk, opening with a long slit at the top, greenish when young, tan or black when spores shed in mid spring to early summer.

Spores. Green, short-lived, released in spring to midsummer.

Habitat. Swamps, wet woods, and stream banks, usually with acidic soils.

Distribution. Eastern United States and Canada.

Chromosome #: $2n = 44$

The fiddleheads of *Osmunda regalis* var. *spectabilis* are smooth, slender, burgundy-colored, and conspicuous in spring.

The closely related *Osmunda regalis* var. *regalis* has a widespread distribution in Europe, Asia, Africa, and India. The anatomic differences between var. *regalis* and var. *spectabils* are minimal.

Recent DNA studies have resulted in placing *O. regalis* var. *spectabilis* in a separate subgenus (*Osmunda*) containing 4 species: *O. regalis, O. spectabilis, O. japonica*, and *O. lancea*.

Another study recognized 3 to 4 varieties of *Osmunda regalis: Osmunda regalis* var. *regalis* (Europe, Africa, and Southwest Asia), *O. regalis* var. *panigrahiiana* (India), and *O. regalis* var. *grasiliensis* (tropical Central and South America, treated as *O.* var. *spectabilis* by some authors). *O. spectabilis* was treated as a separate species.

This distinctive fern has also been called the flowering fern. The lower and middle parts of the fertile fronds are similar to the sterile fronds, with fertile pinnae attached toward the tips by very short stalks. The sterile pinnae and pinnules are the shape and color of black locust tree leaves. The uppermost fertile pinnae, the plants' distinctive "crown" or tassel, are brown and much reduced when mature in early summer.

PELLAEA Link Pteridaceae

Cliff Brakes

See Pteridaceae in appendix for a discussion of the family and a key to the Michigan genera.

Etymology. Greek *pellos*, dusky, referring to the often bluish-gray quality of the fronds of some species in this genus. Brake is a name applied to ferns in the genus *Pteris*, to which many *Pellaea* species were formerly assigned, and to large and course ferns. Probably derived from Middle English *brake* with *braken* (*bracken*) being the plural.

Plants. Monomorphic to somewhat dimorphic, small to medium-sized, terrestrial or epipetric.

Rhizomes. Short to long-creeping, ascending to horizontal; scaly, scales brown to tan, margins dentate or entire.

Fronds. 5–50 cm long, clustered or spread out.

Stipes. Dark, shiny, brown, dark purple to black or straw-colored to tannish; round, glabrous or hairy, usually with a few scales at bases; cross section at base reveals a single vascular bundle.

Blades. 1–2-pinnate, linear to ovate-triangular, leathery; abaxial surface glabrous, hairy, or with hairlike scales scattered along midribs; adaial surfaces usually glabrous, dull.

Pinnules. Elliptic, lanceolate to linear, sometimes more or less rounded, usually more than 4 mm wide, bases rounded, usually stalked.

Veins. Free, obscure, pinnately branched.

Sori. Linear along pinnule margins.

False indusia. Formed by modified in-rolled pinnule margins concealing the sporangia, extending entire length of pinnules.

Habitat. Usually embedded in crevices and cracks of rocky outcrops.

Distribution. Mostly in the southwestern United States and Mexico, with a few found in Asia, southern Africa, the Pacific islands, New Zealand, northeastern Spain, and Australia.

Chromosome #: $x = 29$

Pellaea is a poorly defined genus of with about 35–40 species worldwide, usually found in dry habitats. Fifteen species are treated in *Flora of North America North of Mexico*, vol. 2 (1993). It is represented in Michigan by two species.

KEY TO THE SPECIES OF PELLAEA IN MICHIGAN

1. Rachises densely short-hairy; stipes dark purple to nearly black; undersurface of pinnae with scattered hairlike scales . . . . *P. atropurpurea*
1. Rachises nearly glabrous; stipes dark reddish-brown; undersurface of pinnae smooth with few to no hairlike scales . . . . *P. glabella* subsp. *glabella*

Pellaea atropurpurea (Linnaeus) Link

Pteris atropurpurea Linnaeus

Purple Cliff Brake, Hairy Cliff Brake

Etymology. Latin *atro-*, dark, black, + *purpureus*, purple, alluding to the purplish-black color of the stipe and rachis. The vernacular name is a translation of the Latin name and a description of the color of the stipes. The other common names describe its hairy stipes and preferred habitat and its hairiness.

Plants. Somewhat dimorphic; evergreen.

Rhizomes. 5–10 mm diam., compact, stout, ascending; scales uniformly reddish-brown to tan, tapering to a sharp point from a broader base, 0.1–0.3 mm wide, thin, margins entire to finely toothed.

Fronds. 5–50 × 5–18 cm, usually clustered; *sterile fronds* shorter and less divided than fertile fronds.

Stipes. Circa 1/2 length of blade, **dark purple to nearly black, shiny, round, densely hairy with stiff hairs, bearing scales**.

Blades. Usually 2-pinnate at base, 1-pinnate toward tips, 4–18 cm wide; long-triangular, leathery; sterile blades shorter and less divided than fertile blades; rachises reddish-purple throughout, round, densely hairy on adaxial surface with short, curly, appressed hairs.

Pinnae. 10–12; mostly opposite, perpendicular to rachis or ascending.

Pinnules. 3–14; 1–10 cm long, linear-oblong, tips rounded to slightly pointed, midribs sparsely hairy abaxially; *fertile segment margins* weakly inrolled, margins whitish, crenulate.

Veins. Free, forked, obscure.

Sori. Sporangia borne on inrolled margins of pinnules—false indusia.

Habitat. In Michigan found on ledges, outcrops, low cliffs, and talus slopes. (Usually found on calcareous rocks in most of its range.)

Distribution. Mexico. Guatemala. Canada: Quebec and Ontario. USA: New England south to Florida, west to Arizona, and northeast to Wyoming and Minnesota.

Chromosome #: $2n = 87$, apogamous.

Pellaea atropupurea is probably an autopolypoid arising from a diploid species that has not been found.

Pellaea atropurpurea is a Michigan threatened species. It is considered imperiled in Michigan but secure globally.

Pellaea *atropurpurea* **is larger and much more hairy and scaly than** *Pellaea glabella.*

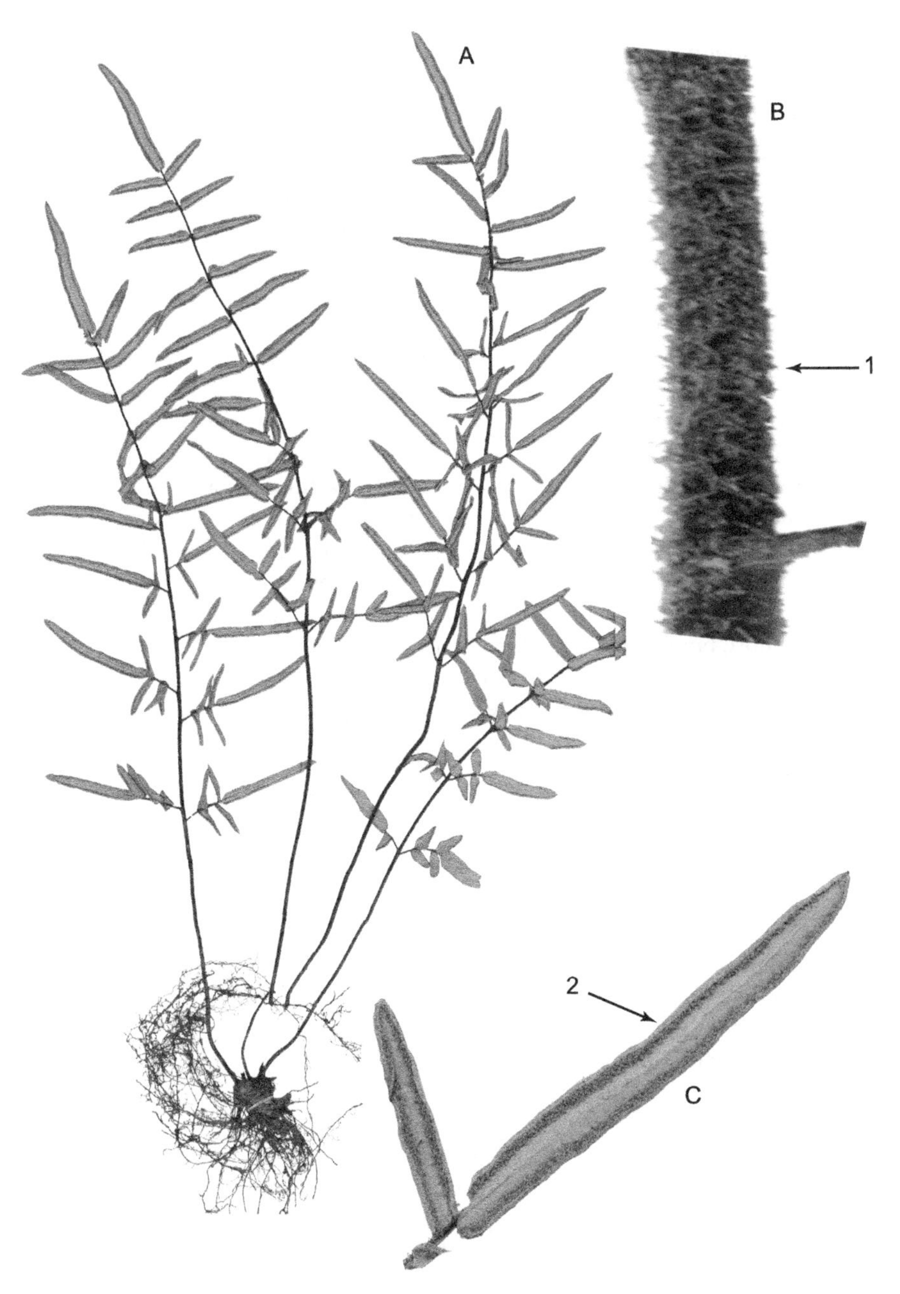

Pellaea atropurpurea: Purple-stemmed cliffbrake: A. Frond cluster. B. Portion of stipe. C. Fertile pinna.

Informative arrows: 1. Stipe densely hairy. 2. Marginal sori covered with in-rolled pinna margin.

Pellaea glabella Mettenius subsp. *glabella*

P. gracilis Hooker

Small Cliff Brake, Smooth Cliff Brake

Etymology. Latin *glabrum*, smooth, hairless, + *-ella*, a diminutive suffix, alluding to the minimal amount of hair and scales on the stipe and rachis of this species. The common name describes its preferred habitat and lack of hairs and scales.

Plants. Monomorphic; evergreen.

Fronds. 10–40 × 2.5–8 cm; clustered, *sterile fronds* somewhat less divided and narrower.

Rhizomes. 5–10 mm diam., ascending, short, branching; scales narrow, 0.1–0.3 mm wide, reddish-brown.

Stipes. About 1/2 as long as blade, round, glabrous, wiry, smooth, shiny, dark reddish-brown.

Blades. Usually 2-pinnate at base, 1-pinnate toward tips, 1–8 cm wide; linear-oblong to ovate-lanceolate, frond tips the same as undivided lateral pinnae; *rachises* **brown, round, nearly glabrous.**

Pinnae. 5–10, somewhat ascending, upper pinnae not divided, lower pinnae usually with 2–5 pinnules, 1–50 mm, sometimes shorter than pinnules.

Pinnules. 0.5–2 cm long; oblong-lanceolate, sometimes lobed, tips obtuse, leathery to herbaceous, nearly glabrous with occasional hairlike scales near midribs on undersurface, **borders whitish**, smooth to finely toothed; fertile pinnae margins recurved, covering less than 1/2 abaxial surface.

Veins. Obscure.

Sori. Partially covered with false indusia formed by inrolled margins of fertile pinnae.

Habitat. Shady vertical cliffs, rocky cliffs, ledges, and rocky slopes.

Distribution. Canada: Ontario and Quebec. USA: Texas, Nebraska, Minnesota, east to Tennessee and Virginia, north to Connecticut and Vermont.

Chromosome #: $2n = 116$, apogamous tetraploid

Four subspecies of *Pellaea glabella* are recognized in *Flora of North America North of Mexico*, vol. 2 (1993). Only the subspecies *glabella* occurs in Michigan.

The subspecies are difficult to separate using morphological characteristics. They have been separated into species and varieties by various authors. Some are diploid and some tetraploid.

Pellaea glabella **is smaller and much less hairy and scaly than** *P. atropurpurea.*

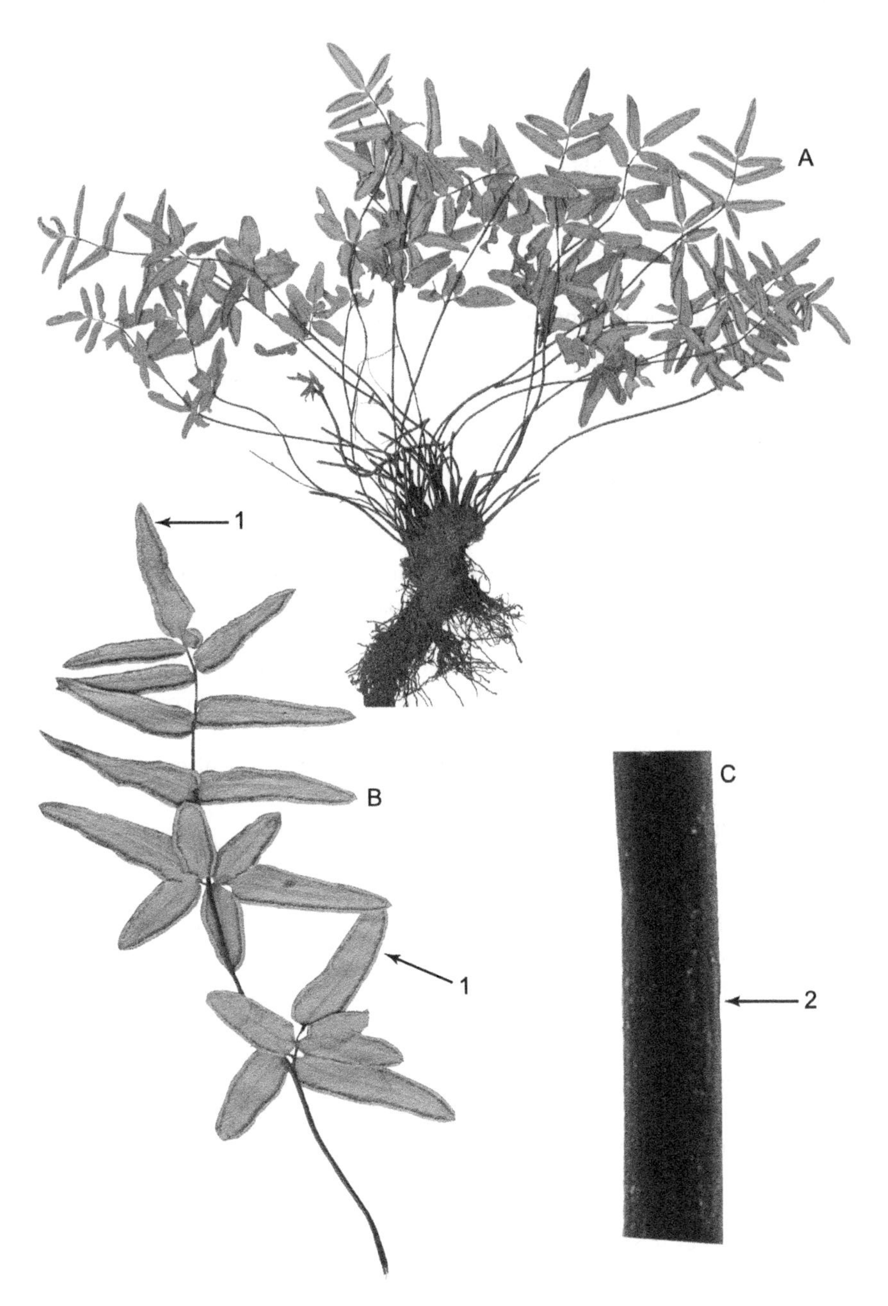

Pellaea glabella var. *glabella:* Smooth cliffbrake A. Frond cluster. B. Fertile pinna. C. Portion of stipe.

Informative arrows: 1. Marginal sori covered with in-rolled margin. 2. Glabrous stipe.

PHEGOPTERIS (C. Presl) Fée Thelypteridaceae

Polypodium Linnaeus sect. *Phegopteris* C. Presl; *Thelypteris* Schmidel subg. *Phegopteris* (C. Presl) Ching

Beech Ferns

See Thelypteridaceae in appendix for a discussion of the family and a key to the Michigan genera.

Etymology. Greek, *phegos*, beech, + *pteris*, fern, alluding to a fern growing under beech trees (although it is not particularly associated with beech trees).

Plants. Monomorphic, deciduous, fronds growing separately in small to large clones with many ramets.

Fronds. Small to medium-sized, arising singly from rhizomes, more or less hairy.

Rhizomes. Stems long-creeping, branching repeatedly, slender, 1–4 mm diam.

Stipes. As long or longer than blades, cross section at bases reveals 2 parallel, linear vascular bundles merging on upper stipes to form a U-shape.

Blades. 1-pinnate-pinnatifid in proximal parts, broadest at bases, tapering to pinnatifid tips; acicular hairs in undersurfaces, rachises, and costae; also with spreading, ovate-lanceolate scales; rachises not grooved.

Pinnae. Deeply lobed, connected by wings along the rachises, wings sometimes forming a lobe with a vein arising from rachis between pinnae; midribs not grooved on upper surface.

Veins. Free, simple, or forked.

Sori. Round, supramedial to submarginal, indusia lacking.

Habitat. Moist woods, usually in full shade, and shaded rock crevices.

Distribution. Northern temperate and boreal East Asia and North America.

Chromosome #: $x = 30$

Phegopteris is a genus of 3 species worldwide. Two are treated in *Flora of North America North of Mexico*, vol. 2 (1993). Both are found in Michigan.

Phegopteris species were formerly included in the genus *Thelypteris*, but they differ in having wings along the rachis and sori lacking indusia.

KEY TO THE SPECIES OF *PHEGOPTERIS* FOUND IN MICHIGAN

1. Rachises not winged between two lowest pinnae; blades usually longer than wide; lowest pinnae angled downward . . . . *P. connectilis*
1. Rachises winged between the 2 lowest pinnae; blades usually about as long as wide; lowest pinnae not or only slightly angled downward . . . . *P. hexagonoptera*

Phegopteris connectilis (Michaux) Watt

Polypodium connectile Michaux

Narrow Beech Fern, Long Beech Fern, Northern Beech Fern

Etymology. Latin *connecto*, join, bind, + *-ilis*, having the quality of, alluding to the pinnae of the upper part of the frond connecting broadly to the winged rachis.

Plants. Monomorphic, deciduous.

Fronds. 15–55 cm long, solitary, 1–2 cm apart.

Rhizomes. Long-creeping, branching, 1–2 mm diam., black, with ovate scales.

Stipes. 1–1.5 blade length, (8–) 15–36 × 0.1–0.3 cm, straw-colored; scales at base tan, lanceolate; sometimes sparingly hairy.

Blades. 1-pinnate-pinnatifid, 9–18 cm wide; triangular, mostly somewhat longer than broad, tapering rapidly to narrow tips, often tilting backward out of plane of stipes nearly 90° from stipes, light green to yellow-green; scales on undersurface of rachis ovate-lanceolate; **rachises winged except between 2 (or sometimes 3) lowest pinnae pairs.**

Pinnae. 12 to 16 pairs, lowest pair 12–20 × 1–3.3 cm, deeply pinnatifid, sessile; subopposite, margins entire or crenate, rarely shallowly lobed, lowest pair of pinnae slightly narrowed near rachis, **lowest pinna pair well separated from second pinna pair, drooping down and forward, more distal pairs strongly adnate and connected to rachis by wings;** scales ovate-lanceolate, to ca. 3 mm long, light tan to shiny brown; translucent, needlelike, 0.3–1 mm (acicular) hairs moderate to dense along costae and veins on undersurface.

Pinna lobes. Entire, those of proximal pinna pair sometimes crenate, uncommonly shallowly lobed.

Veins. Free, pinnately branched and forking again, proximal pair of veins from adjacent segments meeting above sinuses.

Sori. Submarginal, round; mature early to midsummer**;** *indusia* **absent.**

Habitat. Boreal, wet temperate, and cool climates, moist shaded soils, and calcareous rock crevices.

Distribution. Circumboreal. Eurasia. Greenland. Canada. USA: Alaska, Oregon to north central United States and New England, south to Colorado, Missouri, Tennessee, and North Carolina. Uncommon in the southern 2/3 of the Lower Peninsula.

Chromosome #: $2n = 90$, apogamous triploid

A tetraploid hybrid between *Phegopteris connectilis* and another unknown *Phegopteris* species (probably not *P. hexagonopteris*) has been found in New England and should be looked for in Michigan. It has a strong resemblance to *P. connectilis*.

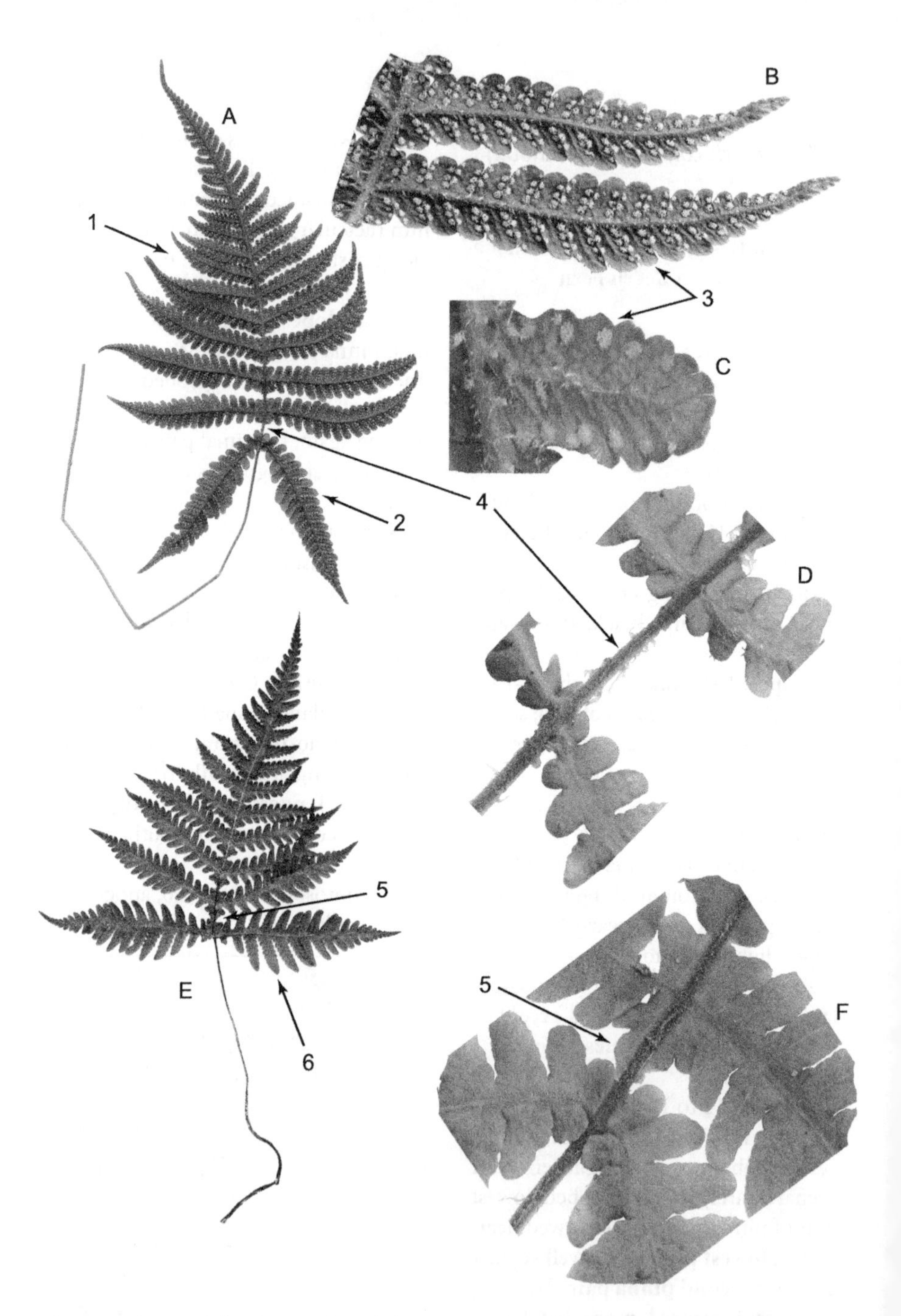

Phegopteris connectilis: Narrow beech fern: A. Frond. B. Pinna. C. Pinnule. D. Rachis between first and second pinnae.

Phegopteris hexagonoptera: Broad beech fern: E. Frond. F. Rachis between first and second pinnae.

Informative arrows: 1. Blade longer than wide. 2. Lowest pinnae pointing downward. 3. Indusia lacking. 4. Wing tissue on rachis between pinnae lacking. 5. Wing tissue on rachis between pinnae. 6. Proximal pinnae longest, spreading or ascending, narrow near rachis.

Phegopteris connectilis **is distinguished from** *P. hexagonoptera* **by its smaller fronds; rachises that are not winged between the 2 lower pinnae; lowest pinnae that are angled downward and outward; and blades that are usually longer than wide.**

Phegopteris hexagonoptera (Michaux) Fée

Polypodium hexagonopterum Michaux

Broad Beech Fern, Southern Beech Fern

Etymology. Greek *hexa*, six, + *gonia*, angled, + *pteron*, wing, referring to the shape of the winged rachises near the base of the pinnae.

Plants. Deciduous, monomorphic.

Fronds. 15–60 cm long to 30 cm wide, well separated in a row, sterile fronds abundant in spring, fertile fronds few and appearing later in the season.

Rhizomes. Long-creeping, branching, 2–4 mm diam., black, scales ovate.

Stipes. As long or longer than blade, 20–42 cm long × 1.5–4 mm diam., dark at base, light green to straw-colored above; scales lanceolate, tan, with hairy margins at base.

Blades. 1-pinnate-pinnatifid, (8–)15–32 cm long, broadly triangular, widest at base, about as wide as long, dull light green, tilted toward horizontal, **proximal pinnae longest, spreading or ascending, narrow near rachis;** acicular hairs 0.1–0.25 mm long, moderate to dense on lower surfaces along costae and veins, yellow glandular hairs 0.1 mm long on veins and blade tissue; **rachises winged,** green.

Pinnae. 12 to 15 pairs, (7–) 10–20 × 2–6 (–8) cm, ovate-lanceolate, narrowed near rachis, more distal pinnae not narrowed; deeply pinnatifid, subopposite; lower pinnae spreading, not drooping down and forward; **proximal pair of pinnae connected to the next pair above by sometimes very narrow wings along rachises;** costae not grooved, scaly abaxially, scales ovate-lanceolate, light tan to shiny brown, with marginal hairs, acicular hairs mostly less than 0.25 mm long, a few sometimes up to 0.5 mm on undersurface.

Ultimate segments. Larger segments lobed.

Veins. Free, pinnately divided in ultimate segments, forked, proximal pair of veins from adjacent segments touch margin above the sinus veins in segments of middle pinnae.

Sori. Small, round, sparse, submarginal and at the base of a sinus at vein tips, mature mid- to late summer; *indusia* **absent.**

Habitat. Humus-rich moist woods and shaded slopes, usually in full shade.

Distribution. Canada: Ontario eastward. USA: eastern states, Minnesota to Texas and east to the Atlantic Ocean. In Michigan found mainly in lower half of the Lower Peninsula.

Chromosome #: $2n = 60$

Phegopteris hexagonoptera **may be distinguished from** ***P. connectilis*** **by rachises that have wings between the 2 lower pinnae, by lowest pinnae that are spreading horizontally or ascending (vs. drooping down and forward on** ***P. connectilis*****), and by blades that are about as wide as long.**

POLYPODIUM Linnaeus Polypodiaceae

Polypody Ferns

See Polypodiaceae in appendix for a discussion of the family and a key to the Michigan genera.

Etymology. Greek *polys*, many, + *podion*, little feet, alluding to short stumps (phyllopodia) left on the rhizome when the fronds abscise. Some say that the name alludes to the highly branched rhizomes, which resemble many little feet.

Plants. Monomorphic, evergreen.

Rhizomes. Usually long-creeping, freely branching, 3–15 mm diam., bearing phyllopodia (outgrowths from the stem to which a stipe is jointed); scaly, scales linear-lanceolate, light brown, peltate, concolored to bicolored, margins entire to denticulate.

Fronds. Erect, 7–30 cm long, close to widely spaced, not conspicuously narrowed at tips.

Stipes. Alternate in 2 rows on upper surfaces of rhizomes, straw-colored; articulate at rhizomes, leaving phyllopodia at base after fronds abscise, scales as on rhizomes.

Blades. Simple, linear to triangular, pinnatifid to 1-pinnate at base, usually with fewer than 25 pinna or lobe pairs; lobes oblong or linear-oblong, tips usually obtuse; rachises glabrous adaxially but sparsely scaly to glabrous abaxially; scales lanceolate-ovate, not clathrate.

Lobes or pinnae. Linear to oblong, margins entire to serrate, tips rounded to attenuate.

Veins. Free, pinnately branched.

Sori. Round or oval; up to 2 mm diam., exindusiate, in 1–3 rows on either side of midribs, on veins tips, frequently confined to distal parts blade, glabrous.

Spores. Bright yellow.

Indusia. Absent.

Habitat. A variety of soils and rocks.

Distribution. *Polypodium* is a genus of perhaps 110 species, mostly in the New World tropics and subtropics, with a few species in temperate areas. Eleven species are treated in *Flora of North America North of Mexico*, vol. 2 (1993). It is represented in Michigan by a single species.

Chromosome #: $x = 35, 36, 37$

Polypodium virginianum Linnaeus
Polypodium americanum Hooker
Rock Polypody, Common Polypody

Etymology. *Virginianum* refers to Virginia, the colony that once covered a large area of what is now the eastern United States and was the source of many of the specimens that Linnaeus described in 1753. The vernacular name derives from the generic name.

Plants. Monomorphic, evergreen.

Fronds. 8–40 × 3–15 cm, forming in lines along the rhizome, longer fronds erect, smaller fronds spreading, glabrous, green.

Rhizomes. Long-creeping, slender, to 6 mm diam., spreading, occasionally exposed, smooth or with lanceolate light brown scales with darker centers.

Stipes. 1/2 to 2/3 length of blades, slender, to 2 mm diam., round, smooth or with light brown scattered scales, withered old stipes often retained; scales thin, lanceolate, twisted, base and margins light brown, sometimes with dark central stripe, margins finely toothed.

Blades. 5–30 × 2.5–5 (–7) cm, **deeply pinnatifid to pinnatisect,** lanceolate to oblong-lanceolate, usually widest at middle, sometimes near base, tips narrowly triangular, pointed, **10–22 pairs of lobes; somewhat leathery,** upper surface bright green lustrous, lower surface dull light green; *rachises* **appear winged because of fused lobe bases**, sparsely scaly to glabrous abaxially, glabrous adaxially, scales lanceolate-ovate.

Lobes. Oblong to lanceolate, bases broadly adnate, tips rounded to broadly acute, less than 8 mm wide, alternate producing a zigzag pattern on the rachis wing; margins entire to crenulate, minutely toothed, **sinuses cut nearly to rachis,** acute-angled (not rounded).

Veins. Free, pinnate, one central vein per lobe, 1-forked, ending in bulbous expansions at minute teeth short of margins, somewhat obscure.

Sori. Large, in single row on both sides of midvein, medial to submarginal, found on more distal parts of the blade.

Indusia. Lacking, replaced with abundant glandular-tipped paraphyses.

Habitat. On a variety of soils in shady areas, cliffs, and rocky slopes, sometimes on trees. In Michigan often found on sandy soils (unusual for this

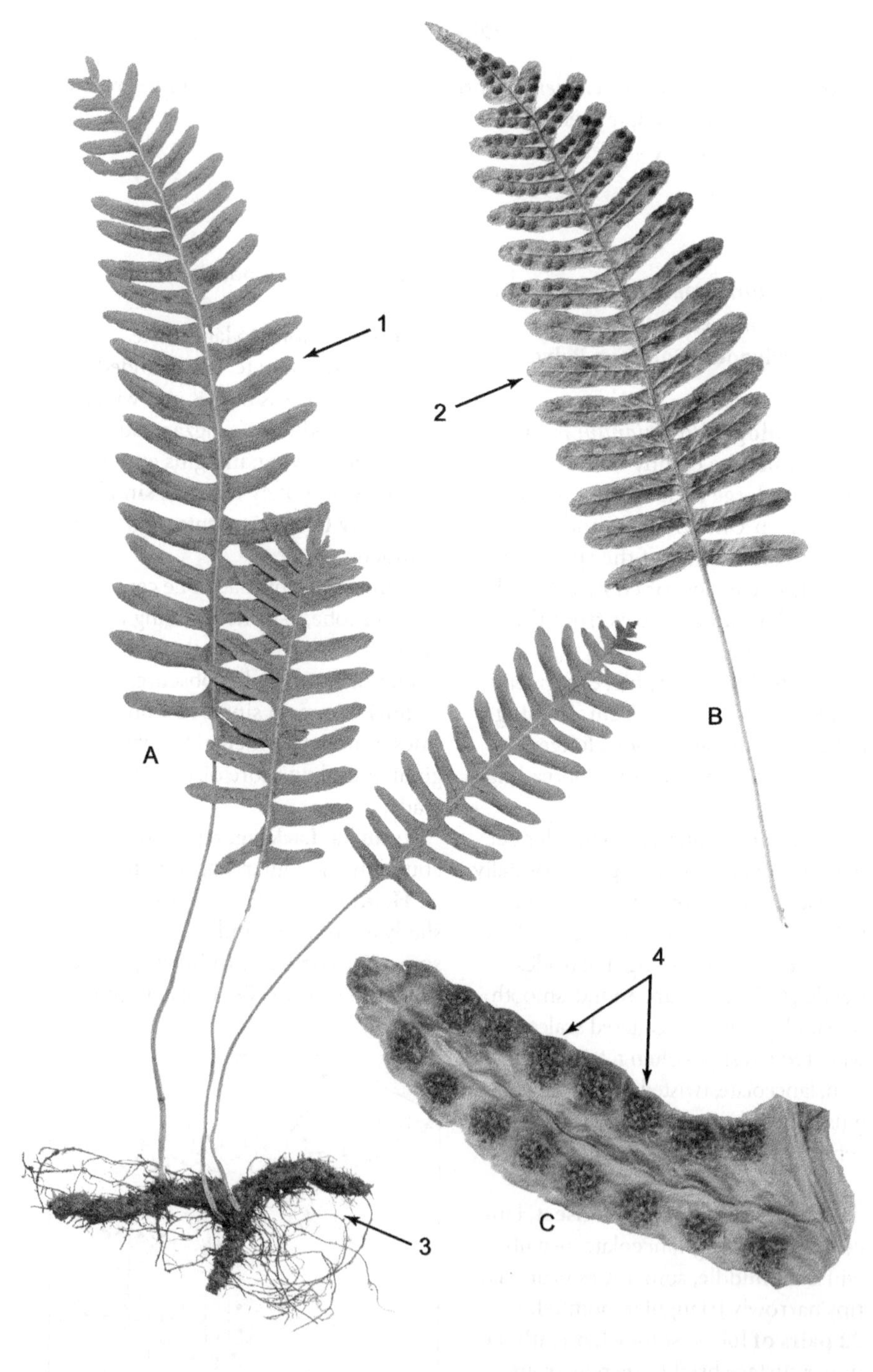

Polypodium virginianum: **Common polypody: A. Frond upper side (adaxial) and rhizome. B. Fertile frond lower side (abaxial). C. Pinna.**

Informative arrows: 1. Blunt-tipped lobes winged at rachis. 2. Sori on upper part of frond. 3. Creeping rhizome. 4. Sori exindusiate.

species when found elsewhere), ridges, and calcareous rocks.

Distribution. Canada and USA: Alaska to Newfoundland, south to South Dakota, Arkansas, Alabama, and Georgia.

Chromosome #: $2n = 148$

Polypodium virginianum varies much in form. The evergreen fronds may shrivel in prolonged dry conditions but recover their original appearance when rehydrated.

Polypodium virginianum **is easily recognized by its evergreen, thick, deeply cut, pinnatifid to pinnatisect fronds. Its mat-forming growth pattern with fronds on long-creeping rhizomes is distinctive. Its sori are conspicuous, round, and lack indusia.**

POLYSTICHUM Roth Dryopteridaceae

Sword Ferns, Christmas Ferns, Holly Ferns

See Dryopteridaceae in appendix for a discussion of the family and a key to the Michigan genera.

Etymology. Greek *poly*, many, + *stichos*, row, alluding to the sori arranged in regular rows on each pinna. The common names Christmas fern and holly fern derive from the evergreen fronds of some species, which were used for Christmas decorations. Sword fern derives from the shape of the blades of some species.

Plants. Monomorphic or dimorphic, evergreen.

Rhizomes. Decumbent to erect.

Fronds. Close, forming a rosette.

Stipes. 1/10–1/3 length of blades, grooved on upper surface, scaly; cross section at base reveals a semicircle of small, round vascular bundles.

Blades. 1–2-pinnate, linear-lanceolate to broadly lanceolate, somewhat leathery to leathery; rachises scaly.

Pinnae. Sessile to short-stalked, bases usually with acroscopic auricle, margins with spiny teeth, adaxially more or less glabrous or similarly scaly; costae grooved on upper surface, grooves continuous with rachis grooves; scales on costae and sometimes between veins abaxially linear to lanceolate.

Pinnules. Marginal teeth, often with spines at tips, auriculate.

Veins. Free, forked.

Sori. Round, between midribs and margins, confluent or not.

Indusia. Peltate, persistent.

Habitat. Forest floors and shaded rocky slopes with cool, moist soils. In Michigan found in second-growth forests on sand dunes.

Chromosome #: $x = 41$

Polystichum is worldwide genus of about 180 species, mostly in temperate areas but also found at higher elevations in tropics and subtropics. Fifteen species are treated in *Flora of North America North of Mexico*, vol. 2 (1993). It is represented in Michigan by 3 species.

KEY TO THE SPECIES OF *POLYSTICHUM* FOUND IN MICHIGAN

1. Fronds 2-pinnate; fertile pinnae not smaller; sori often distinct . . . . *P. braunii*
1. Fronds 1-pinnate; fertile pinnae smaller or same size; sori confluent, completely covering abaxial surface, or sori distinct (2).
2(1). Fertile pinnae on upper 1/4–1/3 of fronds, much smaller than sterile pinnae; sori often confluent, completely covering undersurface of pinnae; stipes 1/4–1/3 length of fronds; blades narrowing very slightly, if at all, toward base . . . . *P. acrostichoides*
2. Fertile pinnae on upper 1/2 of fronds, same size as sterile pinnae; sori distinct, not completely covering undersurface of pinnae; stipes 1/10–1/6 length of fronds; blades narrowing toward base . . . . *P. lonchitis*

Polystichum acrostichoides (Michaux) Schott
Nephrodium acrostichoides Michaux
Christmas Fern

Etymology. The name *acrostichoides* derives from the tropical genus *Acrostichum* + *-oides*, a Greek suffix that indicates resemblance. The mature sori of *Polystichum acrostichoides* appear to cover the entire undersurface of the pinnae like those of the fertile blades of *Acrostichum*. The ommon name, Christmas fern derives from its fronds, which are conspicuously evergreen at Christmastime.

Plants. Dimorphic, evergreen, occurring individually or in small loose colonies.

Rhizomes. Short, thick, creeping to erect, dark, with remnants of leaf stalks (phyllopodia) above; abundant fibrous roots below.

Fronds. 30–80 × 7–12 cm; sterile fronds arching, growing in a fountain-like loose rosette; fertile fronds longer than sterile ones.

Stipes. 1/4–1/3 length of frond, brown at base, green above, densely scaly; scales light brown, large at base, becoming smaller upward.

Blades. 1-pinnate, linear-lanceolate, gradually becoming smaller toward

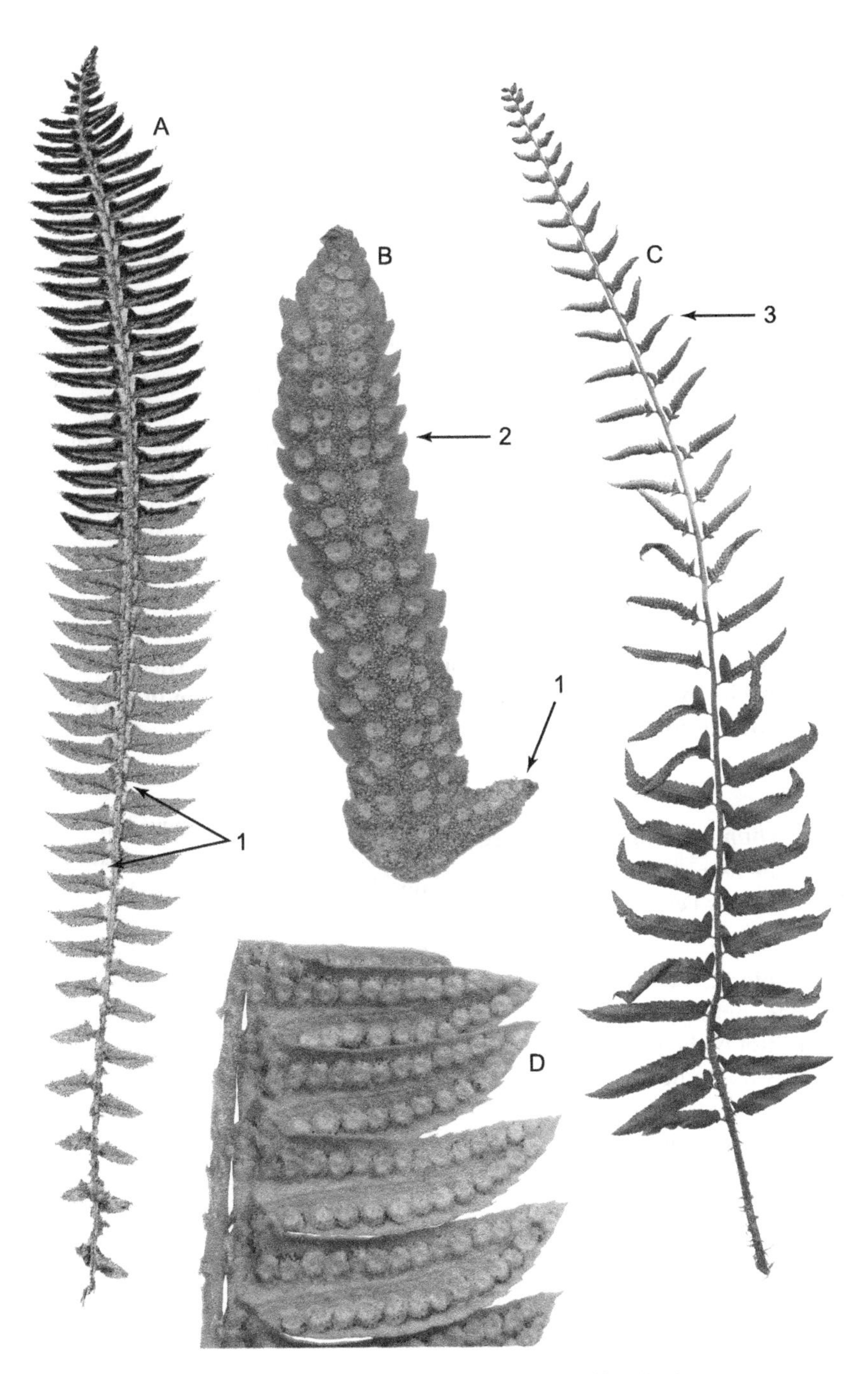

Polystichum lonchitis: Holly fern: A. Fertile frond. D. Detail of fertile pinnae.

Polystichum acrostichoides: Christmas fern: B. Detail of fertile pinna. C. Fertile frond. Informative arrows: 1. "Ears" (auricles) with pointed tips on forward-facing pinnule bases. 2. Crowded sori with peltate indusia. 3. Smaller fertile pinnae toward tip of frond.

pinnatifid tips, leathery, glossy, medium to dark green adaxially, duller lighter green abaxially; rachises with conspicuous dull white to light brown scales.

Pinnae. 20–35 pairs, 2–6 × 0.9–1.1 cm, alternate, several pairs of proximal pinnae usually gradually reduced, **lanceolate to falcate,** not overlapping, tips pointed or blunt, **pointed auricles (ears) on acroscopic bases,** margins notched or finely or minutely toothed or spiny with ascending teeth; minute hairlike scales dense on both surfaces.

Fertile pinnae. Only on upper 1/3 to 1/2 of fertile fronds, much smaller and narrower than sterile pinnae, undersurface covered with sori, costae usually sinuate.

Veins. Free, forked, obscure.

Sori. Round, medial at tips of veins, separate at first, becoming confluent and covering entire undersurface of pinnae; indusia peltate.

Habitat. Moist woodlands, shady slopes, and rocky slopes. In Michigan often found in second-growth forests on sand dunes.

Distribution. Naturalized in Europe. USA and southeastern Canada south to Florida, west to Texas, north to Minnesota.

Chromosome #: $2n = 82$

Polystichum acrostichoides **may be distinguished from the other Michigan** *Polystichum* **species by its much-reduced fertile pinnae found only on the upper 1/3 to 1/2 of the frond.**

Polystichum braunii (Spenner) Fée

Aspidium braunii Spenner

Braun's Holly Fern

Etymology. The name honors Alexander Braun (1805–77), a botanist and director of the Berlin Botanical Garden, who studied *Isoëtes, Marsilea,* and *Equisetum.*

Plants. Monomorphic, evergreen.

Rhizomes. Erect, short, thick, very scaly.

Fronds. 20–70 (–100) × 7–22 cm, arching, rosette-forming; lower stipes, rachis, and fiddlehead densely covered with silvery scales that become tan.

Stipes. 1/8–1/6 frond length, grooved, densely scaly at base, scales silvery at first, then light brown, gradually becoming smaller upward.

Blades. 2-pinnate, elliptic to broadly lanceolate, broadest at middle or above, tapering gradually and conspicuously at base and tips, abaxial surface dull, dark green, leathery, whitish to brown narrow scales on both surfaces; rachises densely covered with light brown linear scales.

Pinnae. 20–40 pairs; 2–10 cm long,

Polystichum braunii: Braun's holly fern: A. Frond. B. Detail of pinnules.
Informative arrows: 1. 2-pinnate blade. 2. Short stipe. 3. Round sori, older sori with small residual indusia. 4. "Ears" (auricles) with pointed tips on forward-facing pinnule bases.

oblong-lanceolate, not or only slightly overlapping, weakly falcate, lower pinnae ± rectangular, tips acute, lowest 3–4 pinnae only 2 cm long, pointing downward; scales linear, narrow, dense abaxially, sparse adaxially.

Pinnules. 6–18 pairs, short-oblong or ovate, close, short-stalked near rachises, sessile farther out, acroscopic auricle prominent on proximal pinnules, margins toothed, tips broadly acute with slender bristle tips.

Veins. Free, pinnately branched.

Sori. Few and separate, in 1 row between midrib and margin, small, well spaced; indusia peltate, ciliate.

Habitat. Cool, moist, humus-rich soils in boreal and northern deciduous forests and on rocky slopes and moist cliffs. In Michigan sometimes found in second-growth forests on sand dunes.

Distribution. Eurasia. North America: In the west Alaska, British Columbia, and Idaho with a large gap at midcontinent; in the east Minnesota, Wisconsin, Michigan, Pennsylvania, New York, and New England to the Atlantic.

Chromosome #: $2n = 164$ (thought to be an ancient tetraploid)

Polysitchum braunii, **while it has the distinctive auricles and spiny-toothed margins on the pinnules, is easily distinguished from the other Michigan** *Polystichum* **species by being 2-pinnate (vs. 1-pinnate for the others).**

Polystichum lonchitis (Linnaeus) Roth

Polypodium lonchitis Linnaeus

Holly Fern, Northern Holly Fern

Etymology. Greek *lonche,* spear, + *-itis,* a suffix indicating a close connection, alluding to the fronds' spearlike shape. The common name holly fern derives from the appearance of its evergreen pinnae.

Plants. Evergreen, monomorphic.

Fronds. 10–60 × 6 cm, erect, not arching except at tips; forming rosettes.

Rhizomes. Short, thick, erect to ascending, surrounded by last year's dead fronds; very scaly, scales variable, ovate to lanceolate, gradually narrowing toward tips, pale brown.

Stipes. 1/8–1/6 length of blade, grooved, green or straw-colored with age; densely scaly, scales light brown, gradually becoming smaller toward tips.

Blades. 1-pinnate, linear-lanceolate, usually widest above middle, narrowing at base, drooping at tips, glossy, stiff; scales small, dense below, glabrous above.

Pinnae. 25–35 pairs, 0.5–3 cm long, well separated, oblong to lanceolate to falcate, pinna tips bending toward

frond tips; tips acute, acroscopic auricle prominent, margins with spinelike teeth; lowest pinnae ± triangular; scales minute, dense on underside only.

Veins. Free, forked.

Sori. Round, medial, conspicuous, only on upper half of blade, indusia peltate, distinct, grayish-white, thin, margins entire or minutely dentate.

Habitat. Alkaline substrates, crevices in rocky cliffs, well-drained sandy soils, cool moist boreal and subalpine coniferous forests. In Michigan found in second-growth forests on older sand dunes.

Distribution. Europe. Siberia. Himalaya. Japan. Greenland. Canada and USA: In the west Alaska and the Yukon south to California; in the Rockies south to Utah and Colorado; in the east Ontario, Iowa, Minnesota, Wisconsin, Michigan, the Gaspé Peninsula, and Newfoundland.

Chromosome #: $2n = 82$

Polystichum lonchitis **may be recognized by its once pinnate fronds and by pinnae with toothed margins and earlike tags at their base. While the sori are found only on the distal half of the fronds, the fertile pinnae are not reduced in size as in** *P. acrostichoides*.

PTERIDIUM Gleditsch Dennstaedtiaceae

Bracken

See Dennstaedtiaceae in appendix for a discussion of the family and a key to the Michigan genera.

Etymology. Greek *pteris*, from *pteron*, fern, wing, feather, + Latin *-idium*, a diminutive suffix (Greek *-idon*), a small fern. An inappropriate name for one of our largest fern species, certainly larger than other ferns in the genus. *Pteris* is an allusion to the fronds, which suggest the spread wings of a bird. The common name bracken is of Old Norse origin, related to the Swedish word *bräken*, meaning "fern."

Plants. Medium-sized, terrestrial, colony-forming.

Rhizomes. Subterranean, 1.3–2.5 cm diam., long-creeping, branching, black, densely hairy, scales absent; roots sparse along rhizomes.

Fronds. ±1 m; remote, broadly deltate, arising at rhizome tips and displaying intermittently along rhizomes; young croziers often very hairy.

Stipes. Shorter than or about as long as blade, grooved abaxially, well separated; cross section at base revealing a complicated-shaped vascular bundle; glabrous to short-hairy; dormant rhizome buds at base.

Blades. 2-pinnate to 4-pinnate, ovate to ovate-triangular, sparsely to densely hairy abaxially, rarely glabous, tilted almost parallel to ground; rachises and costae grooved, grooves connecting; nectaries often present at base of pinnae and pinnules.

Pinnules. Pinnately divided or lobed, ultimate segments ovate to oblong to linear, bases decurrent or surcurrent or short-stalked, margins entire.

Veins. Free, pinnately divided, 2- to 3-forked or joined at margin by commissural vein beneath sori.

Sori. When present, more or less continuous along margins and sinuses, covered with recurved false indusia created by reflexed outer margins of blades; true inner indusia membranous, obscure, hidden by outer indusia, small to nearly absent.

Habitat. Grows on a variety of soils with the exception of heavily waterlogged sites. Grows best on deep, well-drained soils with good water-holding capacity. Tolerates sites too dry for most ferns.

Distribution. Widely distributed on all continents (except Antarctica), avoiding only deserts.

Chromosome #: $x = 52$

As interpreted here, *Pteridium* is a genus of 1 species and several varieties. *Flora of North America North of Mexico*, vol. 2 (1993), treats *P. aquilinum* as having 4 varieties in North America north of Mexico. One variety, *P. aquilinum* var. *latiusculum*, is found in Michigan.

Pteridium is a worldwide genus variously interpreted as containing a single variable species with up to 12 varieties, as divided into up to 11 subspecies, or as containing 2 to 6 species and these divided into varieties. This taxonomic problem is now being studied with modern techniques, but there is no current consensus. Some taxonomists believe several species should be recognized because of morphologic differences and because occasional hybrids between them are sterile as recognized by abortive spores. It is certainly the most studied fern in the world because of its aggressive weedy character and its cancer-causing ability.

Pteridium (bracken), when considered as 1 species with several varieties, is the most common and widely used, as well as the most hated, fern in the world. It is probably the world's most successful fern. It has a 55-million-year-old fossil record and a nearly worldwide distribution. Bracken varieties in some parts of the world may grow up to 2.5 m tall. If considered a single species, the fronds and rhizomes of bracken would have a much greater biomass than any other living terrestrial organism. Its abundance in Michigan requires that more than usual attention must be paid to its various features and uses.

Bracken fronds are often sterile, producing neither sori or spores. Completely or nearly completely sterile clones are frequent. In some years spore production is very low. It is generally lower in shady areas and higher in sun-exposed areas but varies greatly from year to year. The spores that are produced may persist in the soil and remain viable for up to 10 years. A single large bracken frond has been estimated to shed hundreds of millions of spores. New clones of bracken almost invariably begin in open disturbed areas resulting from fire, forest clearing, abandoned fields, and so on. They are almost never established in areas of undisturbed natural vegetation. Spores and sexual reproduction are not the major reasons for its ability to spread. Bracken forms large clones arising from a single spore. When seen in open fields these clones may form nearly perfect circles that arise from a single spore at its center. The rhizomes typically grow to a depth of 50 cm, although in loose soils they may extend to more than 3 m, and they may spread a meter or more underground between fronds. Rhizomes that advance ahead of the main clone may grow more than a meter a year; 2.1 m in one season has been reported. Individual parts of rhizomes may only live a few years, but the rhizome as a whole is persistent. In Finland individual clones of bracken have been dated back 1,500 years. Biochemical tests to determine the extent of spread for individual bracken clones found a maximum dimension of 1,015 m in one plant. A clone that large must be very old.

The thick rhizomes act as storage organs for carbohydrates and water (water comprises 87 percent of the rhizome). They have a diameter about that of a pencil. The main shoots elongate rapidly and have few lateral buds; they do not produce fronds. Slow-growing short shoots are the leaf-bearing lateral branches that arise from the long main shoots. They are closer to the soil surface and produce annual fronds and many dormant frond buds. New rhizome extensions arise from the base of the frond stipes rather than directly from the rhizome.

Bracken invades cultivated fields and disturbed areas, where it vigorously competes for moisture and nutrients. It is allelopathic, releasing toxic substances that prevent the germination of other plants or even their own spores. Residual plant toxins may inhibit plant growth for a year after bracken is removed from a site. Its resistance to many herbicides and various types of mechanical control make it difficult to control. Its thick mats of rhizomes grow in competition with the roots of herbs and shrub or tree seedlings. Broad, closely spaced fronds shade out smaller plants. Other plants are seldom seen growing in bracken clones.

Burning, grazing, and woodland clearance have contributed to the spread of bracken. Heavy grazing reduces the competition from palatable grasses, allowing bracken to spread and compete more aggressively with the remaining grass. In some parts of the world, particularly those with grazing domestic animals, it is a major weed. Bracken is well adapted to fire. While the highly flammable layer of dried fronds may burn, its deeply buried rhizomes are protected and sprout vigorously following fires before most competing vegetation is reestablished.

Bracken fronds are among the earliest to die in the fall, forming a mat of litter that insulates the rhizomes and prevents them from freezing. This litter also slows the rise in soil temperature in the spring and delays the emergence of frost-sensitive fronds. The mat of dead fronds buries and blocks other plants from growing.

Severe disease outbreaks are very rare in bracken. Fossil evidence suggests that it has had at least 55 million years to evolve and perfect antidisease, hormonal, antiherbivore, and allelopathic protective chemicals. These varieties of secondary metabolites act as toxins to natural enemies.

Well over 100 insects, including borers, miners, gall formers, leaf eaters, and sap suckers, attack bracken. In defense the fern produces hormones that stimulate uncontrolled early molting in insects (ecdysones). This interrupts their normal development and kills the insects. Bracken also has the ability to switch off and on its production of prunasin, which releases poisonous hydrocyanide when insects chew on the blades.

Nectaries at the base of *Pteridium* pinna stalks exude a sweet substance containing many sugars and some amino acids that ants and other insects eat. The nectaries are present and active in the spring when the croziers are fresh and soft and need the ants for protection. They harden and darken in the summer after the fronds become mature. In Michigan 5 ant and 10 spider species have been seen attracted to the nectaries, and ants have been observed patrolling plants and biting or stinging unwelcome arthropods. (One study claims that ants do not protect bracken from herbivory.) Bracken nectaries were first reported in 1877 by Francis Darwin (Charles Darwin's son), who believed that the nectaries served as waste glands and that ants attracted to the nectaries played no role in defending the plant because bracken was thought to be free of enemies. He was wrong in assuming that bracken has no natural enemies and probably wrong in believing that the ants do not provide protection against insect parasites.

Bracken fronds have been used in many ways. As a form of herbal remedy the powdered rhizomes were once considered particularly effective against parasitic worms. American Indians ate raw rhizomes as a remedy for bronchitis. The fiddleheads of the bracken fern are eaten in Japan (*warabi*), Korea (*gosari*), China, and Taiwan (*juecai*) either fresh, cooked, or preserved by salting, pickling, or sun drying. The Māori of New Zealand used the prepared rhizomes of bracken (*aruhe*) as a food. Both fronds and rhizomes have been used to brew beer, and the rhizome starch has been used as a substitute for arrowroot. Bread can be made from dried and powered rhizomes alone or with other flours. American Indians cooked the rhizomes, then peeled and ate them or pounded the starchy fibers into flour. In Japan starch from the rhizomes is used to make confections. It has been used for animal bedding, which later breaks down to form a rich mulch that can be used as fertilizer.

Bracken fronds, when ingested in large amounts, are carcinogenic to animals such as mice, rats, guinea pigs, sheep, and, cattle. There may be a connection between eating bracken and the high incidence of stomach cancer in Japan. Bracken spores have also been implicated as a carcinogen. A leached carcinogenic chemical can seep into the soil and water supplies and may explain the increased incidence of stomach cancers in bracken-rich areas.

Uncooked bracken contains the enzyme thiaminase, which breaks down thiamine (Vitamin B_1). Eating excessive quantities of bracken can cause beriberi, especially in creatures with simple stomachs; ruminants are less vulnerable. Bracken poisoning induces vitamin B deficiency in pigs and horses. In cattle bracken poisoning can occur in both an acute and chronic form, acute poisoning being the most common. In livestock it can produce "bracken staggers." Poisoning usually occurs when there is a drought or heavy snowfall and grasses are not available.

Pteridium aquilinum (Linnaeus) Kuhn var. latiusculum (Desvaux) Underwood
Pteris aquilina Linnaeus; *Pteris latiuscula* Desvaux; *Pteridium latiusculum* (Desvaux) Fries
Bracken Fern

Etymology. Latin *aquilinum*, eagle, possibly referring to the transitory eaglelike appearance of the apical pinnae during frond expansion. A possible alternate view of the derivation is that in the Middle Ages scholars and herbalists looking at the dark vascular bundles, seen when the stipes or rhizomes were cut in cross section, believed they saw the outlines of a double-headed eagle or, if turned 180°, an oak tree. (It takes a good imagination to see this.) This view might be the real reason that Linnaeus chose the specific name *aquilinum*: Latin *lati-*, broad, wide, + *-usque*, continuously, all the way to, + *-ulum*, a diminutive suffix, possibly alluding to the tips of the pinnules, which are wide in this variety.

Plants. Monomorphic, forming large clones, deciduous.

Rhizomes. 0.4–0.6 cm in diam., long-creeping, very extensive, subterranean, to 0.5 m deep, dark brown to black, hairs pale to dark, scales absent.

Fronds. 35–180 cm long, widely spaced, forming colonies or thickets; large fronds are spread out along the rhizome, often forming dense stands that shade out plants on the ground beneath.

Stipes. 15–100 cm, ± same length as the blade, hairy when young, smooth and mostly glabrous when mature, rigid, grooved, green turning straw-colored to dark brown later; rhizome buds are found near base of older stipes.

Blades. 3-pinnate to 3-pinnate-pinnatifid, 20–80 × 25–50 cm, broadly triangular, divided into 3 ± equal parts, 1 terminal and 2 opposite, tilting nearly horizontal to ground, dark green, leathery, blade margins and abaxial surface shaggy, rachises and costae glabrous or sparsely hairy on undersurface.

Pinnae. 14 or fewer, 7–50 cm long, 2–5 times as long as wide, opposite to subopposite, lowest broadly triangular, distal pinnae narrowly triangular or oblong, terminal segment of most pinnae circa 2–4 times longer than wide, nectaries producing minute surgary droplets present at abaxial base of pinna stalks, clearly present in early spring, becoming dry and hardened and disappearing in summer.

Pinnules. Ovate to oblong to linear, at 45°–60° angle to costae, circa 3–6 mm wide, terminal segments of pinnules 2–4 times longer than wide, equally decurrent and surcurrent, margins entire; abaxial surface of blade midrib and costae shaggy.

Veins. Free, pinnately 2- to

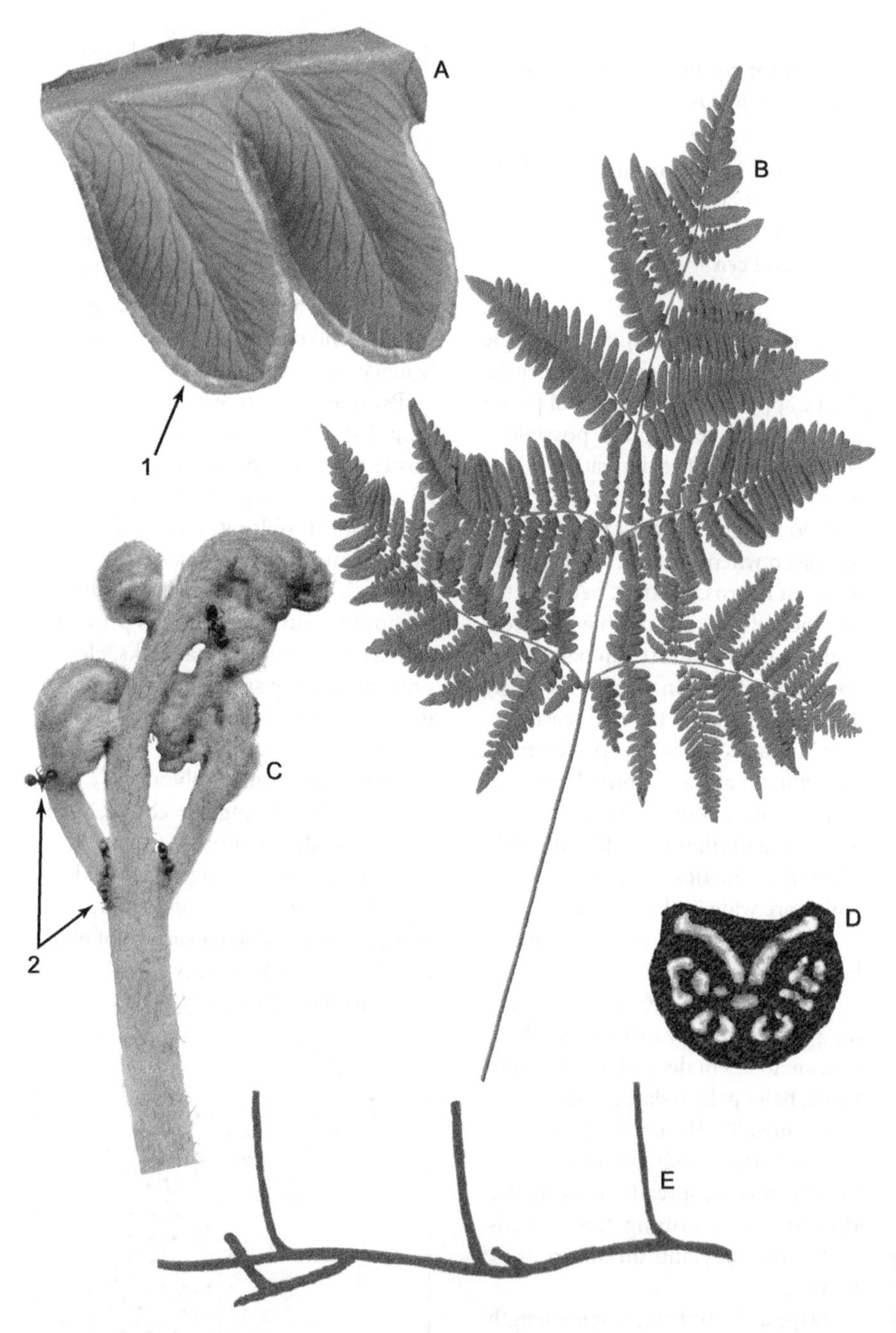

Pteridium aquilinum: Bracken fern: A. Pinnules. B. Small frond. C. Early season crosier. D. Cross section of stipe showing vascular bundles. E. Rhizome with attached stipes. Informative arrows: 1. Marginal sori covered with overlapping leaf margin. 2. Ants feeding at nectaries.

3-forked, joining at margin forming commissural veins beneath sori.

Sori. Often absent, marginal, continuous, covered with a false indusium formed by recurved edges of pinnules, and an obscure inner true indusium; silvery hairs becoming brown and sparse at base of the outer indusium.

Habitat. Very diverse, open fields, barren areas, burned areas, and open woodlands in full sun to partial shade.

Distribution. Canada: Scattered in British Columbia, Ontario to Newfoundland, Nova Scotia, and Prince Edward Island. USA: New England to Florida, west to Louisiana, north to Minnesota. Most frequent in the northern part of the range.

Chromosome #: $2n = 104$

Pteridium aquilinum var. *latiusculum*, the only variety of bracken found in Michigan, is distinguished from other varieties by pinnules set at a 45° to 60° angle to costae, glabrous outer indusia, nearly glabrous or sparsely hairy blades, and terminal segments of pinnules 2–4 times longer than wide. Its pinnule margins and abaxial surfaces of blade midribs and costae are shaggy.

This may be the most common fern seen in Michigan. When traveling along highways and side roads it is seen in large clones that may form a perfect circle or have curved margins if part of the circle is interrupted by woodlands. These clones arise from a single spore and, after producing the first frond, expand each year from the center and often cover large areas. In some locations the plants have expanded and grown together, making it impossible to visually separate the clones.

While bracken is probably the most commonly seen fern in Michigan, it is likely not the most common fern (if each clone is counted as one plant). Some of the woodferns are probably more numerous.

Pteridium aquilinum **var.** *latiusculum*, **Michigan's most commonly seen fern, is easily recognized by its relatively large fronds divided into more or less 3 equal parts and by its large colonies found in mostly open settings.**

THELYPTERIS Schmidel Thelypteridaceae

Lady Ferns

See Thelypteridaceae in the appendix for a description of the family and a key to the Michigan genera.

Etymology. Greek *theyls*, female, + *pteris*, fern. The Greeks applied this name to several delicate ferns, comparing it to the more rugged *Dryopteris filix-mas*, the male fern.

Plants. Monomorphic or dimorphic, deciduous, usually growing in clonal colonies.

Rhizomes. Short to long-creeping, bearing triangular scales at tips.

Stipes. Cross sections at bases reveal ends of 2 ribbon-shaped vascular bundles (appearing as crescent-shaped, somewhat parallel lines) that unite to form a single U-shaped bundle in upper stipes.

Rachises. Grooved adaxially, grooves not continuous with grooves of pinna midribs, with short white to translucent, sharp-tipped hairs (acicular hairs).

Blades. 1-pinnate-pinnatifid.

Pinnae. Costae grooved above, cartilaginous membranes often found at base of sinuses, hairs on adaxial surfaces similar to those on rachises.

Veins. Free and forked, pinnately branched in lobes, sometimes joining below sinuses to form excurrent veins to sinuses.

Sori. Round.

Indusia. Reniform, often hairy.

Chromosome #: $x = 54, 70$

Thelypteridaceae is one of the largest fern families with around 30 genera and about 900 species worldwide. Twenty-one species are treated in *Flora of North America North of Mexico*, vol. 2 (1993). The genus *Thelypteris* has been divided into sections or subgenera by various authors. Two subgenera, each with 1 species, are found in Michigan. These are subg. *Parathelypteris* with *Thelypteris noveboracensis* and subg. *Thelypteris* with *Thelypteris palustris* var. *pubescens*. *Thelypteris simulata* occurs in Wisconsin and southern Ontario and could occur in southern Michigan.

KEY TO THE SPECIES OF *THELYPTERIS* FOUND IN MICHIGAN

1. Fronds tapering at both ends; lowermost pinnae very small, usually less than 2 cm long; stipes shorter than blade ...... *Thelypteris noveboracensis*
1. Fronds tapering toward tips, lowermost pinnae mostly only slightly smaller than middle pinnae; stipes as long as or longer than blades. . . . . *Thelypteris palustris* var. *pubescens*

Thelypteris noveboracensis (Linnaeus) Nieuwland

Polypodium noveboracense Linnaeus

New York Fern

Etymology. Latin *novus*, new, + *Eboracum*, the Roman name for the English town of York, England, + *-ensis*, a suffix indicating origin—the fern from New York. Why Linnaeus chose this name even though the collection

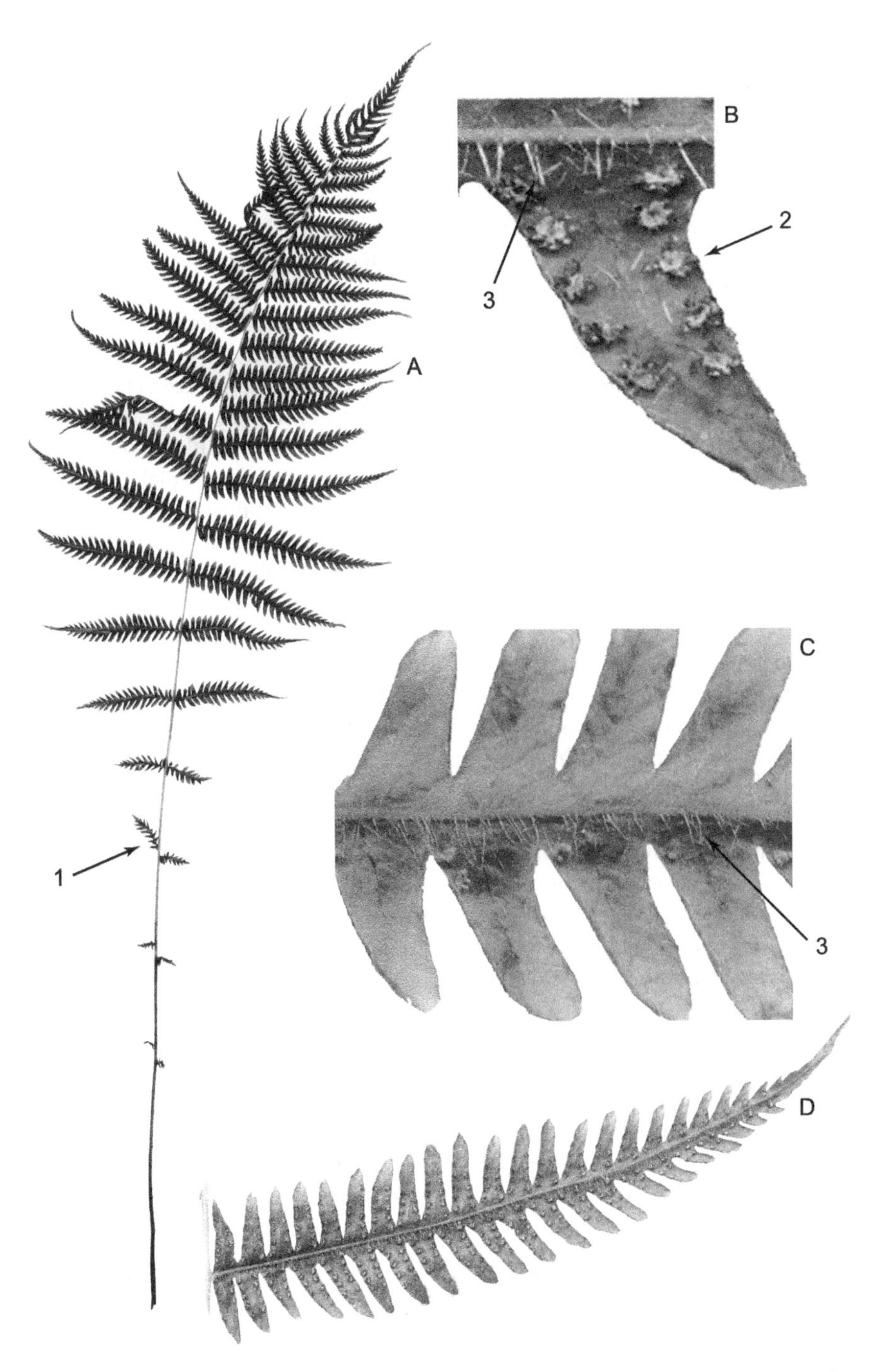

Theylypteris novaboracensis: New York fern: A. Frond. B. Pinnule. C. Enlarged portion of pinna. D. Pinna.

Informative arrows: 1. Fronds tapering at both ends, basal pinnae very small. 2. Round sori. 3. Acicular (needlelike) hairs.

he based the name on was collected in Canada is puzzling.

Plants. Somewhat dimorphic, deciduous, usually growing in small to large colonies, each ramet with fronds arranged as in a shuttlecock.

Fronds. 30–85 × 8–18 cm, often in evenly spaced tufts; *fertile fronds* appearing late in the season are usually longer, narrower, and more stretched out than sterile ones.

Rhizomes. 1.5–2.5 mm diam., dark brown, long-creeping, branching.

Stipes. Short, 1/8–1/6 length of frond, 4–12 cm long × 1–3 mm diam., brown at base, light green to straw-colored above; scales at base few, tan to reddish-brown.

Blades. 1-pinnate-pinnatifid to 1-pinnate-pinnatisect, 15–60 cm long; elliptic, tapering at both ends, lowest 4–10 pinna pairs **gradually becoming smaller toward base, the lowest near base of stipes are tags less than 5 mm long**, blades taper gradually to pinnatifid tips, delicate; **costae and veins moderately to densely covered with acicular hairs (needlelike, short, translucent to white hairs) to 1 mm long;** rachises light green, with scattered or dense acicular hairs and sometimes yellowish sessile glands.

Pinnae. 20 pairs more or less; 3–9 (–13) × 1–2 (–2.5) cm, cut to within 1 mm of midrib, lanceolate, long-tapering, membranous, yellow-green; small cartilaginous membranes found at base of sinuses.

Lobes. Oblong to linear, entire to crenulate, somewhat angled toward pinna tips, margins with scattered acicular (fine, white, needlelike) hairs.

Veins. Free, pinnately divided, first branches of veins from adjacent lobes reach margins above sinus.

Sori. Round; submarginal; indusia kidney-shaped, tan, often slightly hairy.

Habitat. Terrestrial in moist woods, especially in sunny areas near swamps, swamps, streams, and vernal seeps in ravines.

Distribution. Canada: Ontario to the Atlantic Ocean. USA: Michigan, Illinois, Oklahoma, Louisiana east to the Atlantic seashore (except Florida).

Chromosome #: $2n = 54$

Thelypteris noveboracensis, **has a short stipe and is distinguished by fronds that gradually taper toward the tip and the base, with the lowest pinnae very small. It may be distinguished from other ferns exhibiting this pattern by the presence of small, translucent acicular hairs on the costae and veins.**

Small sterile fronds of *Athryium filix-femina* **may resemble** *T. noveboracensis. A. filix-femina* **may be separated by the lack of acicular hairs and by lower pinnae that are somewhat shorter (vs. abundant acicular hairs and very small, taglike lower pinnae on** *T. noveboracensis*)**.**

Sometimes segregated as *Parathelypteris novaboracensis* (Linneaus) Ching.

Thelypteris palustris (Salisbury) Schott var. _pubescens_ (Lawson) Fernald
Polypodium palustre Salisbury; *Lastrea thelypteris* (Linnaeus) Bory var. *pubescens* Lawson
Marsh Fern

Etymology. Latin *palustris*, swampy, marshy, alluding to the preferred habitat of this fern; Latin *pubescens*, downy with short acicular hairs, referring to the hairiness of the blades. The vernacular name refers to a preferred habitat.

Plants. Dimorphic, dying back in late summer or early fall

Fronds. 23–90 × 10–20 cm, fronds isolated, not clustered, sometimes forming large colonies; *fertile fronds* appearing late in season, usually longer, more erect, narrower, and more stretched out than sterile ones.

Rhizomes. Long-creeping, 1–3 mm diam., shallow, branching.

Stipes. 1/3–1/2 as long as blade, 9–50 (–60) cm long, stipes of fertile fronds much longer than those of sterile ones, **black to purple at base, shading to pale green above;** scales tan, ovate, sparse near the base.

Blades. 1-pinnate-pinnatifid to 1-pinnate-pinnatisect, 10–40 (–55) cm long, well separated, thin-textured, lanceolate, blades widest below the middle tapering gradually to pinnatifid tips, lowest pinnae mostly slightly shorter, pale green; **rachises, costae,** green, smooth, slender, **veins covered with acicular (fine, white, translucent needlelike hairs) soon falling hairs**.

Pinnae. 14–20 pairs, 2–10 × 0.5–2 cm, mostly alternate, long-tapering, lanceolate, deeply cut to within 1 mm of midrib, less incised near the tips, perpendicular to the rachises, midribs grooved abaxially, not continuous with the rachis; *fertile pinnae* narrower, folded downward.

Lobes. Oblong, margins crenulate, tips rounded, fine-toothed, with many small acicular hairs; segments of fertile pinnae with leaf edges slightly inrolled over sori, appearing narrower.

Veins. Free, once-forked.

Sori. Round, medial, crowded, almost covering undersurface of pinnae, mature in mid- to late summer; indusia tan, small, hairy, irregular or somewhat reniform, soon shed.

Habitat. Rich, wet soil but not in standing water, swamps, bogs, marshes, riverbanks, wet woods, roadside ditches, and wet meadows.

Distribution. Eurasia. Northern Africa. Bermuda. Cuba. Peru. North America: Manitoba to Newfoundland, south to the Gulf states.

Chromosome #: $2n = 70$

Thelypteris palustris var. *palustris* occurs in Eurasia.

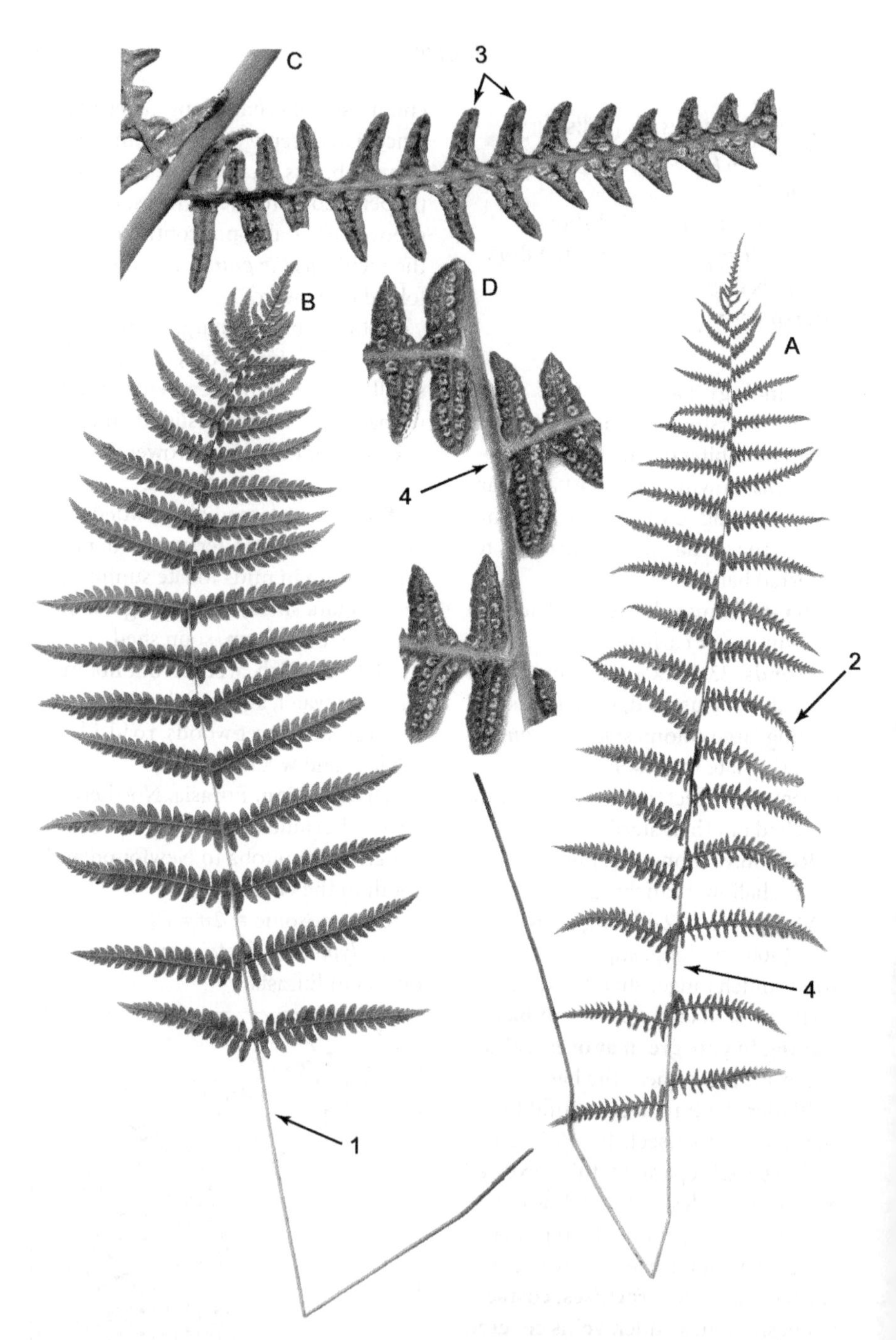

Thelypteris palustris var. *pubescens:* Marsh fern: A. Fertile frond. B. Sterile frond. C. Fertile pinna. D. Enlargement of rachis and costae.

Informative arrows: 1. Stipes smooth, black to purple at base, becoming green above and usually as long as or longer than blades. 2. Narrower fertile blade with narrower pinnae. 3. Marginal sori covered with overlapping pinnule margins. 4. Small acicular (needlelike) hairs along rachises and base of costae.

Thelypteris palustris **var.** *pubescens* **may be distinguished from** *T. noveboracensis* **by its very long stipes with dark bases; by its lower pinnae, which are only slightly shorter than those above (vs. tapering to very small lower pinnae in** *T. noveboracensis*)**; and by its fronds, which emerge singly and scattered from horizontal rhizomes rather than in tufts.**

WOODSIA R. Brown Woodsiaceae

Cliff Ferns

See Woodsiaceae in the appendix for a description of the family and a key to the Michigan genera.

Etymology. Name honors Joseph Woods (1776–1864), an English botanist, architect, and author of *The Tourist's Flora of the British Islands* (1850). The vernacular name, cliff ferns, reflects the common habitat of the ferns.

Plants. Deciduous or persistent over winter, sometimes persistent into the next season; monomorphic.

Rhizomes. Creeping, ascending, or erect, stolons absent.

Fronds. Small, clustered.

Stipes. 1/5–3/4 length of blade, often jointed and easily fractured, bases not conspicuously swollen, cross section at bases revealing 2 round to oblong vascular bundles; scaly, scales uniformly brown, lanceolate.

Blades. Linear, lanceolate or ovate, 1-pinnate-pinnatifid to 2-pinnate-pinnatifid, gradually reduced distally to pinnatifid tips, herbaceous.

Pinnae. Sessile to short-stalked, basal pinnae somewhat smaller, costae often shallowly grooved, grooves more or less continuous with rachis, glandular or nonglandular hairs on both surfaces, rarely absent, segment margins entire to dentate.

Veins. Free, simple or forked, ending in hydathodes, not extending to margins.

Ultimate segments. Margins entire to dentate, not spiny.

Sori. Round, medial, in single row.

Indusia. Arising under sporangia, dissected into several to numerous hairlike or ribbonlike segments surrounding the sori, persistent but usually obscure in mature sori.

Habitat. Rocky substrates.

Chromosome #: $x = 38, 39, 41$

Woodsia has about 30 species worldwide. Ten are treated in *Flora of North America North of Mexico*, vol. 2 (1993). Four are found in Michigan.

The Michigan species of *Woodsia* fall into 2 natural groups. Species in the first group, *W. alpina* and *W. ilvensis*, have jointed stipes, indusial segments that are 1 cell wide, hairs throughout, and composed of cells that are much longer than wide, entire or crenate pinnules, strictly one-colored stipe scales, and chromosome numbers of 39 or 41. They are circumboreal and show clear affinities to Eurasian species.

The other group, represented in Michigan by *W. obtusa* and *W. oregana* subsp. *cathcartiana,* has stipes that are not jointed, indusial segments that are more than 1 cell wide at the base (at the very base in *W. oregana* subsp. *cathcartiana*) and composed of cells as long as wide or slightly longer than wide, dentate pinnules, often 2-colored stipe scales, and a chromosome number of 38. These species are endemic to the New World. Hybridization is common within the 2 groups, but intergroup hybrids such as *Woodsia xabbeae* are very rare.

Woodsia glabella and *W. scopulina* occur in Ontario and could be found in Michigan.

In Michigan *Woodsia* species are identifiable by their fragile stipe bases, jointed or not, which break and leave tufts of the stipe bases on the rhizomes; by their relatively small size for ferns in Michigan; by their affinity for rocky habitats; and by their 2-pinnate fronds.

Cystopteris **is the most closely related genus, and** *Woodsia* **species are often confused with it.** *Woodsia* **may be distinguished by its persistent stipe bases, multisegmented indusia arising beneath the sori (vs. hoodlike, partially covering the sori, in** *Cystopteris*)**, and obscure veins that end in hydathodes before reaching the leaf margins.**

KEY TO THE SPECIES OF *WOODSIA* FOUND IN MICHIGAN

If you find a *Woodsia* species in Michigan the chances are great that it will be *W. ilvensis*; the other species are quite rare.

1. Stipes jointed (seen as small-thickening), all at about the same height along a fracture line above the base; stipe bases of previous season persistent, all nearly the same length, brushlike; indusia composed of narrow hairlike segments, uniseriate throughout; pinnule margins entire or crenate, without marginal pointed teeth; rhizome scales strictly 1-colored (2).

1. Stipes not jointed above base, fracture line absent; stipe bases of previous season persistent but of variable lengths, not brushlike; indusia usually ribbonlike segments or composed of long segments with hairlike tips; pinnule with pointed teeth on margins; rhizome scales often 2-colored (3).

2 (1). Scales common on undersurfaces of pinnae; rachises with abundant hairs and brownish scales; largest pinnae with 4–9 pairs of pinnules .... *W. ilvensis*

2. Scales absent or very rare on undersurfaces of pinnae; rachises with sparse hairs and scales, sometimes nearly glabrous; largest pinnae with 1–4 pairs of pinnules .... *W. alpina*

3(1). Indusia composed of relatively broad, ribbonlike segments many cells wide for most of length, often branched or divided distally; broad-based multicellular scales conspicuous on stipes and extending onto the rachises;

glandular hairs plentiful on both surfaces of blades; mature stipes straw-colored, sometimes with brown base . . . . *W. obtusa*

3. Indusia composed of segments with narrow hairlike tips composed of a single row of long, narrow cells, often obscure; broad-based multicellular scales sparse on the lower stipes, absent on the rachises; glandular hairs on blades sparse, grading to glabrous; stipes reddish-brown, dark purple near bases when mature . . . . *W. oregana* subsp. *cathcartiana*

Some *Woodsia* plants resemble *Cystopteris fragilis* and may be confused with it. The two often grow together. Keys to the *Woodsia* species rely on descriptions of the indusia, which are often lost or damaged as the fronds age. The indusium of *Cystopteris* is attached at one side and arches over the sorus, forming a cup or hood. The indusia of *Woodsia* consist of hairlike or ribbonlike growths that arise under the sorus. The rhizomes of *Woodsia* retain many fractured stipe bases while the rhizomes of *Cystopteris* retain few or none. The veins of *Cystopteris* are more distinct and touch the margins while those of *Woodsia* are less distinct and usually end in hydathodes short of the margins.

Woodsia alpina (Bolton) S. F. Gray

Acrostichum alpinum Bolton; *W. ilvensis* (Linnaeus) R. Brown var. *alpina* (Bolton) Watt

Alpine Cliff Fern, Northern Woodsia, Alpine Woodsia

Etymology. *Alpinus* is botanical Latin for an alpine habitat. The vernacular name describes its preferred habitat in parts of its distribution.

Plants. Deciduous, monomorphic.

Rhizomes. Erect to ascending; scales uniformly brown, lanceolate; roots hairlike, numerous.

Fronds. 2.5–25 × 0.6–2.5 cm, in compact upright tufts.

Stipes. Short, 1/3 or less length of blades, slender, reddish-brown or dark purple, darker at base, smooth and shiny above, **jointed above base at swollen node** 1/3 way up the stipes, **relatively brittle and easily broken, leaving stipe stubble of nearly equal lengths;** lanceolate scales at base monochromatic, fewer upward.

Blades. 1-pinnate-pinnatifid, narrowly lanceolate, bright green, herbaceous, lacking glands; rachises green to straw-colored, grooved, with sparse scattered hairs and scales, the hairs long, fine, red-brown.

Pinnae. 4–15 pairs, ovate to triangular, longer than wide, tips rounded or broadly acute, the shorter ones fan-shaped; upper surface glabrous, undersurface with uncommon isolated

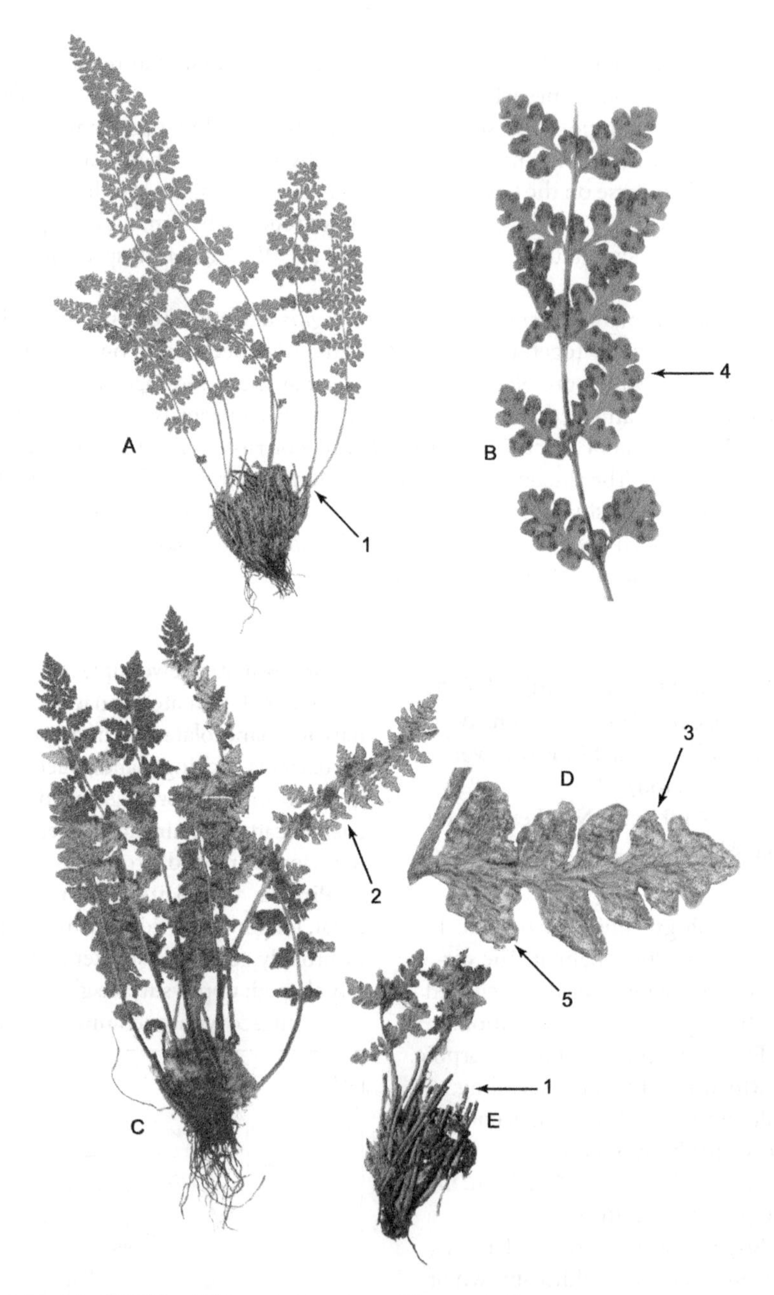

Woodsia alpina: **Alpine cliff fern: A. Fronds on rhizome. B. Pinnae.**

Woodsia ilvensis: **Rusty Woodsia: C. Fronds on rhizome. D. Pinnae. E. Rhizome with attached broken stipes of uniform length.**

Informative arrows: 1. Stipe base stubble uniform in length. 2. Abundant hairs and scales. 3. 4–9 pinnules on pinnae. 4. 1–3 pinnules on pinnae. 5. Indusia consisting of hairs enveloping sori.

hairs and linear scales; costae grooved above, grooves continuous from rachis to costae.

Pinnules. Largest pinnae with 1–3 pairs of entire or broadly crenate, dull, thin lobes, margins with a few scattered hairs.

Veins. Free, simple or forked, not reaching margins, tips with hydathodes visible on the upper surface.

Sori. Small, round, submarginal, **indusia attached under sorus and enveloping it, composed of narrow hairs with cells in 1 row, many times longer than wide, enveloping sorus, untangling when mature.**

Habitat. Talus slopes, crevices and ledges on dry cliffs mostly on slate or limestone, and northern woods in sun to partial shade.

Distribution. Northern Eurasia. Greenland. Canada and USA: Alaska to Newfoundland, south to Minnesota, New York, Vermont, New Hampshire, and Maine. In Michigan it is found at its southern range limit in Keweenaw (including Isle Royale) and Marquette counties.

Chromosome #: Disagreement exists concerning the chromosome number but $2n = 160$ is most probable.

Woodsia alpina is a Michigan endangered species. It is apparently globally secure.

Woodsia alpina **may be distinguished from** *W. ilvensis* **because it is less hairy and scaly and its largest pinnae have only 1–3 pairs of pinnules.** *Woodsia ilvensis* **has abundant hairs and scales, and its largest pinnae have 4–9 pairs of pinnules. In Michigan** *W. ilvensis* **is by far the more likely species to be found.**

Woodsia alpina **may be distinguished from** *W. oregana* **subsp.** *cathcartiana* **by its fragile, jointed stipes, which break off, leaving a stipe-base stubble of uniform length.** *W. oregana* **has unjointed stipes and stipe-base stubble of unequal lengths.**

Woodsia ilvensis (Linnaeus) R. Brown

Acrostichum ilvensis Linnaeus

Rusty Cliff Fern, Resurrection Fern, Rusty Woodsia

Etymology. Linnaeus used the name *ilvensis* to refer to the island of Elba (Napoleon connected) in the Mediterranean (its earlier name was Ilva). Oddly, this species does not grow there. In his description he gave its locality as "Europe, frigidissime" (the coldest parts of Europe, not at all like Elba). The common name alludes to mixed hair and scales on undersurfaces, which become rusty brown. Resurrection fern refers to the dried fronds, which become green and fill out when it rains.

Plants. Deciduous, monomorphic, small.

Fronds. 4.5–25 × 1.2–3.5 cm, 1-pinnate-pinnatifid to 2-pinnate, clustered, light green with silver undersides, turning rusty in fall or dry seasons, fiddleheads silvery white, visible throughout the growing season.

Rhizomes. Erect, **with abundant persistent stipe bases of more or less equal length; scales** abundant, lanceolate, **uniformly brown,** toothed, and pointed.

Stipes. Swollen joint node halfway up the stipe, brittle and easily shattered, brown to dark purple; scales at base lanceolate, with a mixture of scales and hairs above.

Blades. 1-pinnate-pinnatifid to 2-pinnate, narrowly lanceolate, silvery gray, shaggy below when young, rusty brown later; multicellular hairs concentrated along costae; rachises green, with abundant linear-lanceolate scales below, glands lacking.

Pinnae. 10–20 pairs, evenly spaced, ovate-lanceolate to roughly triangular, longer than wide, tapered to a rounded or broadly acute tip, lowest somewhat reduced, sessile or nearly so, subopposite; multicellular hairs and linear-lanceolate scales on abaxial surface, adaxially with multiple hairs mostly along midrib, hairs white on young fronds, turning rusty when dry, costae grooved above, grooves continuous from costae to rachis.

Pinnules. Larger pinnae with 4–9 pairs of pinnules, margins entire or crenate, rarely lobed, fine hairy.

Veins. Free, simple or forked, ending in hydathodes best seen on the adaxial surface.

Sori. Small, round, near margins, partly hidden by hairs; indusia arising under sori composed of hairlike segments with 1 row of long, narrow cells, enveloping sorus, untangling with maturity, persistent but often obscure.

Habitat. Sunny exposed cliffs, thin, dry, or sometimes moist acidic rocky substrates, and boreal forests.

Distribution. Circumboreal. Northern Eurasia. Greenland. Canada and USA: Alaska, Yukon, British Columbia, north central and northeastern Canada, northeastern United States south to West Virginia, west to Iowa and Minnesota.

Chromosome #: $2n = 82$

In dry weather *Woodsia ilvensis* fronds dry out and curl up: they become green again after a rain, thus a "resurrection fern."

The most common *Woodsia* **in Michigan,** *W. ilvensis,* **is distinguished from** *Woodsia alpina* **by its abundant rusty hairs and scales and the greater number of pinnules, 4–9 pairs on each pinna (vs. 1–3). It may be distinguished from** *W. oregana* **and** *W. oregana* **subsp.** *cathcartiana* **by its jointed stipes and even-length stubble on old stipe bases.**

Woodsia obtusa (Sprengel) Torrey subsp. _obtusa_
Polypodium obtusum Sprengel;
Cystopteris obtusa C. Presl
Blunt-Lobed Cliff Fern

Etymology. Latin *obtusus*, blunt, refers to rather obtuse tips of the pinna lobes. The vernacular name derives from the blunt-tipped pinna lobes.

Plants. Monomorphic, small, sterile fronds evergreen, remaining through winter flattened horizontally on substrate, fertile fronds dying back.

Fronds. 8–40 × 2.5–12 cm, sterile fronds shorter and broader than fertile, clustered, green to yellow-green.

Rhizomes. Erect to horizontal, scales, uniformly brown, some with a dark central stripe.

Stipes. Not jointed above base, easily broken, leaving few to many persistent bases of unequal lengths, light brown, straw-colored when mature, sometimes dark at base; conspicuous scales uniformly tan, some with dark centers and light brown margins, narrowly lanceolate.

Blades. 2-pinnate to 2-pinnate-pinnatifid, tapering to both ends, truncated at base, tips pinnatifid, glandular hairs plentiful on both surfaces, herbaceous, gray-green; rachises yellow-green, with glandular hairs and scattered hairlike scales.

Pinnae. 6–15 pairs, subopposite to alternate, sessile or nearly so, ovate-deltate to elliptic, longer than wide, abruptly tapered to rounded or broadly acute pinnatifid tips, 2 lowest pinnule pairs shorter and more triangular than those above; upper and lower surfaces glandular.

Pinnules. 3–8 pairs, usually dentate or shallowly lobed.

Veins. Free, simple or forked, ending before the margin in conspicuous hydathodes visible on the adaxial surfaces.

Sori. Submarginal, round, in 1 row between midrib and margin, brown then black at maturity, mature early to midsummer; *indusia* **arising at base of sori divided into 4 broad ribbon-like segments surrounding sori, persistent, translucent, multicellular throughout,** visible before sporangia mature, obscured by mature sporangia; sporangia brown then black, mature in early to midsummer.

Habitat. Shaded moist ledges near seeps and springs on granite and calcareous substrates and wooded shady hillsides.

Distribution. Midwestern, southern, and northeastern United States.

Chromosome #: $2n = 152$, tetraploid

Two cytotypes of *Woodsia obtusa* have been identified. They are usually treated as subspecies. They show morphologic and ecological distinctions and tend to have different distributions. The tetraploid subspecies *obtusa* occurs in the eastern

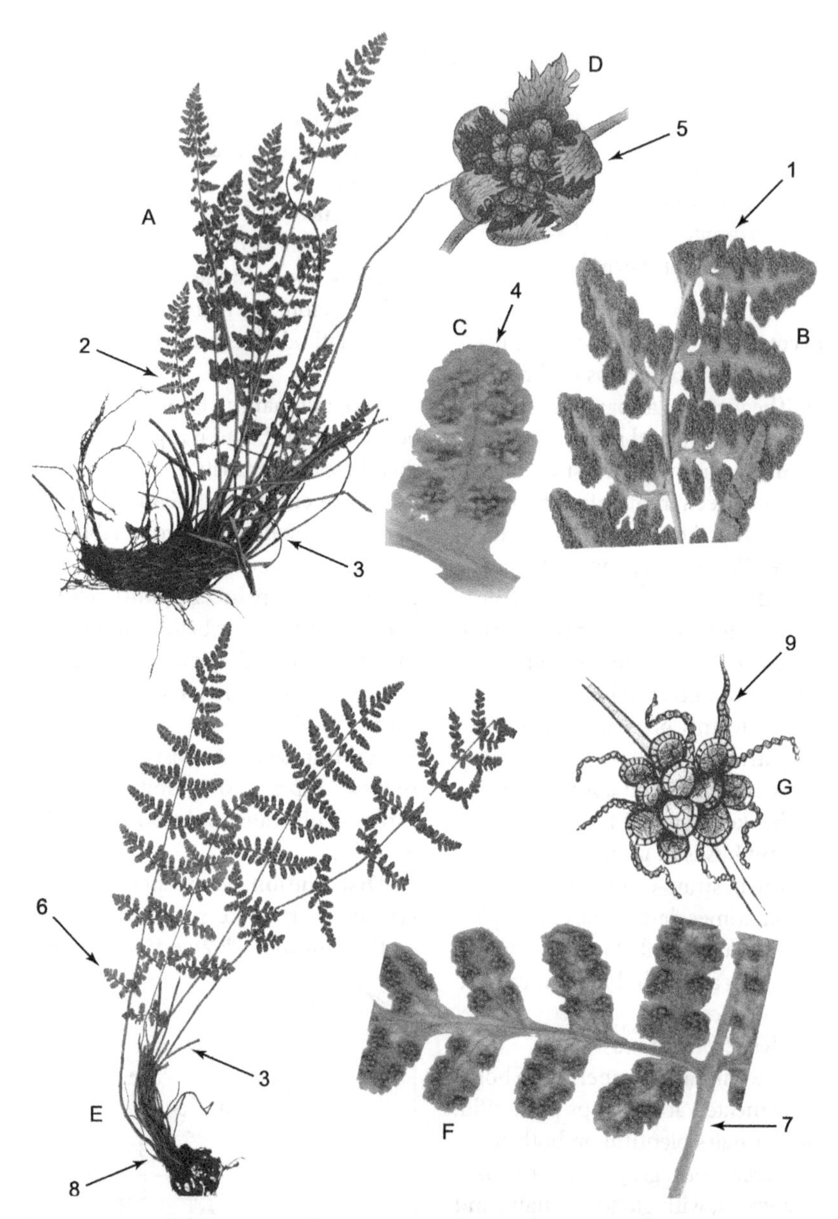

Woodsia obtusa: Blunt-lobed cliff fern: A. Fronds on rhizome. B. Enlargement of portion of pinna. C. Enlargement of pinnule. D. Indusium divided into broad ribbonlike segments surrounding sorus.

Woodsia oregana var. *cathcartiana:* Oregon cliff fern. E. Fronds on rhizome. F. Enlarged pinna section showing pinnules. G. Indusium of linear segments with long threadlike tips.

Informative arrows: 1. Glands plentiful on both surfaces. 2. Hairy rachises and scaly stipes. 3. Broken stipe bases of various lengths. 4. Blunt-tipped pinnules. 5. Indusium composed of broad ribbonlike segments arising at base. 6. Pinna pairs closest to base well separated. 7. Glands sparse to glabrous. 8. Stipes dark purple near base when mature. 9. Indusium composed of long radiating hairs.

United States, usually on limestone. The diploid *Woodsia obtusa* subsp. *occidentalis* is found in the south-central United States, usually on sandstone or granite.

Woodsia obtusa subsp. *obtusa* is a Michigan threatened species. It is considered globally secure.

Woodsia obtusa **subsp.** ***obtusa*** **has unjointed stipes that fracture at irregular intervals, leaving clusters of stipe bases that are irregular in length (vs. jointed and regular in length in** ***W. alpina*** **and** ***W. ilvensis*****). It has glandular hairs that are plentiful on both surfaces. Its mature stipe bases are straw-colored to brown. Its pinnule margins are toothed. It often has 2-colored scales.**

Woodsia obtusa **may be distinguished from** ***W. oregana*** **subsp.** ***cathcartiana*** **by its indusia, which are composed of relatively broad ribbonlike segments (vs. numerous threadlike segments in** ***W oregana*****); by its sparse to absent glandular hairs (vs. glandular in** ***W. oregana*** **subsp.** ***cathcartiana*****); and by its light brown to straw-colored stipes (vs. reddish-brown to dark purple in** ***W. oregana*** **subsp.** ***cathcartiana*****).**

W. obtusa **subsp.** ***obtusa*** **resembles** ***Cystopteris fragilis*** **but may be distinguished by its starlike indusia sitting under the sorus (vs. cup-shaped and partially covering sorus in** ***Cystopteris*****) and by its pinnae with glandular hairs (vs. glands lacking in** ***Cystopteris fragilis*****). It is more easily found in late summer and fall when** ***Cystopteris*** **species have withered.**

Woodsia oregana D. C. Eaton subsp. *cathcartiana* (B. L. Robinson) Windham

Woodsia cathcartiana B. L. Robinson

Oregon Cliff Fern, Cathcart's Cliff Fern

Etymology. *Oregana* is a Latinized name for Oregon, where this species is found. However, the range of the Michigan subspecies, *Woodsia oregana* subsp. *cathcartiana*, does not extend to Oregon. The name *cathcartiana* honors Miss Ellen Cathcart, who collected the type specimen of this species.

Plants. Deciduous; monomorphic; small.

Fronds. 3–25 × 1.3–3 cm

Rhizomes. Erect to ascending, compact; scales lanceolate, brown or some irregularly bicolorous with a dark central stripe and pale brown margins.

Stipes. 2–8 cm long, **few to many persistent stipe bases without joints resulting in stipe base stubble of unequal length,** somewhat bendable and resistant to breaking, reddish-brown, **dark purple near bases when mature.**

Blades. 5.5–17 × 1–4 cm;

1-pinnate-pinnatifid to 2-pinnate; narrowly lanceolate, dull green, sparsely glandular with glandular hairs, grading to glabrous; rachises with scattered glandular hairs and occasional hairlike scales.

Pinnae. 10–14 pinna pairs; lanceolate, abruptly tapered to round or broadly pointed tips, pairs closer to the base well separate, both surfaces smooth to moderately glandular, midribs grooved, grooves continuous with grooves of rachises.

Pinnules. 3–9 pairs on largest pinnae, often shallowly lobed, margins thin, crenate to serrate, toothed, often rolled under; glands sparse; hydathodes inconspicuous.

Veins. Free, simple or forked, ending before the margin.

Sori. Round, submarginal; *indusia* **small, composed of linear segments with long threadlike tips, surrounding and extending above sporangia.**

Habitat. Cliffs, rock ledges, rocky slopes, granite, limestone, and other substrates.

Distribution. Canada and USA: A subspecies of the western and southwestern United States, extending east and north to Montana, Saskatchewan, Manitoba, Minnesota, Wisconsin, and Michigan. A small population is found in New York. (*Woodsia oregana* subsp. *oregana* is found in 5 northwestern states and British Columbia and Alberta in Canada.)

Chromosome #: $2n = 152$, tetraploid

The taxonomy of *Woodsia oregana* needs further study. *Woodsia oregana* subsp. *cathcartiana* is tetraploid, compared to *Woodsia oregana* subsp. *oregana*, which is diploid; their hybrids, produced in a narrow region where they grow together, are sterile. They have different isozyme patterns, spore sizes, and subtle measurements of adaxial epidermal cells.

Woodsia oregana **subsp.** *cathcartiana* **may be distinguished from** *W. alpina* **and** *W. ilvensis* **by its unjointed and uneven length of stipe stubble on its rhizomes (vs. even-lengthed) and by its toothed pinnule edges (vs. rounded). It may be distinguished from** *W. obtusa* **subsp.** *obtusa* **by its indusia of linear segments with threadlike tips (vs. 4 wide ribbonlike segments); by its inconspicuous hydathodes (vs. conspicuous); by its stipes, which are reddish-brown and dark purple near base (vs. light brown to straw-colored); and by its sparse to absent glandular hairs (vs. plentiful).**

WOODSIA HYBRIDS IN MICHIGAN

***Woodsia* ×*abbeae* Butters;** *Woodsia ilvensis* × *Woodsia oregana* D. C. Eaton subsp. *cathcartiana* (B. L. Robinson) Windham; *Woodsia confusa* T. M. C. Taylor; *Woodsia oregana* D. C. Eaton var. *squammosa* B. Boivin

Abbe's Cliff Brake

Etymology. Name honors Mrs. E. D. Abbe, who first noted this plant along the Lake Superior shore of Minnesota.

SUMMARY COMPARISON OF CHARACTERS OF *WOODSIA × ABBEAE*, *W. ILVENSIS*, AND *W. OREGANA* SUBSP. *CATHCARTIANA*

Character	*W. ilvensis*	*W. ×abbeae*	*W. oregana subsp. cathcartiana*
Spacing of pinnae	Basal pinnules ± equally spaced	Basal pinnae remote	Basal pinnae remote
Scale distribution	On all axes plus lamina	On all axes	On rhizome and stipe
Hair structure	With cone-shaped cross walls	With straight cross walls	With straight cross walls
Blade margin	With hairs	With glands and hairs	With glands
Stipes	Jointed, equal length stubble, glands absent	Stipes not jointed, stubble irregular, glands present	Stipes not jointed, stubble irregular, glands present
Spores	Fertile	Abortive	Fertile
Chromosome #	$n = 41$	$3n = 117$ (triploid)	$n = 76$ (tetraploid)

Woodsia ilvensis × *W. oregana* subsp. *cathcarthiana* is a sterile triploid hybrid resulting from a hybrid between species in the groups with stipes fractured at even lengths and those fractured with uneven stipe length.

So far found only in Dickinson and Ontonagon counties in the Upper Peninsula.

Chromosome #: $3n = 117$

WOODWARDIA Smith Blechnaceae

Chain Ferns

Etymology. Name honors Thomas J. Woodward (1745–1820), a British botanist who studied algae. The vernacular name refers to the chainlike rows of sori lying parallel to the midribs of the pinnae.

Plants. Monomorphic to dimorphic, deciduous.

Rhizomes. Long-creeping to ascending, deeply subterranean, slender to stout; scales brown.

Fronds. Clustered or well separated.

Stipes. Thick, cross section reveals 5–9 round vascular bundles in a semicircle at stipe base.

Blades. 1-pinnate to 1-pinnate-pinnatisect; rachises and costae slightly scaly.

Veins. Joining, forming a regular series of areoles (spaces between joining veins) along costae and midribs of ultimate segments, then free and branching to margins.

Sori. Borne in single chainlike lines on either side of the midribs of pinnae lobes on outer arches of areolar veins of pinna.

Indusia. Covering sori, thin to thick, sometimes glandular.

Distribution. Mediterranean regions. Europe. East Asia. North America. Central America.

Chromosome #: $x = 34, 35$

Woodwardia is a genus of about 12–14 species mostly in subtropical and temperate regions of the northern hemisphere. Three species are treated in *Flora of North America North of Mexico*, vol. 2 (1993). One species is found in Michigan.

Woodwardia areolata, not treated here, was reported collected twice in Van Buren County in 1880.

Woodwardia virginica (Linnaeus) J. E. Smith

Blechnum virginicum L; *Anchistea virginica* (Linnaeus) C. Presl

Virginia Chain Fern

Etymology. *Virginica* is the Latinized name for the colony of Virginia, where the plants named by Linnaeus were collected. The vernacular name, chain fern, refers to the chainlike rows of sori parallel to the pinna midribs.

Plants. Monomorphic, deciduous.

Fronds. 50–110 × 14–30 cm, close, arranged in rows along rhizomes, leathery, glossy.

Rhizomes. 8 mm diam., long-creeping, dark brown surface, chalky white in cross section (when fresh); scales sparse, triangular.

Stipes. Long, ca. 1/3 length of blade, adaxial surface with 2 grooves, **dark purple to black proximally, straw-colored distally, base shiny and swollen,** glabrous.

Blades. 28–65 × 14–30 cm, 1-pinnate-pinnatifid to 1-pinnate-pinnatisect, ovate-lanceolate, widest at the middle, abundant minute glandular hairs borne on all surfaces, especially on underside; rachises and midribs smooth, light brown, with scattered small glands and scattered small tan scales.

Pinnae. 12–23+ pairs, alternate, middle pinnae 6–16 × 1–3.5 cm, deeply pinnatifid, ascending, linear to narrowly lanceolate with pointed tips and narrowed bases, often rotated toward blade tip ± 90°.

Pinna lobes. 10–25, oblong, tips rounded.

Veins. Joining to form a single row of areoles (spaces surrounded by veins) near, and on both sides of midribs and costae, then free and branching to margins, embossed on upper surface.

Sori. Short-linear, mostly 2.5–6

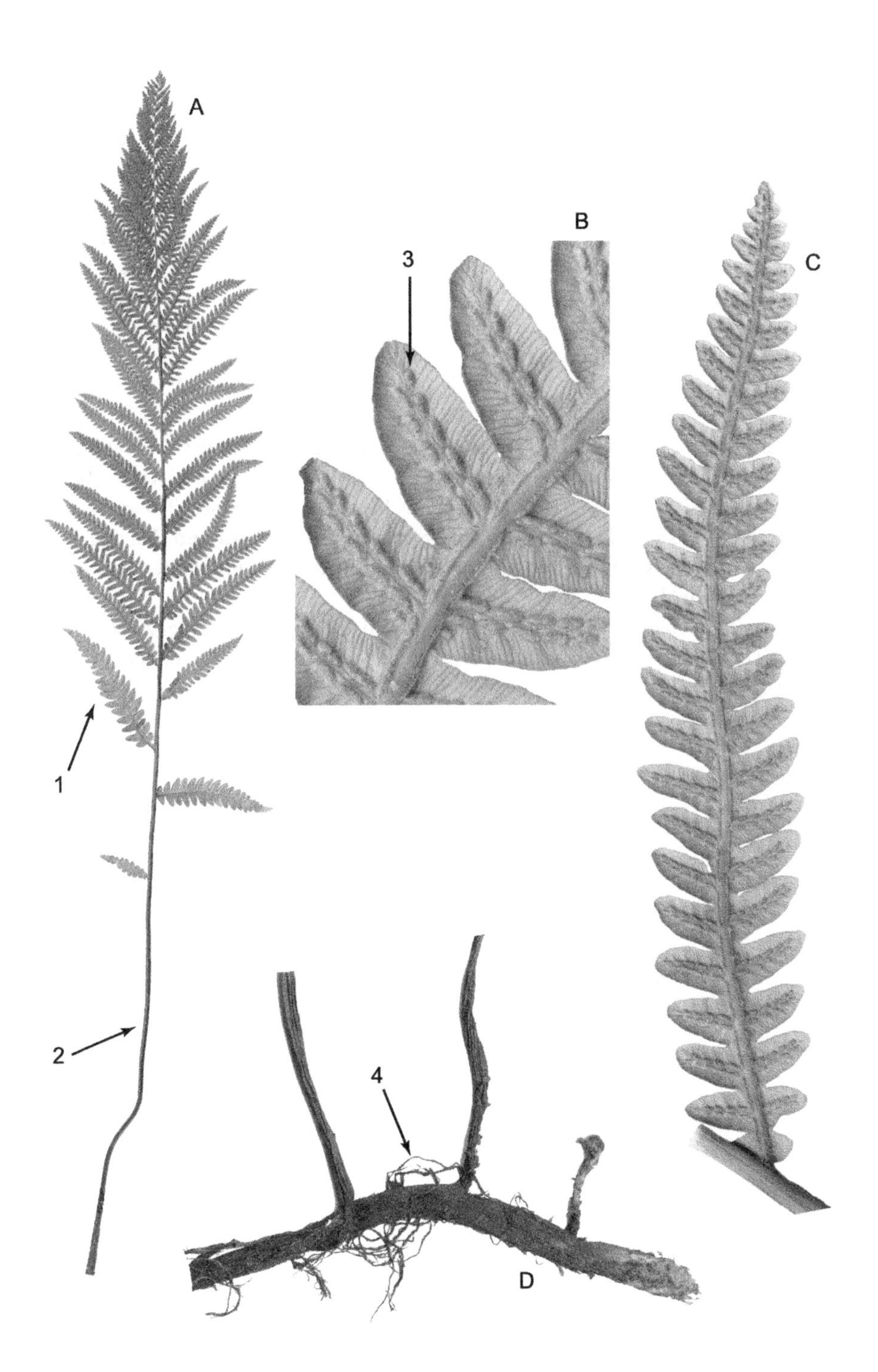

Woodwardia virginica: Virginia chain fern: A. Frond. B. Pinnule. C. Pinna. D. Rhizome. Informative arrows: 1. Pinnae rotated toward blade tip. 2. Long dark stipe with swollen base. 3. Short linear sori parallel to midrib (chainlike). 4. Long-creeping rhizome, usually submerged.

mm long, **in single chainlike rows extending along areolar veins on each side of and parallel to costae and midrib of lobes,** individual sori only as long as single areolar vein, often appearing joined after spores discharged; mature in early summer.

Indusia. Flaplike, leathery, opening toward the midrib, often hidden by discharged sporangia.

Habitat, Mostly in acid wet soil or standing water, swamps, marshes, bogs, and roadside ditches. Grows in full sun but must have its "feet" wet.

Distribution. Bermuda. Canada and USA: Ontario to the Atlantic Ocean, Texas, Arkansas, north to Illinois and Michigan, east to the Atlantic coastal plain.

Chromosome #: $2n = 70$

Woodwardia virginica **has netted veins close to midribs, with free veins farther out, and short-linear sori close to and parallel with the costae midribs of the pinna segments. It is very glandular and has blackish stipe bases. These characters make it easy to distinguish from** *Osmunda cinnamomea* **and** *Thelypteris palustris,* **which have somewhat similar fronds. It grows in marshes, swamps, and wet roadside ditches.**

If the genus *Woodwardia* is split up, our Virginia chain fern becomes *Anchistea virginica* (L.) C. Presl (and *Woodwardia areolata* becomes *Lorinseria areolata* (L.) C. Presl).

PART 3

The Adder's-Tongue Family (Ophioglossaceae)

OPHIOGLOSSACEAE C. Agardh Adder's-Tongues, Moonworts, Grapeferns

See Ophioglossaceae in the appendix for a description of the family and a key to the Michigan genera.

Etymology. Greek, *ophis*, snake, + *glossa*, tongue. The fertile spikes of some genera in this family suggest a snake's tongue.

Ophioglossaceae are represented in Michigan by 4 genera, *Ophioglossum* (2 species), *Botrychium* (11 species), *Botrypus* (1 species), and *Sceptridium* (4 species). *Ophioglossum* has long been recognized as a distinct genus. The other 3 have been treated by various authors as 1 genus (*Botrychium*) divided into subgenera or as the 3 genera recognized here. Hauk, Parks, and Chase (2003) have shown that these 3 genera are clearly distinct when analyzed morphologically and with plastid DNA.

The aerial fronds of Ophioglossaceae are composed of a single leaf divided into a fertile segment (sporophore) and a sterile segment (trophophore). The family is characterized by underground nonphotosynthetic mycorrhizal gametophytes; eusporangiate sporangia opening with a slit; thick, fleshy, hairless roots; and non-circinate vernation. Fronds of plants in Ophioglossaceae develop with erect or nodding tips not by uncoiling and expansion of crosiers or fiddleheads (circinate vernation) as other ferns do.

Their spores lack chlorophyll and are not photosynthetic. They germinate below the soil surface, rely on mycorrhizal fungi for nutrition, and may remain subterranean for several years before sending a single photosynthetic leaf above the soil surface. The green aerial fronds may not rise above the soil in some years.

The family is related to the ferns treated in part 2 of this book. While they have traditionally been treated with the ferns, they are so remarkably different in their morphology and life cycle that they are treated here as a separate group. Gross and microscopic anatomy and chemical (DNA and enzyme) studies have clearly demonstrated this separation.

Botrypus, *Botrychium*, and *Sceptridium* are treated here as separate genera because of clear distinctions in their morphology. *Sceptridium* species, strikingly different from *Botrychium* species, have evergreen, leathery, larger, more divided, dark green sterile leaves that spread horizontally to the ground. *Botrychium* has deciduous, succulent, generally much smaller, more erect, mostly less divided, light or pale green erect fronds. *Sceptridium*'s common stalks divide at or below the soil surface, while those of *Botrychium* and *Botrypus* divide well above the soil surface (except for *Botrychium simplex*, which divides at or below the soil surface).

Botrypus, which is very different from the *Botrychium* species, is a genus of 2 to 3 species with 1 represented in the United States. It is deciduous, generally much larger, has a papery texture, has a single common stalk dividing well above the soil surface into a much-divided trophophore spreading horizontally with the ground, and has an erect sporophore arising at the base of the trophophore.

Because of significant morphologic differences, terms specific to this family used in the treatment of their genera and species are different from those used for most ferns. The small size and simplicity of the leaves limit the number of discrete characters available for identification and classification and make the systematics of the genus *Botrychium* particularly problematic, difficult, and controversial.

Differences between the Genera of Ophioglossaceae in Michigan

Genus Character	*Ophioglossum*	*Botrychium*	*Botrypus*	*Sceptridium*
Evergreen or deciduous	Deciduous or evergreen	Deciduous	Deciduous	Evergreen
Trophophore or sporophore common stalk union	Divides at or below ground level	Divides well above ground level (except sometimes *B. simplex*)	Divides well above ground level	Divides near or below ground level
Trophophore length	To 20 cm, mostly less than 15 cm	To 15 cm (mostly less than 10 cm)	10–45 (–60) cm long	5–25 cm
Trophophore width	3.5–4 cm	Mostly less than 2.5 cm	5–25 cm	More than 5 cm
Trophophore division	Simple, unlobed, margins entire	Simple, lobed to 2-pinnate, seldom 3-pinnate	3-pinnate to 4-pinnate	3-pinnate to 4-pinnate
Trophophore orientation	Erect to prostrate	Ascending to erect	Parallel with ground	Parallel with ground
Trophophore texture	Stiff and papery to herbaceous	Fleshy, delicate	Thin, herbaceous	Leathery
Venation	Netted	Free, pinnate, or fanlike	Free, pinnate	Free, pinnate
Color	Light to dark green	Light to pale green or glaucous	Light yellow-green	Dark green
Presence of sporophore	Often absent	Always present	Often absent	Often absent
Sporangia	Deeply sunken	Sessile to short-stalked	Sessile to short-stalked	Sessile to short-stalked
Sporophores branching	Unbranched	Branched 1 to 3×	Branched 2×	Branched 1 to 4×
Climate preference for species #	Mostly subtropical	Mostly temperate	Mostly temperate	Mostly temperate
Chromosome #	$2n$ up to 1,200+	$2n = 90, 180$	$2n = 92$	$2n = 90$

BOTRYCHIUM Swartz Ophioglossaceae

Moonworts

Etymology. Greek *botry-*, bunch or cluster of grapes, + *-ium*, a diminutive suffix alluding to the small grapelike clusters of sporangia on the fertile segments of the frond leaves. The vernacular name, moonwort, derives from the first described species, *Botrychium lunaria*, which has half-moon-shaped pinnae.

Upright stems. Underground, forming caudices up to 5 mm thick, perennial, herbaceous, occasionally in small clusters, tipped by a single apical bud producing one above-ground leaf per year.

Roots. 0.5–2 mm diam., smooth, succulent, hairless, mostly 10 or fewer, yellowish to brown.

Leaves. Mostly less than 15 cm tall, small to very small, emerging in spring, senescing in summer or fall, consisting of 2 opposing segments, the sporophore (spore bearing) and trophophore (photosynthetic blade) attached to a common stalk.

Common stalks (basal undivided portion of leaf). Divides into 2 segments at (in *B. simplex*) or well above substrate surface (the sporophore and trophophore arise from the top of the common stalk), vascular bundles horseshoe-shaped in cross section.

Trophophores (blade-bearing segments). 1 per plant, **erect, fleshy, simple to 1-pinnate to 3-pinnate, linear, oblong, or deltate** (*B . lanceolatum*)**, mostly less than 2.5 cm wide,** light to pale green, mostly perpendicular to stem, sessile or short-stalked. (Trophophore blades of many species are divided into distinct pinnae. Blades of some species are incompletely divided into lobes or segments.)

Trophophore stalks. Sessile to short to 1/3 of the entire length of the trophophore.

Pinnae and pinnules. Variously shaped, fan-shaped to ovate to lanceolate; margins entire or variously toothed to deeply lobed, pinna tips rounded or acute, spreading to ascending.

Veins. Free, pinnate or arranged like ribs of a fan.

Sporophores (spore-bearing segments). Bearing numerous globose sporangia, normally 1 per frond, 1-pinnate to 3-pinnate, 0.5–2 mm diam., 0.8–3.5 (–8)× the length of trophophores, almost always present. (The sporophores of *Botrychium* continue to grow until they obtain their full length at the time the spores are shed.)

Sporophore stalks (portion below the sporangia). 0.25× to exceeding the length of the entire trophophore.

Sporangia. Clustered, almost completely exposed, borne in 2 rows.

Useful characters for describing *Botrychium* **species.**

Pinna span. That portion of a circle that is "spanned" by the outer circumference of the pinna. (See illustration, p. 239.)

Pinna stalk. The attachment of the pinna to the rachis, that is, the "petiole" of a pinna.

Pinna outer margins. The outer margins of fan-shaped pinnae; often crenulate, dentate, or lobed.

Habitat. Very diverse; dense woods, open sand dunes, open fields, and shady second-growth woods. Most species prefer habitats dominated by perennial, herbaceous, ground-level vegetation.

Chromosome #: $x = 45, 90$

Botrychium is a genus of around 32 species worldwide but mainly found in north temperate areas. Twenty-four species of moonworts are treated in the *Flora of North America North of Mexico*, vol. 2 (1993). They are represented in Michigan by 11 species (some of which were not yet published when the *Flora of North America* appeared) and possibly more not yet recognized in the state.

After fertilization occurs on the subterranean gametophyte, *Botrychium* plants spend up to several years in the dark below the soil surface. During that time they rely on nutrition from mycorrhizal fungi and are not photosynthetic. After beginning to produce above-ground leaves the plant may go a year or two without producing another green frond. In some known localities there may be hundreds of plants above ground one year and very few the next.

Only 5 species of *Botrychium* (as recognized in this book) were recognized in North America in 1927. These were *B. angustisegmentum* (as *B. lanceolatum* subsp. *angustisegmentum*), 1854; *B. neolunaria* (as *B. lunaria*), 1801; *B. matricariifolium*, 1847; *B. minganense*, 1927; and *B. simplex*, 1823. These species have long been accepted and are relatively easy to identify.

In more recent decades, description of new species has increased this number to about 30 species. Warren (Herb) Wagner, Florence Wagner, Don Farrar, and others have thoroughly studied the genus and were able to find subtle distinguishing characters, allowing them to describe these many new species. The ability to recognize these species was aided by chromosome counts and molecular (isozyme) data. Six of these newly described species are found in Michigan, including *B. campestre* (1986), *B. mormo* (1981), *B. pallidum* (1990), *B. spathulatum* (1990), and *B. michiganense* (A. V. Gilman, Zika, and Farrar).

Some of the recently recognized *Botrychium* species are difficult to recognize even with considerable field experience. Many were once considered to be minor morphological variants within a single previously recognized species. They are as distinct genetically as are species of flowering plants within a genus.

The life-span of individual *Botrychium* plants is unknown, but ages from 10 to 50 years have been documented by counting annual leaf scars (produced 1 per year) on underground stems. Spores may remain viable for a long time, possibly decades, helping the species to survive long periods of unfavorable conditions. Their spores may arrive at a habitat much different from the one in which the sporophytes will mature. After germination several years are required to complete maturation of the underground gametophyte and several more years to produce a mature sporophyte. It takes about 15 years following a ground-clearing disturbance for the first above-ground sporophytes to appear, about the time it takes for succession from weedy annuals to perennial herbaceous vegetation and early hardwood stands. Further succession to closed-canopy forest takes about 50 years.

The subterranean gametophytes of *Botrychium* plants spend several years in the dark below the soil surface. During that time they rely on nutrition from mycorrhizal fungi. After beginning to produce the first above-ground leaves, the plant may go a year or two without producing another green frond. In some localities there may be hundreds of plants above ground one year and very few the next.

The long period of underground growth before the first leaf appears above the ground, the absence of any above-ground growth in some years, the light green color of the fronds, and the short life-span of the green above-ground fronds certainly represent a very different life pattern from those of other groups of ferns.

Botrychium species are quite variable in morphology and have subtle and difficult to recognize characters. Their small size and simple morphology make differences between species slight. In addition these differences tend to be statistical rather than absolute. Further complicating the problem, some species are allopolyploids and, because of shared chromosomes from 2 diploid parental species, have an intermediate morphology. There are good morphological characters that define every species. The problem is that the ranges of variability within species tend to cause some degree of overlap between species. This being the case, there are a number of plants (around 10 percent) that cannot be identified by morphology alone. This emphasizes the importance of collecting more than one plant to compare.

Many of the *Botrychium* species require multiple well-collected and preserved specimens to show the morphologic variation within a population. Limited samples often are not adequate. Small plants may resemble small plants of other species. Both small and unusually large plants often give insufficient or incorrect information for correct identification.

A single silhouette or illustration will not show the range of forms found within these species. However, comparing known verified silhouettes with the frond in question is helpful. Multiple similar *Botrychium* species often occur in the same location. Using the keys helps, but these do not always get you to the correct identification. It is often necessary to consult a *Botrychium* expert to arrive at a precise diagnosis.

When collected, the fronds must be carefully pressed with all the pinnae flat. (A telephone directory makes a fine field press for collections.) Shriveled or folded specimens are seldom helpful. The underground parts of *Botrychium* species are not useful diagnostically (except when searching for gemmae) and should not be collected. Above-ground leaves bearing all characters necessary for identification may be collected by cutting them at ground level, allowing the below-ground parts to continue producing new fronds in later years. Observable field data describing color, luster, fleshiness, and stature are helpful and should also be recorded.

The sporophore stalks of *Botrychium* continue to elongate until they achieve their full length at the time the spores are shed. Collections at that time yield the best information.

Morphology and genetic studies continue to find new species within currently recognized species and new varieties within species. New areas of distribution are being found. *Botrychium ascendens*, *B. crenulatum*, *B. lineare*, and *B. pseudopinnatum* have recently been found in Minnesota and Ontario, but, while they may well be present, they have not yet been found in Michigan. These species, while not treated in this book, should be sought.

In mediaeval times it was thought that the "Fern-seed" of moonworts imparted to its owner the power to become invisible and resist magical charms and incantations. The absence of visible seeds made the mode of fern propagation mysterious. This led to interesting theories. Ferns were unmistakably plants, and it was thought that they must have seeds, a very reasonable conjecture according to Joseph Pitton de Tournefort, a noted French botanist in the late seventeenth century.

The mysterious regeneration of ferns meant that their seeds were invisible. Hence, after the doctrine of signatures, it was concluded that persons who possessed the secret of wearing this seed about them would become invisible.

This belief is illustrated in literary form by Shakespeare's *Henry IV*, Part 4, scene 1:

> GADSHILL: We steal as in a castle, cock-sure, we have the receipt of Fern-seed—we walk invisible.
>
> CHAMBERLAIN: Now, by my faith I think you are more beholding to the night than to the Fern-seed for your walking invisible.

To be effective the seeds had to be collected at midnight on Midsummer Night, or St. John's Eve, the shortest night of the year, June 23, at the precise moment at which the saint was born.

In order to catch invisible fern seed you had to stack 12 pewter plates beneath a fern frond; the seeds would fall through the first 11 plates and rest on the twelfth. The seed seeker, when collecting the seeds, had to be barefoot, in his shirt, and in a religious state of mind. Roaming fairies were blamed if no fern seed was collected.

It was also believed that moonworts would open or break locks if placed in the keyhole and turned. They possessed the power to unshoe horses, as well as humans, who stepped on them, and the power to loosen iron nails and break chains by simply touching them. The moonwort even allowed woodpeckers to acquire the strength to pierce iron if they rubbed their beaks on a leaf.

From Botrychium Illustrated glossary from Farrar Moonwort Systematics Hayden Arboretum Iowa State University

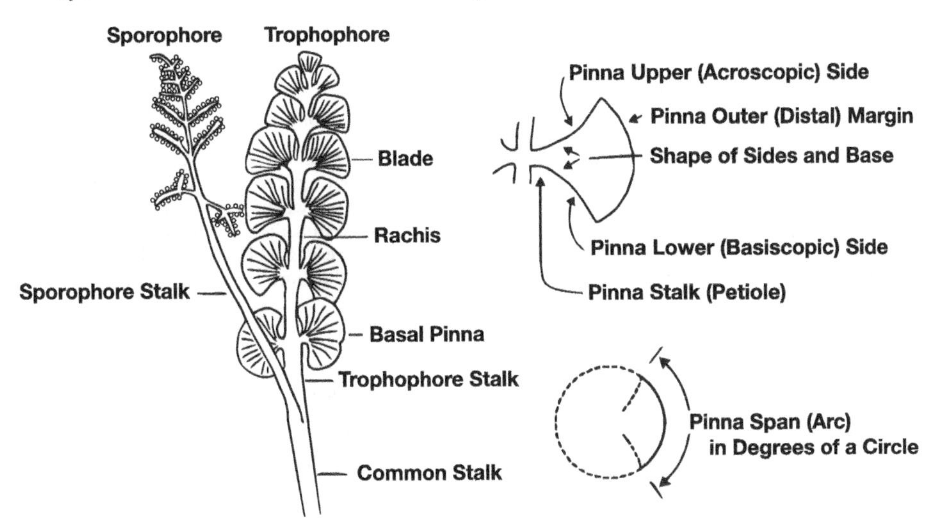

Upright stem. Underground stem tipped by a single apical bud, producing 1 above-ground leaf per year.

Common stalk. Stipe, extending from underground upright stem tips and dividing above ground into the trophophore and sporophore.

Trophophore. One of 2 divisions extending from the common stalk, bearing the expanded photosynthetic, sterile blade. (Greek, *trophe*, food, nourishment, + *phore*, carrier, alluding to the photosynthetic function of this blade.)

Trophophore stalk. The stalk between the common stalk and the trophophore blade: the trophophore "petiole."

Sporophore. One of 2 divisions extending from the common stalk, bearing numerous globose sporangia: the fertile blade.

Sporophore stalk. The stalk between the common stalk and the branching, sporangia-bearing portion of the sporophore. It continues to lengthen until spores mature.

Pinna span. That portion of a circle that is "spanned" by the outer circumference of the pinna. (See illustration)

Pinna stalk. The attachment of the pinna to the rachis, that is, the "petiole" of a pinna.

Pinna side margins. The sides of fan-shaped pinnae (acroscopic = upper side, basiscopic = lower side); usually entire.

Pinna outer margins. The outer margins of fan-shaped pinnae; often crenulate, dentate or lobed.

Characters Differentiating Once Pinnate Michigan *Botrychium* Species, Some with Fan-Shaped Pinnae

(The first 5 characters will almost always narrow the choice to 1 or 2 species. Then it becomes relatively easy to go to the species descriptions and confirm the correct ID.)

Species Character	*Botrychium campestre*	*Botrychium minganense*	*Botrychium mormo*	*Botrychium neolunaria*
Sporophore stalk length relative to the entire length of the trophophore when sporophore is fully elongated at maturity (when sporangia begin to turn color)*	Sporophore nearly sessile, stalk not reaching more than 1/4 of the entire length of the trophophore	Sporophore strongly stalked, stalk equal to or exceeding the entire length of the trophophore	Sporophore stalk about equal to or somewhat exceeding the entire length of the trophophore	Sporophore strongly stalked, stalk equal to or exceeding the entire length of the trophophore
Basal pinna span	10–45°	60°–129°	Variable	(120°–) 180°
Span of third pinna	10°–45°	60–90°	Not available	60–90°
Trophophore stalk length	Usually sessile to 10% of trophophore length	0–2 cm, stalk length exceeds the distance between the first 2 pinna pairs	0.2–2.5 cm	Usually sessile, 0–1 mm, stalk length never equaling or exceeding the distance between the first 2 pinna pairs
Pinna shape	Linear to spathulate	Flabellate or ovate, pinna sides concave near rachises and more or less straight along pinna margins	Linear to fan-shaped	Fan-shaped to lunate
Trophophore stalk length relative to average pinna spacing	Shorter	Equal or longer	0.3–6×	Shorter
Blade color	Dull, whitish-green	Dull green	Deep green to nearly white depending on degree of light exposure	Dark green to dark green, lustrous to dull, never glaucous
Pinna pairs	4–5	5–7 (–10)	(1–) 2 (–3)	Up to 9
Basal pinna size compared to next pair	Approximately equal in size	Usually smaller	Approximately equal in size	Same size
Outer pinna margins	Dentate to lobed	Nearly entire, shallowly crenate	Entire to slightly crenate	Mainly entire to undulate, rarely dentate
Basal pinnae cleft	Often	Rarely	Often	Rarely

Botrychium pallidum	*Botrychium simplex*	*Botrychium spathulatum*	*Botrychium tenebrosum*
Sporophore stalk equal to or somewhat exceeding the entire length of the trophophore	Sporophore strongly stalked, stalk equal to or exceeding the entire length of the trophophore	Sporophore stalk short, reaching about 1/2 the entire length of the trophophore	Sporophore stalk short, reaching about 1/2 the entire length of the trophophore
Ca. 90° (60–100°)	Variable, 60–180°	Variable	Variable
60–90°	60–90°	Not avail.	Not avail.
0–2 cm, 0–1/5 of trophophore rachises	0–3 cm, 0–1.5× length of trophophores	Sessile	Very short
Mushroom-shaped, larger pinna with 2 lobes, upper lobes larger and cleft	Fan-shaped to wedge-shaped	Spatulate to flabellate, mostly cleft with wide sinuses	Short, elongate, rounded
Equal or longer	Longer	Sessile	Shorter
Glaucous blue-green fresh or living	Dull to bright green to pale green	Yellowish-green	Whitish-green to yellow-green
3–5	To 5 (–7)	(2–) 4–5 (–7)	1–3
Approximately equal in size	Mostly larger and often more complex	Larger, commonly folded over rachis	Mostly larger and often more complex
Entire to irregularly crenulate-denticulate	Usually entire or shallowly crenulate	Rounded to entire or cleft with wide sinuses and lobes rounded to entire	Almost entire
Often	Often divided into 2 unequal parts	Often, with wide sinuses	Very short

Note: Unusually large forms of some fan-leaved *botrychiums* may show enlarged basal pinnae, making them appear to be 1-pinnate-pinnatifid.

See note at bottom of *Botrychium* key below.

Characters of the 1-Pinnate-Pinnatifid to 2-Pinnate Species of *Botrychium*

Species Character	*B. angustisegmentum*	*B. matricariifolium*	*B. michiganense*
Common stalk color	Green or red	Pale green, pink stripe down 1 side of the common stalk, especially in open-grown plants	Green, pink stripe down 1 side of the common stalk, especially in open-grown plants
Trophophore shape	Triangular	Ovate-oblong, oblong, narrowly elliptic	Long-triangular (because of long basal pinnae, lanceolate above basal pinnae)
Trophophore surface	Lustrous	Glaucous	Glaucous
Trophophore stalk	0–1 mm, short to sessile, shorter than average pinna spacing	0–10 (–15) mm, usually stalked, not shorter than average pinna spacing	0–2 (–3) mm, usually sessile, shorter than average pinna spacing
Basal pinnae disproportionately large compared to 2nd pinna pair	Basal usually somewhat larger, blade often appearing ternate	Usually similar in size (basal pinnae often larger on large sun-exposed plants)	Usually much larger, mostly abrupt in transition, 2nd pinna pair much smaller (basal pinnae often smaller on shade-grown plants)
Symmetry in 1st and 2nd pinna pair	Both sides similarly cut	Both sides similarly cut	Upper margin crenulate to entire, much less dissected than basal margin
Dissection of basal pinnae	Distinct lobes or segments with broad sinuses	Sinuses, if present, broad and rounded	Sinuses narrow and angular
Suprabasal pinna outline	Linear to broadly lanceolate, entire to divided to tips	Broadly ovate-elliptic, suprabasal pinnae broader, sessile to short-stalked base	Oblong to narrowly ovate-elliptic, suprabasal pinnae often little broadened beyond sessile base
Dissection of basal pinnae (division between pinnules)	Sinuses broad and rounded	Sinuses broad and rounded	Sinuses narrow and angular
Pinna tips	Acute	Rounded	Rounded to acute

Note: Unusually large forms on some fan-leaved *botrychiums* may show enlarged basal pinnae, making them appear to belong in this category.

KEY TO THE *BOTRYCHIUM* SPECIES OF MICHIGAN

See the discussion of sporophore stalk length versus total trophophore length following the key.

1. Trophophores (sterile blade-bearing segment) 1-pinnate-pinnatifid to 2-pinnate or ternate or 1-pinnate-pinnatifid with elongate basal pinnae; pinna veins pinnately branched (2).
1. Trophophores (sterile blade-bearing segment) simple or 1-pinnate, basal pinnae elongate or not, pinna veins mostly arranged in a fan shape (4).
2. Trophophores ovate or oblong to linear, approximately equal in size and cutting to adjacent pairs; stalks not sessile .... *B. matricarifolium*
2. Trophophores tripartite, basal pinna pair larger than those above, pinnae and pinnules tips pointed or rounded; stalks sessile (3).
3. Basal pinna pair usually much longer than second pinna pair, second pinna pair much less dissected than basal pinnae, especially on the upper margin, upper margin of suprabasal pinnae crenulate to entire .... *B. michiganense*

3(1). Basal pinna pair only slightly longer than second pinna pair, second pinna pair equal in dissection to basal pinnae, upper margin of suprabasal pinnae with narrow pointed segments .... *B. angustisegmentum*

4(1). Common stalk often not extending above ground level, shorter than the trophophore; trophophore linear with rounded tip or basal pinnae sometimes enlarged and pinnately dissected; sporophore stalk thin and flexible, very long, usually greatly exceeding the entire length of the trophophore .... *B. simplex*

4. Common stalk extending above ground level, longer or same length as trophophore, basal pinnae not enlarged or pinnately dissected, sporophores stalks thin or thick, not greatly exceeding the entire length of the trophophore (5).

5(4). Pinnae not fan-shaped, squarish or rounded, span not exceeding 45° (6).

5. Pinnae variously fan-shaped, span of basal pinnae between 60° and 180° plus (8).

6(5). Trophophores mostly sessile, sporophore stalks less than 1/3 the length of the entire trophophore, pinnae narrow mostly linear, rachises broad (in Michigan), pinna margins dentate to shallowly lobed, occasionally cleft into 2 somewhat spreading lobes, found in open fields and sand dunes... .... *Botrychium campestre*

6. Trophophores not sessile, sporophore stalks usually longer than trophophore, pinnae squarish or broad with rounded margins; found in dense forest leaf litter or open meadows and sandy areas (small etiolated plants grow in and around marshes) (7).

7(6). Completely to partially buried in hardwood leaf litter; ranges in color from nearly white below leaf litter to deep green above leaf litter; very succulent, pinnae squarish, outer pinna margin coarsely dentate, tip of trophophore clearly divided into blunt segments, succulent; sporophores thick with embedded sporangia; found in shady, mature hardwood understory .... *B. mormo*

7. Typically emerges completely above the leaf litter, mostly uniformly yellow-green, herbaceous, pinnae typically round, outer margins somewhat rounded, outer margin smooth; tip of trophophore is typically entire; sporophores thin, sporangia not embedded; found in forests and open meadows and dunes (small etiolated plants grow in and around marshes) .... *B. tenebrosum*

8. Sporophore stalks 1/2 to longer than the entire length of the trophophore; trophophore rachises not unusually wide; pinnae variously fan-shaped with span of basal pinnae span exceeding 60° (9).

9(8). Sporophore stalks about 1/2 the entire length of the trophophores (10).

9. Sporophore stalks 1/2 to longer than the entire length of the trophophore; pinnae variously fan-shaped with span of basal pinnae exceeding 60° (7). the trophophores (11).

10(9). Trophophores sessile or nearly so, pinnae linear to spatulate........... .... *B. spathulatum*

10. Trophophores conspicuously stalked, pinnae fan-shaped, rounded or rhombic .... *B. tenebrosum*

11(9). Basal pinnae broadly fan-shaped (almost perfect half-moons); occasionally incised; semicircular with arcs of span greater than 150°; pinnae usually overlapping..*B. neolunaria*

11. Basal pinnae fan-shaped to more or less rounded; pinnae often cleft or lobed; pinnae usually not overlapping; with span of less than 120°; pinnae usually not overlapping (12).

12(11). Trophophore whitish-green (changing to green when wet); usually 4 or fewer pinna pairs, largest pinnae with an even number of lobes, often cleft unequally into 2–4 lobes with the upper lobes larger than the lower; outer pinna margins entire to irregularly crenulate-denticulate..... *B. pallidum*

12. Trophophores green to yellow-green, (not changing when wet); 5–7 (–10) pinna pairs; largest pinnae, if lobed, with an odd number of more or less equal lobes; outer pinna margins nearly entire, rounded, shallowly crenate, (occasionally lobed) ....*B. minganense*

The sporophore stalk lengths (the portion of the sporophores below the sporangia) strictly apply only when the sporangia are mature (have turned at least somewhat yellow or some color different from that of the trophophore). With practice this method of identification can work at any time, except very early in the season (when sporangia are still tightly clustered and the same color as the trophophore). With immature plants an educated guess can be made concerning what the stalk length will be when the sporangia are mature. For example, when *B. minganense* is still clearly immature, its sporophore stalk length will still be at least 1/2 of the trophophore length, whereas very immature plants of *B. spathulatum* or *B. campestre* will appear nearly sessile.

Botrychium angustisegmentum (Pease & A. H. Moore) Fernald

Botrychium lanceolatum (S. G. Gmelin) Angström subsp. *angustisegmentum* (Pease & A. H. Moore) R. T. Clausen

Lance-Leaved Moonwort, Little Triangle Moonwort, Narrow Triangle Moonwort

Etymology. Latin, *angusti-*, narrow, + *segmentum*, segmented, alluding to the linear, somewhat pointed pinnae and pinnules of this species. Common name, from the narrow, pointed shape of the pinnae or the shape of the sterile blades.

Plants. Deciduous, emerge in late spring to early summer, dying as late as October.

Common stalks. To 10 cm long, relatively long, succulent, slender, pale green to reddish green.

Trophophores. To 6–7 cm long, 1–2-pinnate, broadly triangular, divided into 3 main axes; blades thin but firm, smooth, dull, to shiny green, to dark green.

Trophophore stalks. 0–1 mm long; short to sessile.

Pinnae. To 5 pairs, linear to broadly lanceolate, entire to divided to tips, margins with distinct lobes or segments, tips acute to rounded; ultimate segments 1–2 mm wide, narrowly linear-oblong; pinnae and lobes ascending, mostly sharply pointed.

Veins. Free. pinnately branching.

Sporophores. 1–3-pinnate, ternately divided into very long basal branches more or less equal to the distal portion of the pinnately branching sporophore; spores released mid-July.

Sporophore stalk. Short, about ¼ the entire length of the trophophore.

Habitat. Mesic closed canopy deciduous woodlands, or mesic coniferous forests.

Distribution. Canada: boreal forests of Alberta to Newfoundland. USA: Great Lakes region east to Maine and south to Ohio and Tennessee.

Chromosome #: $2n = 90$ diploid

Until recently two subspecies of *Botrychium lanceolatum* had been recognized by most authors. These were subsp. *angustisegmentum* found in Michigan, and subsp. *lanceolatum*, a larger and more robust plant found in western USA, Alaska, and Canada. They are now recognized as separate species.

Distinguished from other small *Botrychium* species by its nearly sessile and broadly triangular trophophore with 3 main axes. The trophophores are thin but stiff, and often glossy dark green. The sporophore is usually divided into three main branches of nearly equal length. Most other botrychiums have a pinnately branching sporophore or one with the basal branches elongated but shorter than the distal portion of the sporophore. In deep shade trophophores sometimes become pinnate rather than ternate, but these can be distinguished by their glossy green surface.

Botrychium campestre W. H. Wagner & Farrar
Dune Moonwort, Prairie Moonwort

Etymology. Latin *campestre*, of fields, plains, prairies, alluding to a preferred habitat of this fern. The vernacular names refer to common habitats.

Plants. 6 (–12) cm tall, emerging in early spring, dying back in late spring or early summer.

Common stalks. Short, sometimes to 5 (–10) cm long.

Trophophores. 1-pinnate; 1.5–3 (–6) × 1.3 cm, deeply troughed, especially at base when alive, oblong to linear-oblong, usually widest above middle, fleshy, tips rounded to acute, dull whitish-green, glaucous, central rachises wide, 2–4 mm (in Michigan a wide rachis is a prominent character).

Trophophore stalks. Sessile, rarely to 10 percent of trophophore length, 2–4 mm diam., fleshy.

Pinnae. 4–6 (–9) pairs, well separated, mostly linear to linear-spatulate, ascending mostly at 30–50° from rachises, spanning an arc of less than 45°, often notched or cleft into 2 or more lobes; outer margins of pinnae and pinna lobes crenulate to dentate; lowest pinnae generally narrower and less divided than the distal 1–3 pairs; largest pinnae reaching 4 × 7 mm, strongly asymmetrical and typically cut into 2 lobes with the lower lobe 1/3–2/3 the length of the upper lobe; distal pinnae often irregularly fused.

Veins. Free, spread in a fan shape; midrib absent; immersed.

Sporophores. 1-pinnate (rarely 2-pinnate), stubby, fleshy and flattened.

Sporophore stalks. Less than 1/3 the entire length of the trophophores.

Habitat. Prairies, dunes, grassy railroad sidings, and fields over limestone; in Michigan mostly on lightly vegetated sand dunes, often under shrubs of *Juniperus communis*.

Distribution. Canada: Alberta, Saskatchewan, and Ontario. USA: Great Lakes, New England, Iowa, Nebraska, Black Hills of South Dakota, and Wyoming.

Chromosome #: $2n = 90$, diploid

Botrychium campestre is a Michigan threatened species and legally protected.

The closely related *B. lineare*, with a range extending from Alaska and Yukon south to California and New Mexico and also found in Minnesota and Quebec, may be expected to be found in Michigan. In time *B. lineare* may be treated as a variety of *B. campestre*.

Botrychium campestre is the earliest moonwort to emerge in the spring. Its sporangia mature in late May and early June, and it dies back in mid-June. This may allow the plants to adapt to drier habitats and avoid the summer heat and drought

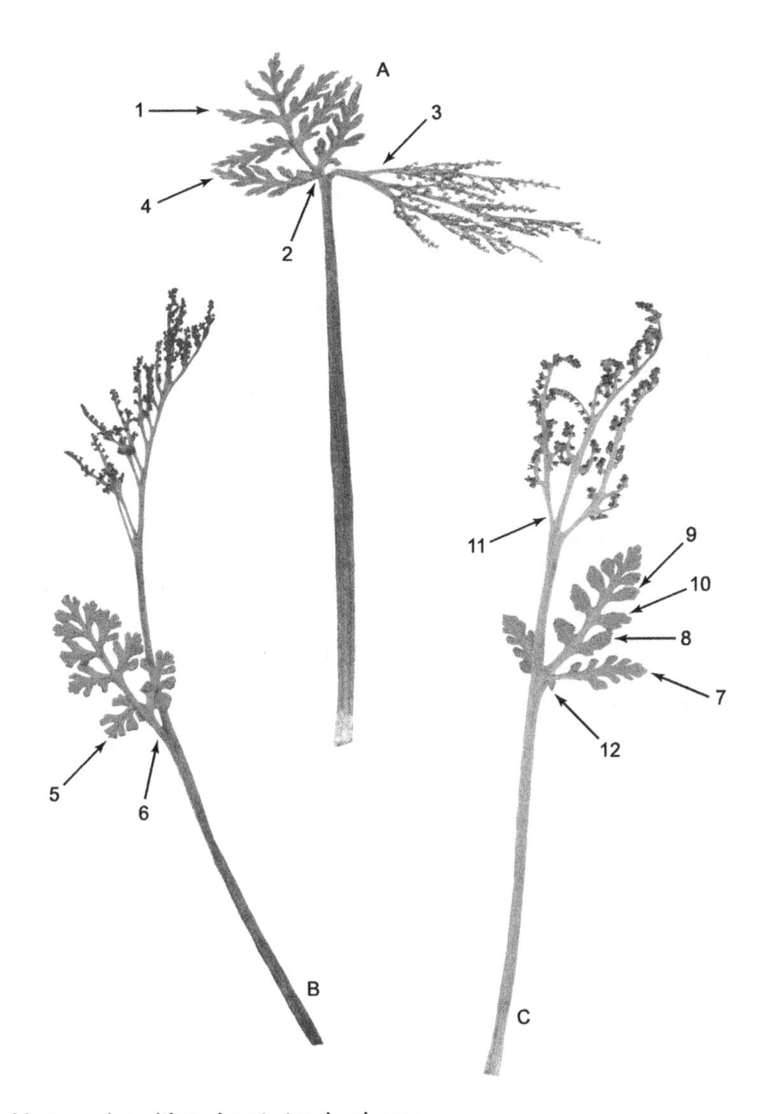

Botrychium species with 2-pinnate trophophores

A. *Botrychium angustisegmentum:* Lance-leaved moonwort.

Informative arrows:

1. Ternate with 3 main axes, broadly triangular, thin, stiff usually glossy dark green.
2. Trophophores nearly sessile. 3. Sporophores usually divided into 3 main branches.
4. Lobes and pinna tips sharply pointed.

B. *Botrychium matricariifolium:* Daisy-leaf moonwort.

5. Basal pinnae usually same size or smaller than the next pair out. 6. Trophophore distinctly stalked.

C. *Botrychium michiganense:* Michigan moonwort.

7. Basal pinnae usually much larger than the next pinna pair. 8. Dissection of second pinna pair none to minimal. 9. Sinuses narrow and angular. 10. Pinnae beyond basal pinnae are narrowly ovate-elliptic and minimally wider than their sessile base.
11. Sporophores divided into 3 main axes with stiff dense upper sporophore branches. 12. Trophophores usually sessile or very short.

associated with its habitat. It may well be more common and widely distributed than is currently recognized because its above-ground presence is very short and in grassy habitats it is inconspicuous and difficult to find.

Botrychium campestre, among the smallest of the moonworts, is inconspicuous and difficult to find. It has oblong, whitish-green, usually sessile trophophores, and its sporophore stalks are the shortest of all moonworts, usually less than 1/3 the length of the entire trophophore. It has narrow pinnae with outer margins spanning less than a 45° arc. Its 1–5 pinna pairs angle strongly upward, especially the lower pair, and are attached broadly to usually broad rachises. Pinna margins are dentate to shallowly lobed, occasionally cleft into 2 somewhat spreading lobes.

Botrychium matricariifolium (Döll) A. Braun

Botrychium lunaria (Linnaeus) Swartz var. *martricariifolium* Döll; *B. acuminatum* W. H. Wagner

Daisy-Leaf Moonwort, Daisy-Leaf Grape Fern

Etymology. *Matricaria*, from the Latin name for *Matricaria recutita* (German chamomile), + *-folium*, *matricariifolium* alluding to the resemblance of the sterile blades to the blades of this plant in the daisy family. The common name is an anglicized version of the Latin species name.

Plants. Appearing in spring, dying in late summer.

Common stalks. Up to 10 cm long, slender to stout, succulent, pale green often with a pink stripe down 1 side.

Trophophores. 1-pinnate-pinnatifid to 2-pinnate-pinnatifid, up to 10 × 9 cm, usually oblong to ovate, but variable in shape, firm, pale dull green.

Trophophore stalks. 5 mm, up to 1/6 the length of the trophophores.

Pinnae. Up to 7 pairs, somewhat ascending, close to slightly remote, basal pinna pairs approximately equal in size and cutting to adjacent pairs (occasionally longer in very large shade-grown plants), spatulate-ovate to narrowly ovate, divided to tips; ultimate segments rounded to square to acute at tips, margins entire to lobed to fully dissected, tips rounded to acute.

Veins. Free, pinnately branching.

Sporophores. 1-pinnate to 2-pinnate, in large plants sometimes with much elongated basal branches.

Sporophore stalks. 1/2 as long to longer than the entire length of the trophophore.

Spores. Released in late June and July.

Habitat. Diverse, woodland edges, early second-growth woods, thickets,

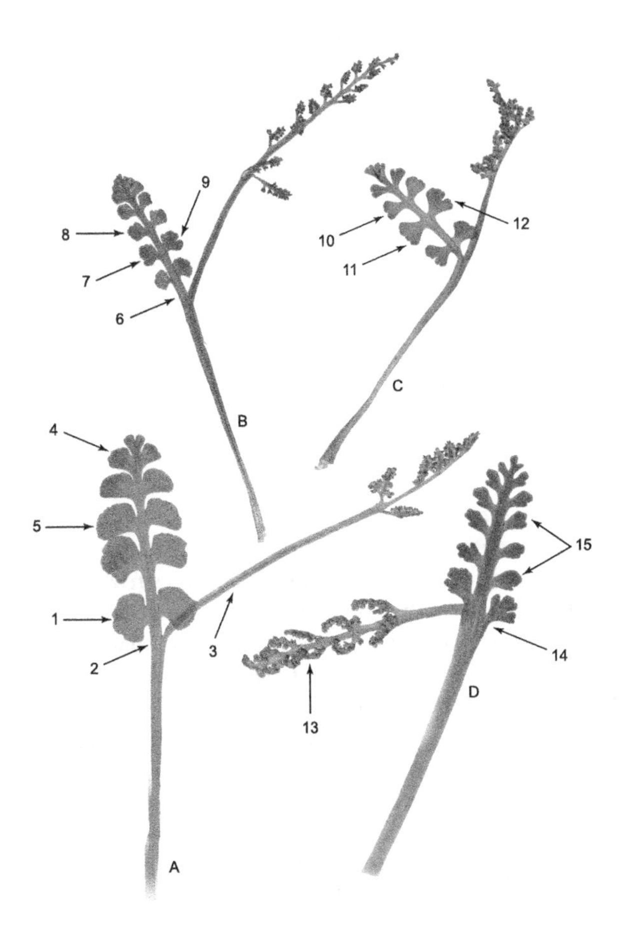

Botrychium species with fan-shaped pinnae (1-pinnate)

A. *Botrychium neolunaria:* New World moonwort.

Informative arrows: 1. Basal pinnae half-moon-shaped, span arc of nearly 180º. 2. Trophophore stalk typically sessile to short. 3. Sporophore long-stalked at spore release. 4. Upper pinna angle upward. 5. Pinna margin with rachis: upper side nearly parallel, lower side at nearly 90º angle.

B. *Botrychium minganense:* Mingan moonwort.

Informative arrows: 6. Trophophores usually stalked. 7. 5–7 well-separated pinna pairs. 8. Pinnae span a quarter circle, usually less than 120º. 9. If pinnae lobed at all, usually 3-lobed.

C. *Botrychium pallidum:* Pale moonwort.

Informative arrows: 10. Look for trophophores that are a blue-green pallid color when living. 11. Usually 3–5 pinna pairs, pinna span circa 90º, margins recurved, giving pinnae a mushroom shape. 12. Largest pinnae often split into 2 distinct lobes, the upper lobe larger.

D. *Botrychium spathulatum:* Spatulate moonwort.

Informative arrows: 13. Sporophore length 1.2–1.8 length of trophophore. 14. Trophophores sessile or nearly so, flat, thick, fleshy. 15. Pinnae narrow spathulate to fan-shaped, broadly attached, lowest pinnae the largest, outer pinna margins varying from entire to crenate with rounded teeth to shallowly lobed, sometimes cleft with wide sinuses, somewhat ascending.

grassy meadows, old fields, roadsides, old beds, and utility corridors.

Distribution. Europe. Greenland. Canada: boreal regions of Alberta to Newfoundland. USA: South Dakota, Minnesota, Wisconsin, Illinois, Indiana, New England, Kentucky, Tennessee, North Carolina, Virginia, Maryland, and New Jersey. Found throughout Michigan but more common in the Upper Peninsula and northern areas of the Lower Peninsula.

Chromosome #: $2n = 180$, tetraploid

Botrychium acuminatum is here treated as a variant of *B. matricariifolium*. It is genetically undifferentiated from *B. matricariifolium*, and morphological intermediates exist.

Botrychium matricariifolium is the most common moonwort in the Great Lakes region.

***Botrychium matricariifolium*, *B. angustisegmentum*, and *B. michiganense* are the only regularly 2-pinnate moonworts found in the Michigan. *Botrychium matricariifolium* and *B. michiganense* have a dull, glaucous surface (shiny in *B. angustisegmentum*), trophophores that are oblong and pinnately divided (*B. angustisegmentum* is ternate and triangular). *B. matricariifolium* usually differs from both *B. angustisegmentum* and *B. michiganense* in having a distinctly stalked trophophore.**

***B. michiganense* also differs from *B. matricariifolium* in usually having basal pinnae that are much longer than the second pinna pair. The basal pinnae of *B. michiganense* are dissected similarly to those of *B. matricariifolium*, but its second pinna pair is much less dissected and often hardly, or not at all, dissected on the upper margin. *B. matricariifolium* may have much enlarged basal pinnae in very large shade-grown plants, but the second pair remains dissected similarly to the basal pair. (Very small plants of *B. matricariifolium* and *B. michiganense* may be impossible to distinguish morphologically.)**

Botrychium michiganense A.V. Gilman, P. Zika, & Farrar
Michigan Moonwort

Etymology. Michigan,+ *-ense*, a Latin suffix indicating place of origin—the Michigan moonwort.

Plants. Deciduous, emerging in midspring, dying back in early fall.

Common stalks. 3–13 cm long; pale green, often with a pink stripe down 1 side.

Trophophores. 1-pinnate-pinnatifid to 2-pinnate-pinnatifid, 5–20 cm long, long-triangular (because of long basal pinnae, lanceolate above basal pinnae),

sessile to short-stalked, firm, dull gray-green.

Trophophore stalks. Sessile to very short-stalked, up to 2 (–3) mm.

Pinnae. Up to 8 pairs, oblong to narrowly ovate-elliptic, ascending, usually close, oblong to oblong-lanceolate, margins lobed to tips; **basal pinna pair usually much longer and more divided than adjacent pair; the second pair of pinnae with greatly reduced cutting relative to the basal pair, especially on the upper margin;** sinuses narrow and angular, distal pinnae much smaller and often entire, ovate to lanceolate with rounded tips, broadly attached to a relatively wide rachis.

Veins. Free, pinnately branching.

Sporophores. 1-pinnate to 2-pinnate, 3–10 cm tall, **usually with basal branches strongly elongated.**

Sporophore stalks. 1/2 to about as long as the entire length of the trophophore.

Spores. Released in June and July.

Habitat. Sparsely vegetated sand dunes and open-canopy woodlands.

Distribution. Canada: Alberta, Saskatchewan, Ontario, and Quebec. USA: Montana, eastern Washington and Oregon, Wyoming, South Dakota, Minnesota, and Wisconsin in the Lake Superior region.

Chromosome #: $2n = 180$ Tetraploid

Botrychium michiganense was formerly thought to be identical with western *B. hesperium*, and it was treated as such in *Flora of North America North of Mexico*, vol. 2 (1993). It has since been recognized as a distinct species, differing from *B. hesperium* in having usually unstalked or short-stalked trophophores and an abrupt transition from longer and deeply dissected basal pinnae to a distinctly smaller and minimally dissected second pinna pair.

Botrychium michiganense (as *B. hesperium*) is a Michigan threatened and legally protected species.

Botrychium michiganense **is a stout, dull green plant with sessile to short-stalked, 1-pinnate-pinnatifid trophophores and a sporophore usually divided into 3 main branches.**

Botrychium michiganense **may be distinguished from** *B. matricariifolium* **by basal pinnae that are usually much larger than the next pinna pair (this character is sometimes present in** *B. matricariifolium*)**, by sinuses that are narrow and angular (vs. broad and rounded in** *B. matricariifolium*)**, by dissection of the pinna pair above the basal pair that is none to minimal (vs. similar to the basal pair in** *B. matricariifolium*)**, by pinnae that are narrowly ovate-elliptic and minimally wider than their sessile base (vs. broadly ovate-elliptic and suprabasal pinnae broader than their sessile to short-stalked base), that are usually sessile or very short-stalked (vs. usually stalked), or that have stiff, dense, upper sporophore branches, (vs. lax and open).**

***Botrychium minganense* Victorin**
Botrychium lunaria (Linnaeus) Swartz var. *minganense* (Victorin) Dole
Mingan Moonwort, Mingan Island Grapefern

Etymology. *Mingan* refers to the Mingan Islands in the Gulf of Saint Lawrence where the type was collected, + *-ense*, a Latin suffix indicating place of origin. The vernacular name derives from the Latin name.

Plants. Deciduous, emerging in spring, dying back in late summer, green to yellow-green.

Trophophores. 1-pinnate, to 10 × 2.5 cm, oblong to linear, usually distinctly stalked, lobed to tips of trophophores, firm to herbaceous, dull green to yellowish-green.

Trophophore stalk. Usually distinct, 0–2 cm long, 0 to 1/5 length of trophophore rachis; stalk length usually equal to or greater than the distance between pinnae.

Pinnae. 5–7 (–10) pairs, narrow, wedge-shaped to ovate to usually fan-shaped, well separated, horizontal to ascending, sides somewhat concave, margins nearly entire to shallowly crenate; occasionally 3-lobed; pinna tips rounded, nearly circular; span about a quarter of a circle (60–120°); basal pinna pair slightly reduced in size.

Veins. Free, spread in a fan shape with short midrib.

Sporophores. 1-pinnate, 2-pinnate in very large, robust plants.

Sporophore stalks. Equal to or longer than the entire length of the trophophore.

Sporangia. Spores released July to October.

Habitat. Diverse, dense forest to open meadows, dry to permanently saturated soils, roadsides, borrow pits, and rocky areas.

Distribution. Iceland. Throughout Canada. USA: Alaska south to California, Arizona, eastward along the northern tier of states, and New England.

Chromosome #: $2n = 180$, tetraploid

Botrychium minganense is one of the more widespread and common Michigan moonworts. It is an allotetraploid with *B. neolunaria* as one of its parents. Because *B. neolunaria* is one of the parents, and because the morphology of both species varies greatly with size and habitat, there is much overlap in their morphology. This is especially true with some shade forms of *B. neolunaria*.

While *Botrychium minganense* **varies greatly in size and habitat, the combination of distinctly stalked trophophore and sporophore and fan-shaped pinnae that span about a quarter of a circle is distinctive. It may be distinguished from** *B. neolunaria* **by its narrower basal pinnae with a span of less than 120°**

(vs. 120–180° for *B. neolunaria*) **and pinnae that are well separated and do not overlap (mostly overlapping in** *B. lunaria*).

Botrychium minganense **is usually larger than** *B. pallidum* **and has more pinnae (5–7 vs. 3–5). Its pinnae are longer; they have a narrower attachment to the rachis; and the side margins are not as recurved.** *Botrychium minganense* **pinnae are symmetrical and, if lobed at all, are 3-lobed, whereas** *B. pallidum* **often has 2 distinct lobes, well spread, the upper lobe being larger. In live plants** *B. pallidum* **is easily distinguished by its light gray-green color. The basal pinnae are slightly smaller in** *B. minganense* **but are the largest pinnae in** *B. neolunaria, B. spathulatum,* **and** *B. pallidum.*

Botrychium mormo W. H. Wagner
Little Goblin Moonwort

Etymology. In Greek mythology *Mormo* was a hideous spirit that bit bad children—a goblin. The Latin and vernacular names suggest a pale little goblin peeping through leaf litter in dark forests.

Plants. Deciduous, fronds appearing in late spring to fall, plants very small, often hidden in leaf litter, sometimes so small that they don't even rise much above the leaf litter; up to 8 cm tall, fleshy, shiny, pale yellow-green.

Common stalks. 50 (20–70) percent of the total frond length, often whitish to pale green.

Trophophore stalks. 0.5–2.5 cm long, usually 0.3–0.6 times length of trophophores.

Trophophore stalks. 0.5–2.5 cm long, usually 0.3–0.6 times length of trophophores (usually less than 1/3).

Trophophores. 1-pinnate; 2 (1.3–4.5) cm long, 5 (3–7) mm wide, linear to linear-spatulate, very succulent, quite variable, in larger plants the blade may bear 2–3 pairs of small, blunt pinnae, in smaller plants the pinnae may be nearly absent, tips with 2–4 angular triangular or squarish lobes, shiny, yellow-green to green.

Pinnae. Up to 3 pairs; ascending; mostly widely separated, often closely spaced in distal half of trophophore; upper pinnae broadly attached to rachis; pinna outline extremely variable, linear to fan-shaped to rhomboidal with straight outer margins that are entire to shallowly crenate to toothed.

Veins. Spread in a fan shape, midrib absent.

Sporophores. 1-pinnate, 0.2–3 cm; succulent.

Sporophore stalks. Equal to or longer than the entire length of the trophophore.

Sporangia. Few, large, sunken to embedded in stalk and branches;

spores released from late summer to October.

Habitat. Found in rich, moist, dark, mature hardwood forests of basswood (*Tilia americana*), sugar maple (*Acer saccharum*), beech (*Fagus americana*), and black ash (*Fraxinus nigra*), and in the rich humus layer of leaf litter in mature and second-growth mesic northern hardwood forests, where small plants may remain covered by a layer of dead leaves.

Distribution. Michigan, Minnesota, and Wisconsin.

Chromosome #: $2n = 90$ diploid

The visual portions of the fronds above the leaf litter are often 4 cm long or less. Some small plants may remain covered with a layer of dead leaves.

The sporangia of *B. mormo* may not open when they remain under leaf litter; on emerged plants the sporangia open.

Botrychium mormo is extremely sporadic in distribution. In wet years leaves may easily be found in known localities; in dry years plants remain dormant and do not produce above-ground leaves.

Botrychium mormo is a Michigan threatened, legally protected fern. It is considered imperiled in the state and is globally vulnerable.

Botrychium mormo **is quite different from all other** *Botrychium* **species except** *B. tenebrosum.* **The tips of** *B. mormo* **are clearly divided into blunt segments while those of** *B. tenebrosum* **are mostly entire.** *Botrychium tenebrosum* **may also have some squarish outer pinna margins that are coarsely toothed, but it usually has some or all pinnae that are more fan-shaped with rounded outer margins.**

B. mormo **is a more succulent plant with a thicker sporophore and deeply embedded sporangia. The sporophore stalk of** *B. mormo* **equals or exceeds the length of the trophophore, whereas that of** *B. tenebrosum* **is about half the length of the trophophore.** *Botrychium mormo* **is a deep forest plant that may be nearly white to deep green depending on how much of it is exposed above the leaf litter.** *Botrychium tenebrosum* **is primarily a plant of open habitats and is fully exposed and green.** *B. tenebrosum* **frequently produces sporangia on its basal pinnae.**

Botrychium neolunaria Stensvold & Farrar

Osmunda lunaria sensu auct. non Linnaeus; *Botrychium lunaria* sensu auct. non (Linnaeus) Swartz

New World Moonwort

Etymology. Latin *neo-*, new, (+ *luna*, moon, + *-aria*, a suffix indicating connection, alluding to the half-moon shape of the pinnae) and *neo-* referring to this species' newly recognized status as distinct from *Botrychium lunaria.* (*Osmunda lunaria*, the basionym of *Botrychium lunaria,* was first described and published in volume 2 of Linnaeus's *Species Plantarum* [1753].) In Michigan this species has been recognized as *Botrychium lunaria* (Linneaus) Swartz. That species as defined by Stensvold & Farrar is now under-

stood not to occur in Michigan but instead is found in Canada, northern Europe, Asia, South America, Australia, and New Zealand. The common name alludes to the half-moon-shaped pinnae of this species.

Plants. Deciduous, appearing in spring, dying back in late summer; up to 10 × 4 cm.

Common stalks. About 5 cm long.

Trophophore. 1-pinnate; up to 10 × 4 cm; oblong, fleshy, subdivided into 4 to 9 (sometimes more) closely spaced or overlapping smooth-edged and commonly concave pinna pairs; trophophore tips usually cuneate to spatulate, notched, approximate to adjacent lobes; green to dark green with a surface that is lustrous to dull, never glaucous.

Trophophore stalks. Usually sessile, 0–1 mm, if stalked the stalk length is shorter than the distance between the basal and second lobes.

Pinnae. Up to 9 pairs, broadly fan-shaped, midribs absent, undivided, margins mainly entire to undulate, rarely dentate, usually the basal pinnae with a span of nearly 180° and the third pinna pair with a span of approximately 90° degrees, basal pinnae pair approximately equal in length and cutting to the adjacent pair, and much less ascending, upper pinnae ascending, with basiscopic inner margin creating a large angle (nearly 90°) with the rachis, the acroscopic inner margin nearly parallel to the rachis; pinnae mostly overlapping, in deep shade pinnae sometimes more distant and not overlapping.

Veins. Spread in a fan shape, midribs absent.

Sporophores. 1-pinnate to 2-pinnate, 0.8–2× length of trophophores.

Sporophore stalks. Usually long-stalked, mostly exceeding the entire length of the trophophore.

Sporangia. Spores released in June and July.

Habitat. At high latitudes and altitudes often found in open to lightly wooded fields and sparsely vegetated scree slopes. At lower elevations and latitudes found in deep woods, as well as meadows and moist, sparsely vegetated but well-drained sandy soils and sand dunes. In the far north, *B. neolunaria* is pretty much confined to bottomlands and coastal plains.

Distribution. Mostly found throughout the boreal and high-elevation areas of North America: Pennsylvania north to Labrador and west across Minnesota and South Dakota to the high mountains of all southwestern states and across all Canadian provinces and coastal and lowland Alaska.

Chromosome #: $2n = 90$, diploid

Botrychium species: 1-pinnate species with non-fan-shaped pinna

A. ***Botrychium campestre:*** **Dune moonwort (or prairie moonwort)**

Informative arrows: 1. Sporophore stalks less than 1/3 length of trophophore. 2. Pinnae outer margins narrow, spanning less than 45º. 3. 1–5 pinna pairs, angling upward. 4. Broad rachis.

B. ***Botrychium mormo:*** **Little goblin moonwort.**

Informative arrows: 5. Tips clearly divided into blunt segments. 6. Sporophore stalk equals or exceeds the length of the trophophore. 7. Coarsely toothed squarish outer pinna margins.

C. ***Botrychium tenebrosum:*** **Shade-loving moonwort.**

Informative arrows: 8. Trophophores usually conspicuously stalked, 1/2 to 2/3 the entire length of the trophophore. 9. Common stalk long. 10. 2–4 (–6) pinnae pairs.

Botrychium neolunaria is endemic to North America and one of the most abundant and widely distributed moonwort species. Throughout much of its range it is the most common moonwort in its habitat, which includes open fields, stabilized sand dunes, mountain meadows, roadsides, and less frequently mesic forests.

In 2008 Mary Clay Stensvold, using genetic analysis, as well as morphology, discovered that the circumboreal species previously called *B. lunaria* contains more genetic variability than was previously recognized. In the contiguous United States, much of southern Canada, and the coastal and lowland interior of Alaska, the plants were found to greatly differ genetically from the European *B. lunaria*, resulting in recognition of these plants as *B. neolunaria*, a species different from *B. lunaria*.

Botrychium neolunaria differs morphologically from *B. lunaria* in having a sporophore stalk that greatly exceeds the entire length of the trophophore, comprising about 2/3 of the entire length of the sporophore, and has strongly ascending sporophore branches. The sporophore stalk of *B. lunaria* is about equal to the entire length of the trophophore, comprises about 1/2 of the entire length of the sporophore, and has branches that spread more or less horizontally to the rachis. Trophophores of the two species are very similar.

Botrychium neolunaria is remarkably uniform in morphology. It often grows with other species of the genus.

Botrychium lunaria, a circumboreal species, has been found in North America across northern Canada into Alaska and from Newfoundland to Quebec (along the St. Lawrence Seaway), James Bay, Hudson Bay, and Yukon. It has not been recorded in the western Great Lakes area or the western St. Lawrence Valley and has not been found in the contiguous 48 states of the United States.

Plants of high mountain habitats in Alaska and northern Canada are most commonly *B. lunaria*. Where *B. lunaria* is found in proximity with *B. neolunaria* it tends to grow on higher, better drained sites.

The distributions of *B. lunaria* and *B. neolunaria* are still being explored.

***Botrychium neolunaria* is distinguished from other *Botrychium* species by its broad and mostly overlapping pinnae. The basal pinnae are half-moon-shaped, spanning an arc of nearly 180°. The upper pinnae angle strongly upward, the lower side margin creating a large angle (nearly 90°) with the rachis and the upper side margin nearly parallel to the rachis. Although it is occasionally short-stalked, the trophophore of *B. neolunaria* is typically sessile, the stalk length seldom equaling or exceeding the distance between the stalks of the first pinna pair as it usually does in *B. minganense*. Plants are green to dark green with a surface that is lustrous to dull, never glaucous.**

The sporophore is long-stalked, with the stalk at spore release exceeding the length of the trophophore.

Botrychium pallidum W. H. Wagner
Pale Moonwort

Etymology. Latin *pallidus*, pale, alluding to the pale green to whitish sterile frond. The common name derives from the Latin name and the plant's pale green to whitish fronds.

Plants. Deciduous, emerging in late spring and early summer; glaucous, pale green to blue-green, becoming green when moist.

Common stalk. 2–4 (–8) cm., strongly glaucous.

Trophophore stalks. 2–8 mm, 0–1/5 length of trophophore rachises.

Trophophores. 1-pinnate, up to 4 × 1 cm, narrowly oblong, more or less folded lengthwise, forming a trough on living fronds, tips of trophophore rounded.

Pinnae. 3–5 pairs; fan-shaped to mushroom-shaped with recurved side margins; closely spaced; small, up to 6 mm long; ascending; basal pinna pair approximately equal in size and cutting to next distal pair; broadly attached to rachis; the larger pinnae ascending and strongly asymmetrical; largest pinnae often split into 2 unequal lobes, the upper 1 larger and more cleft than the lower, in very large plants primary lobes subdivided into additional lobes; outer margins entire to irregular crenulate-denticulate; pinna margins spanning an arc of 60–100°.

Veins. Free; spreading fanlike; midribs absent.

Sporophores. 1-pinnate to 2-pinnate; basal sporophyte branches are divided into a longer upper and shorter lower branch that often extends downward.

Sporophore stalk. As long as or exceeding the length of the trophophore.

Sporangia. Small; spherical; yellow; releasing spores in June.

Habitat. Mainly in meadows and fields with sparse to dense herbaceous vegetation, occasionally in woodlands.

Distribution. Canada: central Alberta through Quebec. USA: Minnesota, Michigan, Maine, Black Hills of South Dakota, and Wyoming. In Michigan it is found in very local and small populations.

Chromosome #: $2n = 90$

Botrychium pallidum is a Michigan species of special concern and is not legally protected. It is considered vulnerable in Michigan and globally.

B. pallidum reproduces both sexually by means of spores and asexually by means of minute gemmae produced on the underground stems. It is often found growing with other *Botrychium* species.

A usually small plant (large specimens are occasionally found), *Botrychium pallidum* suggests a very pale dwarf and narrow sun form of *B. minganense.* It may be distinguished, when living, by its pale green to whitish-blue color and by usually having some pinnae that are unequally bilobed.

The pallid color is caused by a mat of tubular wax excretions on the leaf surfaces. When dry this mat holds air spaces that reflect light. When moisture replaces the air (as when it is placed in a moist plastic bag) the plant doesn't reflect light and appears green.

The trophophores of *Botrychium pallidum* are stalked, oblong in outline, and once pinnate, with pinnae spanning an arc of about 90° (1/4 of a circle), characters closely resembling *B. minganense*.

In live plants *B. pallidum* is easily distinguished by its silvery gray-green or blue-green "pallid" color, which quickly changes to green when moist, especially if placed in a moist plastic bag. *B. pallidum* is usually smaller than *B. minganense* and has fewer pinnae (3–5 vs. 5–7). Its pinnae are shorter, they have a wider attachment to the rachis, and the side margins are recurved, giving the pinnae a mushroom shape. (The pinna sides of *B. minganense* are concave near the rachis and are more or less straight along the pinna margins.) The pinnae of *B. pallidum* often have 2 distinct, well-spread lobes, the upper lobe being larger (*B. minganense,* if lobed at all, has 3 or 5 more or less symmetrical lobes). This unequal lobing is also seen in the basal branches of the sporophores, those of *B. minganense* being symmetrically branched whereas those of *B. pallidum* are divided into longer upper and shorter lower branches.

Botrychium simplex E. Hitchcock
Dwarf Moonwort, Least Moonwort

Etymology. Latin *simplex*, simple, undivided, unbranched, alluding to the simple shape of the sterile and fertile blades of very small plants. The vernacular names derive from their very small size, usually the smallest of the *Botrychium* species.

Plants. Emerging from mid-spring to early fall, usually very small, highly variable, inconspicuous, and scattered.

Common stalk. Very short, usually ending near ground level, may be 1–2 cm in plants growing in dense vegetation.

Trophophores. Simple to 2-pinnate to 3-pinnate (in the basal pinnae), 2.5–7 cm, linear to ovate-oblong, to oblong, to fully triangular with ternately arranged pinnae in large plants with enlarged and pinnately dissected basal pinnae, herbaceous to thin-papery, trophophore tips rounded, usually undivided and somewhat folded (forming a "boat hull" shape) to strongly divided and flat, often somewhat clasping the

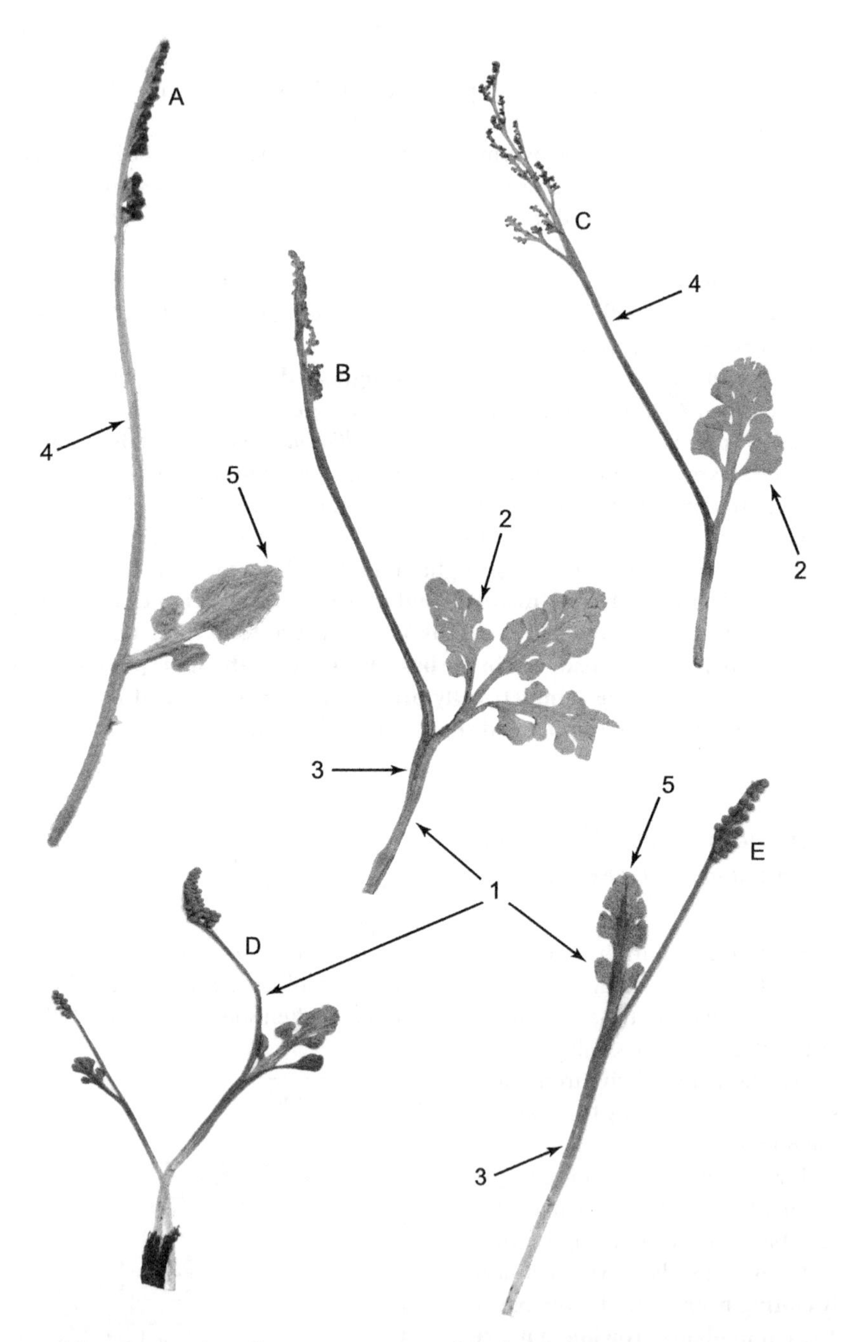

Botrychium simplex: **Dwarf moonwort: A–E. Various frond shapes.**

Informative arrows: 1. Generally small size, simple to 1-pinnate, undivided pinnae with rounded margins and decurrent bases. 2. Basal pinnae on larger plants sometimes larger and 1-pinnate. 3. Sporophore and trophophore usually diverging near the ground. 4. Sporophore usually long-stalked. 5. Look for living blade tips, which are often semifolded, forming a "boat hull."

sporophore stalk, dull to bright green to pale green.

Trophophore stalks. 0–3 cm, slender, 0–1.5× length of trophophores, usually longer than the distance between the first 2 pairs of pinnae, equal to or exceeding the length of the common stalk.

Pinnae. Undivided (in very small plants) to 7 pairs, ascending, widely separated proximally to close distally, distance between first and second pinnae frequently greater than between second and third pairs, basal pinna pair mostly undivided but sometimes much larger and more divided than adjacent pair, often divided into 2 unequal parts, pinna wedge-shaped to fan-shaped, strongly asymmetrical with upper portion larger, basiscopic margins usually ± perpendicular to rachis and somewhat to strongly decurrent, acroscopic margins strongly ascending, outer margins usually entire or shallowly sinuate.

Veins. Free, pinnate or spread fan-like, enlarged basal pinnae appearing to have a midrib.

Sporophores. Mostly 1-pinnate, usually long, lax, and drooping.

Sporophore stalk. Very long, usually greatly exceeding the entire length of the trophophore.

Habitat. Diverse, open habitats, summer dry fields, orchards, and roadside ditches, often in densely vegetated settings.

Distribution: Europe. Iceland. Greenland. USA and Canada: southern Ontario south to Iowa, east to West Virginia and along the Appalachians to South Carolina, from eastern North and South Dakota to the Atlantic Ocean.

Chromosome #: $2n = 90$, diploid

Two varieties of *Botrychium simplex* have been recognized in North America: var. *simplex* and var. *compositum*. *Botrychium simplex* var. *simplex* has been recognized as the only variety that occurs in Michigan and the Great Lakes region. It is also found in southeastern Washington and northeastern Oregon and is the most common variety in Southern California. *Botrychium simplex* var. *compositum* is a western plant found largely in mountainous areas. The differences are subtle and have to do with the division of the common stalk at ground level or farther out and secondarily divided basal pinnae and trophophore stalks equal to or exceeding the length of the common stalk. There is enzymatic support for the separation.

Botrychium simplex, Michigan's smallest fern, has many environmental forms and juvenile stages that have caused the naming of many infraspecific taxa that are not treated separately here. The taxon recognized as *Botrychium simplex* var. *tenebrosum* by some is recognized in this book as *B. tenebrosum* as originally described.

Botrychium simplex, **while quite variable, may be recognized by its very small size; simple to 1-pinnate, frequently clasping trophophore blades (sometimes with enlarged basal pinnae); undivided pinnae with rounded margins; basiscopic margins usually ± perpendicular to rachis and somewhat to strongly decurrent; acroscopic margins strongly ascending (sometimes large basal pinnae on larger plants are 1-pinnate); entire, boat-hull-shaped apices; and the junction of the sporophores and trophophore usually near ground level.**

Botrychium simplex **resembles** *B. tenebrosum* **but can be separated due to its enlarged basal pinnae (not enlarged in** *B. tenebrosum*)**, sporophore and trophophore stalks that are longer than in** *B. tenebrosum*, **and a trophophore-sporophore junction near ground level.** (*B. tenebrosum* **sporophore stalks diverge from the trophophore well above the ground.)**

Botrychium spathulatum W. H. Wagner
Spatulate Moonwort

Etymology. Latin *spathulatus*, spatula-shaped, referring to this plant's spatula-shaped pinnae. The vernacular name derives from Latin name.

Plants. Deciduous, appearing late spring, dying back in late summer; relatively large and fleshy.

Common stalk. Stout, 3–6 cm, dark green.

Trophophores. 1-pinnate, to 8 × 2.5 cm, oblong to narrowly triangular, flat, thick, leathery, lowest pinnae the largest, shiny yellow-green to dark green.

Trophophore stalks. Sessile (or nearly sessile).

Pinnae. (2–) 4–5 (–7) pairs, mostly narrowly spatulate to fan-shaped, somewhat ascending, well separated, basal pinnae largest, commonly folding over the rachises, the junction of lower side margin with the outer margin rounded, narrowly adnate 1/4 to 1/3 of the pinna width, outer margins entire to crenate with rounded teeth to shallowly lobed, tips sometimes cleft with wide sinuses.

Veins. Free, spread in a fan shape, midrib absent.

Sporophores. 1-pinnate to 2-pinnate; branches twisted, angling away from rachis.

Sporophore stalk. About 1/2 the length of the trophophore.

Sporangia. Spores released by late June.

Habitat. Stabilized sparsely vegetated sand dunes, open woods, grassy meadows, and railroad sidings.

Distribution. Widespread in North America from New Brunswick and New England to Alaska and south in the Rocky Mountains to Colorado.

Chromosome #: $2n = 180$, tetraploid

Allozyme studies have shown that *Botrychium spathulatum* is an allotetraploid between *B. neolunaria* and *B. campestre*.

Botrychium spathulatum closely resembles *B. ascendens*, a species that should be sought in Michigan. Small plants resemble *B. campestre* but can be differentiated from that species by having the longest basal pinnae.

Botrychium spathulatum **can be distinguished from** *B. neolunaria*, *B. mingan-*

ense, **and** *B. pallidum* **by its much shorter sporophore stalk (usually less than 1/2 or less the length of the trophophore); by its narrower, spatulate, broadly attached pinnae; and by its nearly sessile trophophore.**

The pinnae of *B. spathulatum* **tend to be less broadly fan-shaped than** *B. minganense,* **and sporophores of** *B. spathulatum* **tend to be more extensively divided than those of** *B. minganense.*

The trophophores of *B. spathulatum* **are sessile or nearly so, while those of** *B. tenebrosum* **are conspicuously stalked.**

Botrychium tenebrosum A. A. Eaton

Botrychium simplex E. Hitchcock var. *tenebrosum* (A. A. Eaton) R. T. Clausen

Shade-Loving Moonwort, Swamp Moonwort (Misnamed Moonwort)

Etymology. Latin *tenebrosus*, dark, gloomy very shady, alluding, incorrectly, to this species' formerly perceived preferred habitat. The vernacular name, misnamed moonwort (formerly called shade-loving or swamp moonwort), derives from the fact that when the plant was originally named it was thought to grow in gloomy shady habitats (hence the specific name *tenebrosum*) and is now known to prefer open meadows and dunes.

Plants. 3–10 cm tall, slender.

Common stalks. 2.5–7.5 cm long, relatively long.

Trophophores. 1.4–3.8 cm. long; herbaceous, light green to yellow-green, slender with undivided basal pinnae.

Trophophore stalks. Usually 1/3 the length of the entire trophophore, occasionally sessile.

Pinnae. 2–4 (–6) pairs, rounded to rhombic, undivided basal pinnae somewhat elongate, margins mostly entire, basal pinnae in open habitats often with sporangia on the margins.

Veins. Free, spread in a fan shape, midrib absent.

Sporophores. 1-pinnate to 2-pinnate, with short, strongly ascending branches.

Sporophore stalks. 1/2 to 2/3 the entire length of the sporophore.

Sprorangia. Open-grown plants of *B. tenebrosum* often produce sporangia on lower pinna margins, spores released in mid- to late June.

Habitat. Open meadows and dunes. Small etiolated plants grow in and around marshes.

Distribution. Iceland. Greenland. Canada: Ontario and Quebec. USA: New Jersey and Pennsylvania west to the eastern Dakotas.

Chromosome #: $2n = 90$, diploid

Botrychium tenebrosum has been a problem species that resembles and has often been treated as a variety (var. *tenebrosum*) of *B. simplex* based on its small, etiolated plants growing in and around marshes. Recent studies have shown it to be a genetically and morphologically distinct species with a primary habitat of open meadows and dunes.

Botrychium tenebrosum **resembles** *B. simplex* **but can be separated by basal pinnae that are seldom enlarged (enlarged in** *B. simplex*), **by sporophore and trophophore stalks that are shorter than in** *B. simplex*, **and by a usually longer common stalk (***B. simplex* **sporophore and trophophore stalks diverge from the common stalk close to the ground). Open-grown plants of** *B. tenebrosum* **often produce sporangia on lower pinna margins of the trophophores.**

Botrychium tenebrosum **is more likely to be confused with** *B. mormo.* **The latter may be distinguished by its squarish outer pinna margins with blunt teeth, its more succulent texture with a thicker sporophore and deeply embedded sporangia, and its habitat of hardwood forests where it is often nearly covered with leaf litter.**

The trophophores of *B. tenebrosum* **are usually conspicuously stalked, while those of** *B. spathulatum* **are sessile or nearly so.**

BOTRYPUS Michaux Ophioglossaceae

Botrychium sect. *Osmundopteris* (J. Milde) R. T. Clausen; *Botrychium* subg.

Rattlesnake Fern

Etymology. Greek *botry-*, bunch or cluster of grapes, + *-pus*, footed, in reference to the clusters of sporangia on the fertile stalk. The common name, rattlesnake fern derives from the fact that when it first opens the sporangia-bearing tip of the sporophore resembles a rattlesnake's rattle.

Plants. Perennial, deciduous, producing a single frond each year, 10–45 cm tall.

Rhizomes. Short, erect, fleshy rhizome, 15 or fewer fleshy yellow to brown roots produced at base of rhizomes.

Common stalks. Dividing well above substrate surface.

Trophophores. 5–25 cm wide when mature, usually 1 per rhizome, deltate, to 3-pinnate-pinnatifid, arising from and sessile to middle to distal portion of common stalks well above ground level; blade horizontal to the ground, thin, herbaceous, light yellow-green.

Pinnae. Slightly narrower at base, tapering toward tips.

Pinnules. First pinnule on basal pinnae usually borne on acroscopic side.

Sporophores. 2-pinnate, long-stalked, commonly absent.

Sporophore stalk. Arising from base of trophophore blade high on common stalk, stiff.

Sporangia. Beadlike, bright yellow, appearing in early summer, soon withering.

Veins. Free, pinnately divided.

Habitat. Shaded forests and shrubby second growth.

Distribution. Eurasia. Central and South America. USA: every state except Hawaii.

Chromosome #: $x = 92$

A genus of 2 to 3 species worldwide with 1 species treated in *Flora of North America North of Mexico*, vol. 2 (1993) (treated there as *Botrychium virginianum*).

This single species is found in Michigan.

***Botrypus virginianus* (Linnaeus) Swartz**

Osmunda virginiana Linnaeus; *Botrychium virginianum* (Linnaeus) Swartz; *Japanobotrychium virginianum* (Linnaeus) M. Nishida; *Osmundopteris virginiana* (Linnaeus) Small.

Rattlesnake Fern, Virginia Grape Fern

Etymology. Latin *Virginianus*, of Virginia, alluding to the colony of Virginia where early collections of this fern were made.

Plants. Monomorphic, somewhat dimorphic (fronds often lacking sporophores); deciduous.

Fronds. 10–45 (–60) cm tall.

Rhizomes. Erect, roots arising 5 cm or more below ground, fleshy, thick, branching, tangled, with numerous vertical, budlike growths.

Common stalks. 10–30 cm long to where sterile blade and fertile blade diverge, 1/2 to 2/3 length of frond, erect, slender, smooth, succulent, pink at base.

Trophophores. 10–30 (–40) × 10–33 (–45) cm wide to 3-pinnate, broadly triangular, reflexed to almost horizontal with ground, thin-textured, bright pale green.

Pinnae. 5–12 pairs, stalks short, subopposite, slightly ascending; ovate to lanceolate, divided to tips, well separated in shady areas, close to overlapping in sunny areas.

Pinnules. Lanceolate, deeply lobed.

Lobes. Elliptic, lanceolate to ovate, tips pointed and toothed, serrate, midvein present.

Veins. Free, pinnately divided.

Sporophores. Often absent; 2-pinnate; 0.5–1.5 (–2)× length of fertile blade; arising at base of sterile blade at junction of stipe and rachis well above the ground.

Botrypus virginianum: Rattlesnake fern. A. Frond with sporophore and trophophore. B. sporophore enlargement. C. Enlargement of sporangia.

Informative arrows: 1. Look for blade parallel with ground. 2. Sporophore erect and long stalked. 3. Sporophore arising at base of blade.

Sporangia. Numerous, beadlike, sessile or short-stalked; borne in 2 rows; bright yellow, appearing in early summer, soon withering.

Habitat. Wide variety of forests (shaded, moist or dry rich woods, and shrubby second growth), less vigorous in open sun, commonly found in groups.

Distribution. Eurasia. Mexico. Central America. South America. Canada: all provinces except Nunavut. USA: All states except Hawaii.

Chromosome #: $2n$ =184

Distinguished from *Sceptridium* and *Botrychium* species by its much larger size and lacy, thin-textured blades. Distinguished from *Sceptridium* by sporophore stalks arising at the base of the blade (arising at ground level in *Sceptridium*). It is very distinctive and easily identified.

OPHIOGLOSSUM Linnaeus Ophioglossaceae

Adder's-Tongue Ferns

See Ophioglossaceae in appendix for a discussion of the family and a key to the Michigan genera.

Etymology. Greek *ophis*, snake, + *glossa*, tongue, in reference to the fertile spike resembling a snake's tongue.

Plants. Perennial or evergreen, terrestrial, small, herbaceous.

Stems. Erect, forming underground caudex, up to 1.6 cm thick, short, gemmae absent.

Roots. 0.1–1.5 mm diam., up to 20 per plant, unbranched, fleshy, smooth, hairless, yellow to tan to black.

Trophophores. Blades nearly sessile or stalked, erect to prostrate, 4–10 cm × 3–4.5 cm, simple, lanceolate to ovate, margins entire, unlobed, erect, tips rounded, acute, or apiculate, without midvein, glabrous.

Veins of sterile blades. Netted, often with included veins, main areoles up to 15 × 4 mm but mostly less than 5 × 3 mm.

Sporophores. 1 per leaf but commonly absent, unbranched, stalked, borne from ground at base of sterile blades, tips usually ± apiculate.

Sporangia. 2 vertical rows with up to 80 pairs, large, deeply embedded in simple, linear, fleshy sporophore tips containing thousands of spores opening by horizontal slits.

Habitat. Open, often disturbed grassy habitats.

Distribution. Mainly tropical and subtropical. A genus of about 25–30 species worldwide, found mostly in the tropics and subtropics. Seven species were treated in *Flora of North America North of Mexico*, vol. 2 (1993). Represented in Michigan by 2 species.

Chromosome #: $2n$ = 120 up to 1,200+

Ophioglossum species are inconspicuous and difficult to find even at known locations, and they are probably more common than collections would suggest. One or more fronds usually emerge each year, but sometimes years are skipped. They produce subterranean gametophytes adapted to drought and fire.

The plants may be confused with small, superficially similar monocotyledonous plants but may be distinguished by their single, simple, oval leaf with netted veins (monocots have parallel venation) and its fertile stalk with 2 rows of beadlike sporangia.

The genus *Ophioglossum* has the highest known chromosome number of any vascular plant.

KEY TO THE SPECIES OF *OPHIOGLOSSUM* IN MICHIGAN

If you find an *Ophioglossum* in Michigan, it will probably be *O. pusillum.*

1. Blades pale green, dull when alive, elliptic or ovate (widest near middle), herbaceous, bases gradually tapering; open pastures, meadows, marshes, ditches (rare on both peninsulas) . . . . *O. pusillum*
1. Blades dark green, somewhat shiny when alive, ovate to trowel-shaped (widest below middle), firm, abruptly tapering to base; damp forest habitat (Lenawee County only) . . . .*O. vulgatum*

Ophioglossum pusillum Rafinesque
Northern Adder's-Tongue

Etymology. Latin *pusillus*, very small, insignificant, probably referring to the inconspicuous character of this fern in nature. The common name derives from the resemblance of the fertile spikes to a snake's tongue.

Plants. Deciduous, emerging mid-spring in single flush.

Stems. 1–2 cm long, 1–5 mm diam., upright, expanding gradually into blade; short-lived membranous basal sheath present.

Sterile blades. 5–10 cm × ca. 3.5 cm, erect or spreading, herbaceous, one leaf per stem, smooth, flat, dull pale grass green when living, eliptic (widest near middle) to ovate, base gradually tapering, tips rounded.

Fertile spikes (sporophores). 20–45 mm long × 1–4 mm wide, 2.5–4.5× length of sterile blades, arising at base of sterile fronds, topped with 2 vertical

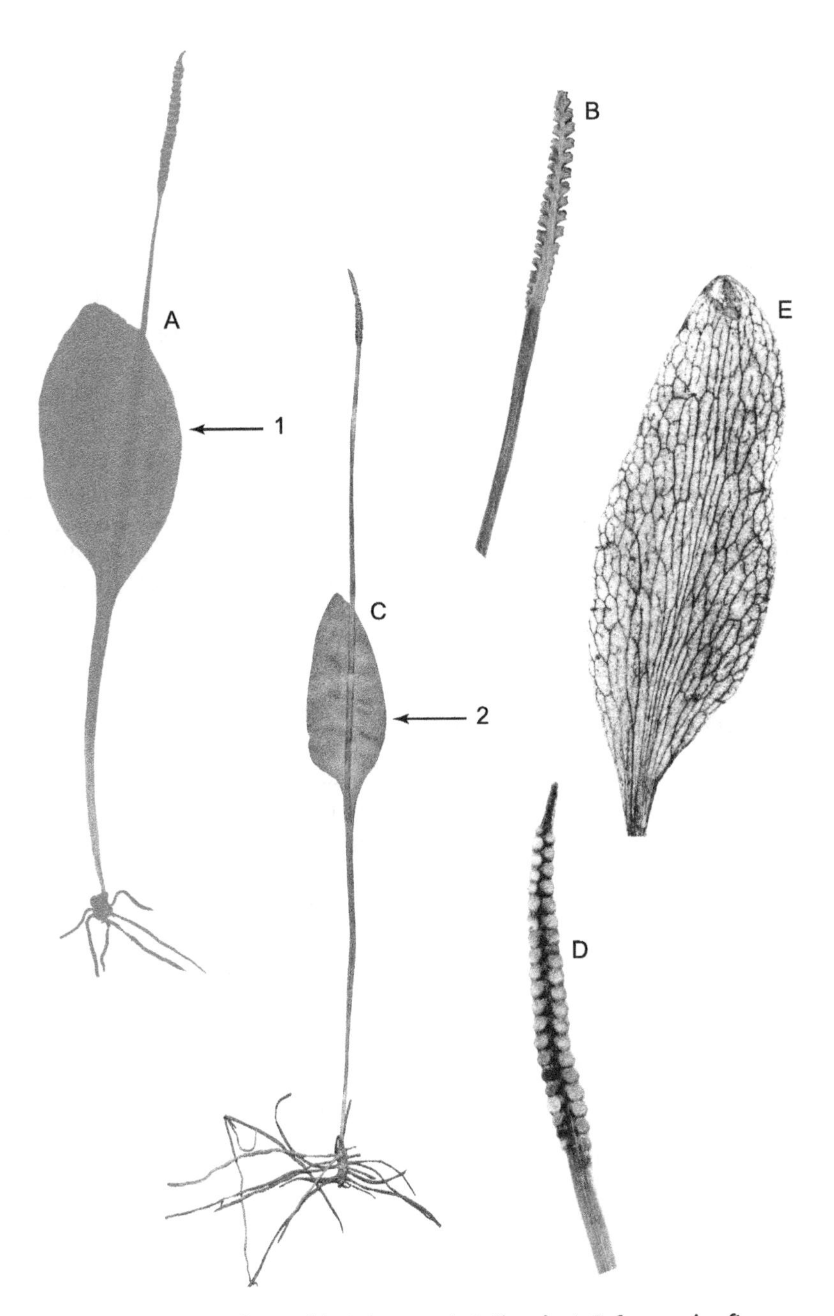

Ophioglossum pusillum: Northern adder's-tongue. A. Entire plant. B. Sporangia after spores discharged.

Ophioglossum vulgatum: Southern adder's-tongue: C. Entire plant. D. Sporophore with embedded sporangial.

Ophioglossum: E. Generic leaf bleached to show reticulated (netted) veins.
Informative arrows: 1. Blade widest at middle (look for dull pale green color).
2. Blade widest in lower half (look for shiny dark green color).

rows of embedded sporangia, tips sterile with 1–2 mm apiculum.

Sporangia. 2 rows of embedded beadlike structures with horizontal slits, 10–40 pairs in linear cluster.

Habitat. Moist sandy fields, often in sandy soil, marsh edges, old fields, pastures, grassy shores, and roadside ditches, not in heavily shaded sites.

Distribution. Canada: British Columbia, skipping Alberta and Saskatchewan, from Manitoba east to Nova Scotia. USA: New England south to Virginia, west to Wyoming and Montana, also Alaska, Washington, Oregon, and California. In Michigan it is scattered in both peninsulas, but because it is inconspicuous it is rarely found.

Chromosome #: $2n = 960$

The simple ovate fronds of *O. pusillum* are dull pale green when alive and have a gradually tapering base. The separate fertile branch (sporophore) arises near the base of the blade. *O. vulgatum,* currently found only in Lenawee County, has obovate blades that are dark green and shiny and have an abruptly tapering base. A short-lived membranous basal sheath is present; it is leathery and persistent in *Ophioglossum vulgatum*.

Ophioglossum vulgatum Linnaeus
Southern Adder's-Tongue

Etymology. Latin, *vulgatus*, common or ordinary, common only because of its widespread distribution. The common name derives from its southern distribution in the United States.

Plants. Deciduous, appearing in spring to early summer.

Fronds. 1 per stem, appearing in a single flush in spring to early summer.

Stems. Up to 1 cm long, 3 mm diam., upright, leathery basal sheath persistent (ephemeral in *Ophioglossum pusillum*).

Sterile blades (trophophylls). Up to 10–20 × 4 cm, mostly ovate to trowel-shaped, base tapering abruptly to stalk, widest in lower half, tips rounded; arising at or near ground level, erect to spreading, usually flat in one plane, somewhat shiny (when living), firm, dark green.

Fertile spike (sporophylls). 2.0–4.0 × 0.1–0.4 cm, 2–4× length of sterile blades, erect, arising from base of fertile blades, cluster of 10–40 pairs of sporangia near tips.

Sporangia. 10–35 pairs of sporangia in 2 rows of beadlike structures with horizontal slits, tips apiculate.

Habitat. Mostly southern rich forests, forested bottomlands, floodplain forests, and shaded secondary woods.

Distribution. Eurasia. Mexico.

USA: southeastern and east-central United States from Georgia to Texas, north to Kansas, Wisconsin, and Michigan, also Massachusetts and Arizona. The sole Michigan locality is in a rich, somewhat disturbed woods in Lenawee County.

Chromosome #: $2n = 480$ (probable)

Ophioglossum vulgatum is a Michigan endangered species. It is considered critically imperiled in Michigan but is globally secure.

It has smaller spores, 35–45µm, compared to 50–60 µm in *O. pusillum* (reflecting the difference in chromosome numbers).

***Ophioglossum vulgatum*, so far found only in Lenawee County, has ovate blades that are dark green and shiny when alive and have abruptly tapering bases. *Ophioglossum pusillum* has obovate fronds that are dull pale green, when alive and have gradually tapering bases. *Ophioglossum vulgatum* differs from *O. pusillum* in having an unusually persistent, leathery basal leaf sheath rather than a membranous ephemeral one.**

SCEPTRIDIUM Lyon Ophioglossaceae

Botrychium Swartz subg. *Sceptridium* (Lyon) T. Clausen

Grapeferns

Etymology. Latin *sceptrum*, scepter, wand, baton, staff, rod, + *-idium*, a diminutive suffix. The fertile spike resembles a small staff. The vernacular name alludes to the appearance of the "grapelike" clusters of sporangia on the fertile fronds.

Plants. Evergreen, appearing in spring, dying the following spring; arising as individual plants.

Roots. 2–4 mm diam. at 1 cm from stem, usually 10 or fewer, spreading, 7–8 cm below surface, thick, fleshy, fibrous, dark gray-brown, pale gray to tan, sometimes large roots darker close to stems, root hairs lacking; mycorrhizal.

Upright stems. Erect, thick, short, subterranean.

Common stalks. Stalks of trophophores and sporophores up to 15 cm long, joined together at or near substrate surface; horseshoe-shaped vascular bundle seen in cross section.

Trophophores. Mostly more than 12 cm long and 5–25 cm wide, 2-pinnate to 4-pinnate, usually 1 per plant, **deltate, spreading horizontally**, leathery, long-stalked, **dark green**.

Sporophores. Erect **(when present), 2-pinnate to -3-pinnate**, stalk 0.5–0.8 mm diam., **arising near substrate surface, long-stalked,** often absent.

Sporangia. Globular, 0.6–1.0 mm diam., in 2 rows along fertile portions, opening with transverse slits, producing abundant yellow spores.

Pinnae. First pinnule away from rachis on basal pinnae basiscopic.

Veins. Free, pinnately branched.

Gametophytes. Mycorrhizal, lacking chlorophyll; rarely seen.

Habitat. Variety of habitats, open grassy areas, sandy soils, moist, shady secondary forests, and swamps.

Chromosome #: $x = 45$

A genus of 13 species worldwide, 6 are treated in *Flora of North America North of Mexico*, vol. 2 (1993). It is represented in Michigan by 4 species.

Sceptridium spores germinate underground in the dark. After germination, and a few cell divisions, their nonphotosynthetic gametophytes form a mycorrhizal relationship with fungi in the soil. A large part of a *Sceptridium* species' life cycle is spent underground. After many years above-ground fronds appear. The fronds are divided into a sterile trophophore and a fertile sporophore, which diverge from a common stalk at or near ground level.

Usually one above-ground frond is produced per year. A frond is not always produced, and in some years neither trophophores nor sporophores are produced. In some years, sometimes many years, sites that produced plentiful above-ground fronds one year will produce very few, while in other years when above-ground fronds are sparse many will only have the trophophore. The leathery trophophores often persist over winter.

The combination of few, difficult, and variable species-specific characteristics and substantial infraspecific morphological variation makes *Sceptridium* species difficult to distinguish. Identification may be difficult, even for experienced botanists. The dissected forms of *S. dissectum* are the most easily recognized. Separating *S. multifidum, S. oneidense,* and undissected *S. dissectum* is more difficult.

The 4 species of *Sceptridium* found in Michigan have dark green, somewhat leathery, broadly triangular sterile blades usually over 8 cm wide (sometimes up to 15 cm), which are horizontal and parallel and close to the ground, and fertile and sterile branches diverging at or near ground level.

KEY TO THE SPECIES OF *SCEPTRIDIUM* FOUND IN MICHIGAN

1. Trophophore blade pinnules coarsely lacerate and deeply cut—more than halfway to the midvein, entire blade lacerate . . . . *S. dissectum* (in part)

1. Trophophore blade pinnule margins entire or finely to coarsely toothed or shallowly cut, entire blade not lacerate (2).

2(1). Terminal pinnules of blade and pinnae much larger than lateral pinnules; pinnae undivided except in proximal 1/2–3/4; vegetative blade segments coarsely and ± irregularly toothed or cut; dissection of blade into segments stopping at circa 1–2.5 cm from tips of blades (3).

2. Terminal pinnules similar to or only slightly larger than lateral pinnules; pinnae divided to tips; vegetative blade segments finely toothed to ± entire; dissection of blade into segments extending to within 1 cm of apex at tips of blades (4).

3(2). Pinnules few; pinnules obliquely ovate, usually more or less rounded at tips, terminal pinnules of vegetative blades narrowly to broadly ovate, large, obtuse to rounded at apex, more or less symmetrical at base; margins finely toothed to crenulate but never lacerate, tips rounded to acute; trophophore blades green where exposed to sunlight in winter; habitat uniform in acid, low woodlands. . . . . *S. oneidense*

3. Pinnules numerous; pinnules obliquely trowel-shaped or linear, usually pointed at tips, terminal pinnules of vegetative blades lanceolate, acute, and strongly asymmetric at base, margins variable, mostly finely toothed to irregularly cut, tips acute; trophophore blades bronze in winter if exposed to sunlight; habitats diverse . . . . *S. dissectum* (in part)

4(2). Pinnules of vegetative blade rounded, larger pinnules mostly 9–17 mm long; symmetrically tapered to an often more or less blunt or rounded tip; margins usually entire or nearly so; flat in living state; texture leathery; plants robust and large . . . . *S. multifidum*

4. Pinnules sharply angular, asymmetrical, wedge-shaped to the apex, pointed; larger segments mostly 4–9 mm long; margins clearly finely dentate; somewhat channeled and concave adaxially in living state; texture semi-herbaceous; plant usually slender and smaller . . . . *S. rugulosum*

Sceptridium dissectum (Sprengel) Lyon
Botrychium dissectum Sprengel; *B. dissectum* f. *confusum* (Wherry) Bartholomew; *B. obliquum* Muhlenberg ex Willdenow.
Cut-Leaved Grapefern, Dissected Grapefern

Etymology. Latin *dissectus*, deeply divided or cut into numerous segments, alluding to the dissected sterile blade of some forms of this species. The vernacular name reflects the same meaning as the Latin one.

Plants. Evergreen, blades appearing in early June and persisting until the following spring.

Trophophores. 8–14 (-20) × 8–15 (-20) cm, 2-pinnate-pinnatifid to 3-pinnate-pinnatifid, triangular (divided into 3 major divisions, each division with a triangular shape), leathery, shiny green, often bronze in winter, varying extensively in blade shape and color; var. *dissectum* pinnules deeply dissected with coarse, sharp teeth along the margins; var. *obliqum* blade less deeply dissected.

Trophophore stalks. 3–20 cm, 1–2.5× length of trophophore rachis.

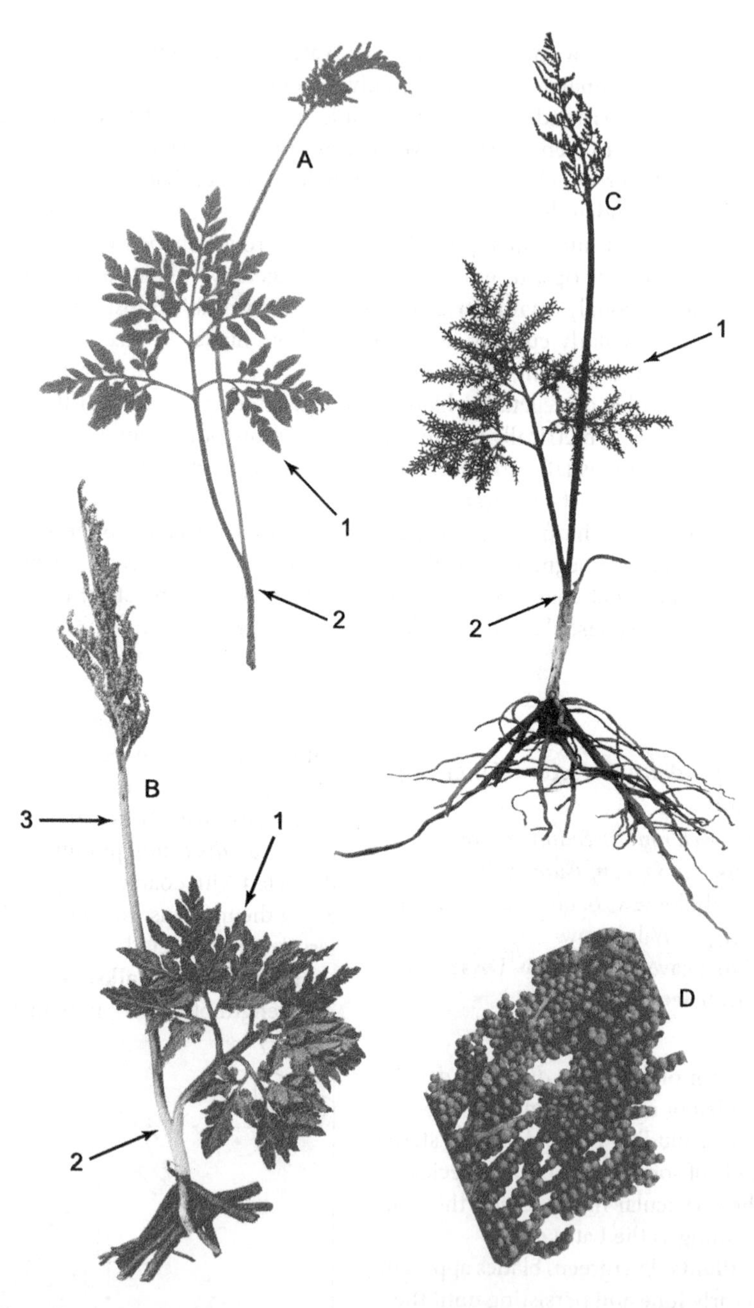

Sceptridium dissectum: Cut-leaved grapefern: A–B. Variations of undissected forms. C. Dissected form. D. Enlargement of sporangia showing sori.

Informative arrows: 1. Sharp-pointed segments that are longer than wide. 2. Common stalk divided close to ground. 3. Sporophore erect.

Pinnae. Up to 10 pairs, approximate to remote, slightly ascending, undivided except in proximal 2/3–3/4.

Pinnules. Proximal pinnae with up to 7 pairs of pinnules, proximal pinnules stalked, the distal sessile, **longer than wide; sharp-pointed at tips**, margins finely serrate, finely dentate or in some fronds lacerate to coarsely and irregularly cut halfway to midribs, terminal pinnules of pinnae usually more elongate than the lateral pinnules.

Sporophores. 5–20 cm long, 1.5–2.5× length of trophophores, 2-pinnate to 3-pinnate.

Sori. Spores shed late summer or fall.

Habitat. Variety of habitats, often in disturbed areas, swamps, old abandoned fields or pastures, open grassy areas, or deep forest.

Distribution. Mexico. Central America. Jamaica and Hispaniola. West Indies in the Antilles. Canada: Prince Edward Island to Ontario. USA: New England south to Florida, west to Minnesota, Iowa, and Kansas, south to Texas, and east to Florida.

Chromosome #: $2n = 90$

Sceptridium dissectum, the most variable North American grapefern species, has much variation in the degree of blade dissection, blade color, pinnule shape, and pinnule margins, even within the same population. It has historically been recognized as having up to 5 infraspecific taxa, the most common being var. *obliquum* and var. *dissectum*. DNA studies have shown no variation between these infraspecific taxa, and it is treated here as a single variable species.

Sceptridium dissectum **may be distinguished by its sharp-pointed segments that are longer than wide.** *S. multifidum* **and** *S. oneidense* **have blunt ultimate segments.** *S. dissectum* **may be confused with** *B. oneidense*, **which has larger, longer, blunter pinnules and fine teeth.**

Sceptridium multifidum (S. G. Gmelin) M. Nishida

Osmunda multifida S. G. Gmelin; *B. multifidum* (S. G. Gmelin) Ruprech

Leathery Grapefern

Etymology. Latin *multi-*, many, + *-fidus* divided, alluding to the much-divided sterile blade. The vernacular name, leathery grapefern, alludes to the texture of the leaf.

Plants. Evergreen, robust.

Trophophores. 6–30 × 3.5–20 cm, 2-pinnate-pinnatifid to 3-pinnate-pinnatifid, broadly triangular, ternate, **leathery, shiny green,** sometimes reddish or copper tinged.

Trophophore stalks. 0.3–1.2× length of trophophore rachis, 1.5–6 cm long.

Pinnae. Up to 10 pairs, close and

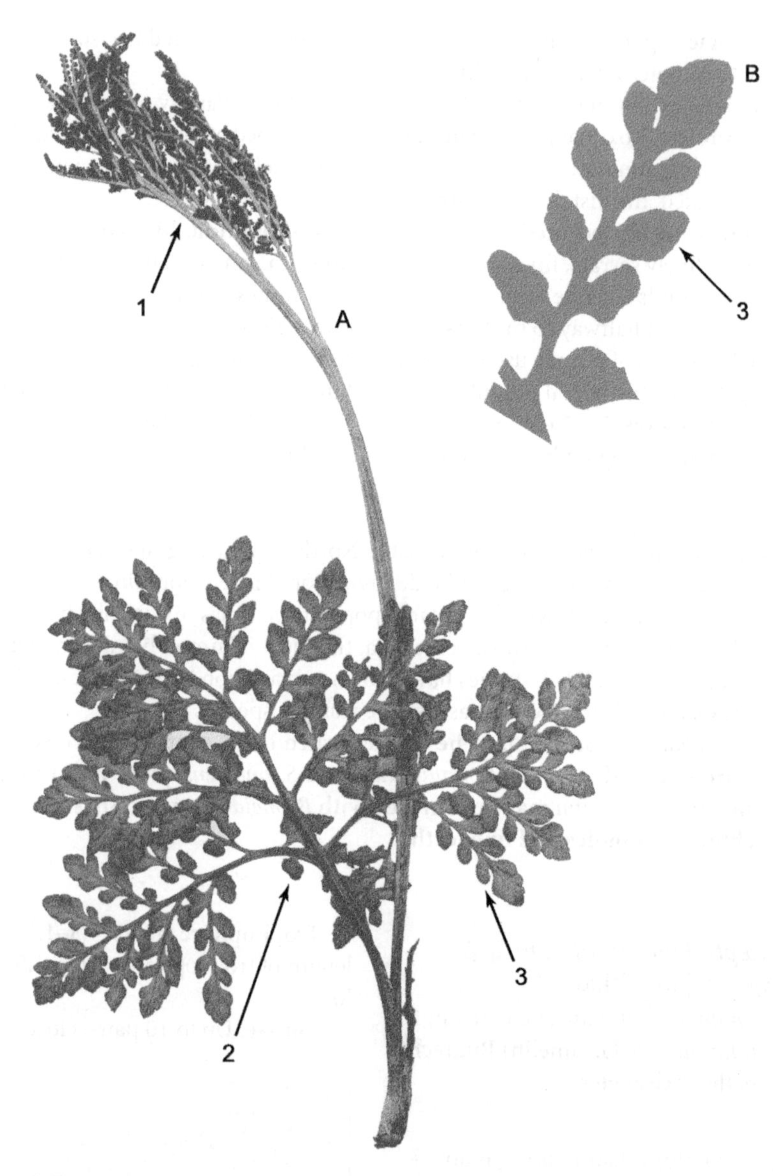

Sceptridium multifidum: Leathery grapefern: A. Plant. B. Pinnule.
Informative arrows: 1. Sporophore large. 2. Blade triangular, much divided. 3. Pinnae and ultimate segments all about the same size with short blunt-tips.

overlapping to well separated, horizontal to ascending, divided to entire, variable with acute, obtuse, or rarely rounded tips, margins crenate.

Ultimate segments. Ovate to rounded, short, tips rounded, oblique, margins entire to shallowly crenulate, sometimes slightly denticulate, flat when living.

Sporophores. If present, 8–40 cm long, 2-pinnate to 3-pinnate, 1.2×length of trophophores.

Sori. Mature in July through September.

Habitat. Fields, forest clearings, and open grassy areas, sandy soils.

Distribution. Eurasia. Greenland. Canada: all provinces except Nunavut. USA: Alaska, northern 2 tiers of states south to California, Arizona, and Colorado, New England south to South Carolina.

Chromosome #: $2n = 90$.

Sceptridium multifidum is quite variable, and this variability has led to the description of several varieties.

The previous year's *S. multifidum* fronds are retained over winter, new fronds appear in early spring, and old yellowed and shriveled fronds may persist through summer. Fertile fronds may not be produced under a closed canopy in deciduous forests.

Sceptridium multifidum is the largest member of this genus in Michigan, although there is considerable overlap in size with the other species. It is most easily seen in the fall when the erect sporophores turn from green to yellow. It can be quite common in some sunny sites, such as old abandoned pastures or fields, forest clearings, and forest edges.

Sceptridium multifidum **differs from** *S. oneidense* **by its highly divided, sterile blades (***S. oneidense* **has less divided sterile blades), and from** *S. dissectum* **by its short, blunt-tipped segments (long, narrow, and pointed segments in** *S. dissectum***). The pinnules of** *S. multifidum* **are all about the same size and shape.**

Sceptridium oneidense (Gilbert) Holub

Botrychium dissectum Sprengel var. *oneidense* (Gilbert) Farwell; *B. obliquum* var. *oneidense* (Gilbert) Waters

Blunt-Lobed Grapefern

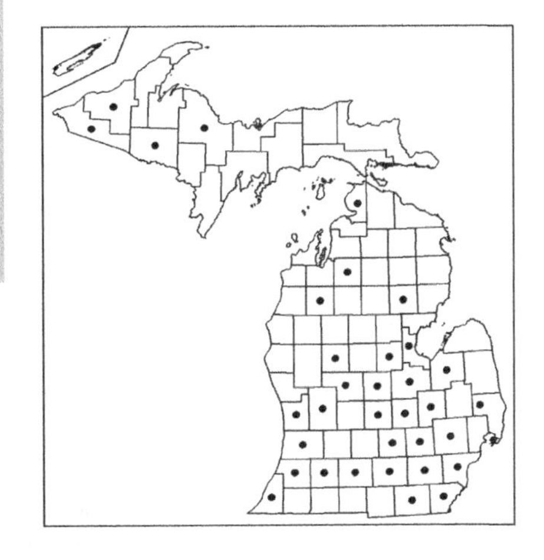

Etymology. *Oneidense* refers to Oneida County, New York, where the type specimen was collected. The vernacular name derives from the round-tipped lobes.

Plants. Evergreen.

Sceptridium oneidense: Blunt-lobed grapefern; A. Plant. B. Enlargement of pinna.

Sceptridium rugulosum: St. Lawrence grapefern: C. Plant. D. Enlargement of pinna. E. Enlargement of pinnule showing toothed margins.

Informative arrows: 1. Fewer, larger, less divided and blunter pinnules than *S. dissectum*. 2. Margins finely toothed. 3. Look for terminal pinnae divided into smaller segments that are convex and somewhat rugulose. 4. Pinnae long-stalked. 5. Basal pinnae pair well separated from the next distal pair.

Trophophores. 2-pinnate to 3-pinnate, dull bluish-green, up to 14 × 23 cm, broadly triangular, near to and parallel with the ground, leathery, mature blades with reddish pigment extending up to 3–15 percent of the midrib, fronds persisting through winter and remaining mostly green where exposed.

Trophophore stalks. 2–15 cm, 1.5–2.5× length of blade.

Pinnae. Up to 5 pairs, usually remote, horizontal to ascending, **divided in proximal 2/3–3/4, undivided nearer tips, blunt at the tips, margins finely or bluntly toothed.**

Pinnules. Usually few, **terminal pinnae large as if segments are fused**; more or less rounded at tips, margins usually finely toothed, surface somewhat dull, smooth; pinnules toward pinna tips relatively undivided, obliquely arranged, tips rounded to acute.

Sporophores. 1.5–2.5× length of trophophores, up to 22 cm long, 2-pinnate to 3-pinnate, sporangia-bearing portion up to 14 cm long.

Habitat. Moist, shady woods with full or nearly full canopy cover, intermittently periodically wet conifer and deciduous swamps, and acidic low woodlands.

Distribution. Canada: Prince Edward Island to Ontario. USA: New England west to Minnesota, Wisconsin, and Illinois, south to Tennessee and North Carolina.

Chromosome #: $2n = 90$

The trophophores of *Sceptridium oneidense* begin to appear in May, and the sporangia mature in late September. The trophophores persist through the winter, usually remaining green where exposed to sunlight.

Sceptridium oneidense is difficult to separate from the other Michigan *Sceptridium* species. It is usually local and rare in most of its distribution. It often occurs with the more common *S. dissectum*, which is found in more diverse habitats. Young plants of *S. dissectum* and *S. multifidum* may resemble *S. oneidense.*

Sceptridium oneidense **is often confused with** *S. dissectum* **and** *S. multifidum.*

Sceptridium oneidense **has fewer, larger, and less divided pinnules than** *S. dissectum.* **The more distal pinnules appear more fused at the base and are larger and blunter than in** *S. dissectum.* **Its pinnules usually have uniformly and finely denticulate to crenulate margins (those of** *S. multifidum* **are entire to crenulate). The pinnules of** *S. oneidense* **are more or less rounded at the tips compared to the more acute tips of** *S. dissectum.* **It is found in a uniform habitat of acidic low woodlands.**

Sceptridium multifidum **differs from** *S. oneidense* **by its highly divided, sterile blades (***S. oneidense* **has less divided sterile blades).** *S. multifidum* **pinnules are about the same size and shape, while those of** *S. oneidense* **are variable.**

***Sceptridium rugulosum* (W. H. Wagner) Skoda & Holub**
Botrychium rugulosum W. H. Wagner
St. Lawrence Grapefern

Etymology. Latin *rugulosus*, somewhat wrinkled, referring to the tendency of the segments to become more or less wrinkled and convex. The vernacular name, St. Lawrence grapefern, alludes to a large part of its distribution.

Plants. Evergreen, new leaves appearing in early May.

Fronds. 8–16 cm long (3 cm in sun to 30 cm in shade), somewhat herbaceous.

Trophophores. (2–) 4–8 (–16) cm, 2-pinnate to 4-pinnate, triangular, divided into 2 or 3 broadly triangular pinnae, convex adaxially, more or less rugulose toward tips, leathery, green.

Trophophore stalks. 2–15 cm long, 1–2.5× length of trophophore (shorter in sun forms, longer in shade forms).

Pinnae. Up to 9 pairs, usually close, horizontal to ascending, lateral and basal pinnae ovate-triangular, surface in the living state convex above and more or less coarsely dentate, **divided to the pinna tips, ultimate segments of pinna with small sinuses.**

Pinnules. Small, 0.2–0.5 cm wide, obliquely and angularly trowel-shaped, rhomboidal or oblong, tips acute, **margins usually coarsely to finely dentate** with wide teeth.

Sporophores. Up to 22 cm long, 2-pinnate, 1–2× length of trophophore, sporangia-bearing part of sporophore up to 14 cm long.

Habitat. Open fields and early secondary forests.

Distribution. Over a wide range in the vicinity of the St. Lawrence Seaway. Canada: New Brunswick, Ontario, and Quebec. USA: Connecticut, Michigan, Minnesota, New York, Vermont, and Wisconsin.

Chromosome #: $2n = 90$

Sceptridium rugulosum resembles *S. dissectum* and *S. multifidum*, with which it usually grows. Its fronds emerge in the spring before the former and after the latter. It is usually found in groups of only 5 to 10 individuals but occasionally in populations of over 100.

Sceptridium rugulosum **may be distinguished by its terminal pinnae, which are divided into smaller segments to the tips and are convex and somewhat rugulose. The terminal pinnae of** *S. dissectum* **and** *S. oneidense* **are larger than the lateral pinnae, and the pinnules of the terminal pinnae are more or less undivided at the tips and are flat and smooth.** *Sceptridium rugulosum* **is found in open habitats while** *S. oneidense* **is usually found only in low shaded forests and swamps.**

Sceptridium rugulosum has long-stalked pinnae and rather large separation between the basal pair of pinnae and next upper pair.

The pinnae of *S. multifidum* are usually close, and the pinnules are larger, rounded, and mostly 4–8 mm wide. Its leaf surface is flat and without wrinkles. The edges of the ultimate segments of the pinnules have rounded teeth or are entire.

PART 4

Lycophytes (Lycopodiopsida)

There are three lycophophyte families found in Michigan. The families, the genera in the families, and the species within the genera are treated alphabetically.

ISOËTACEAE Quillwort Family

ISOËTES Linnaeus Isoëtaceae

Quillworts

Etymology. Greek *isos*, equal, + *etes*, year, alluding to the evergreen leaves of some species that persist throughout the year.

Plants. Aquatic or terrestrial, evergreen or deciduous, tufted, grasslike; heterosporous.

Rootstock. 2-lobed, cormlike, nearly globose; bearing a cluster of spirally arranged leaves above and roots below; corky.

Roots. Tubular with central air cavity and a strand of conductive tissue along the tube wall; regularly forked.

Leaves. Linear; 1–100 cm long × 0.5–3 mm diam., several to many, arising from the corm apex, tufted erect to spreading, straight to recurved, grasslike with tubular leaves tapering toward tips, hollow and quill-like; bases flared with overlapping membranous margins, with small, triangular to cordate, 1–6 mm flaps (velum) covering the sporangia on inner surfaces where leaves narrow.

Sporangia. 3–15 mm, solitary, large, with either megasporangia or microsporangia, variable in size and shape, ovoid to ellipsoid or oblong, borne in cavities on inner sides of leaf bases below ligules; walls unpigmented or brown-streaked to completely brown.

Megaspores. Mostly 300–800 μm diam., globose, trilete, each with an equatorial ridge and 3 converging proximal ridges, smooth or with spines, tubercles, or ridges; white, gray.

Microspores. Mostly 20–50 μm, monolete, smooth or ornamented with spines or tubercles; grayish or brownish.

Habitat. Emergent in clear ponds and slow-moving streams or on wet ground that dries out in the summer when the plants become dormant.

Chromosome #: $x = 11$

Isoëtes is a genus of possibly 200 species found worldwide. Twenty-four are treated in *Flora of North America North of Mexico*, vol. 2 (1993). It is represented in Michigan by 3 species.

The growth habit of *Isoëtes* species is very uniform. They have a short tuberous stem with rows of roots on the undersurface and a dense rosette of spirally arranged, linear, quill-like leaves on a flattened or concave top. Throughout the plant's life new leaves are produced from the center of the top. New roots arise from a single or multiple radial furrows around the stem. A growth layer bearing new leaves and roots produces new stem tissue around the stem's central conducting layer. The older stem tissue is pushed outward from the center until it dies and decays. The stem becomes wider but not longer with age.

Quillworts are difficult to distinguish by general appearance. Examining the megaspore ornamentation is most helpful in distinguishing the species. Megaspore ornamentation and diameter provide characteristics that separate the Michigan species. A 10× hand lens is adequate to see the dried megaspore surface ornamentation of the Michigan species, but other species may require higher magnifications. Accurate spore diameter measurements require a dissecting microscope fitted with an ocular micrometer, but for Michigan species a good estimate of spore size can be obtained with the use of a ruler scaled in millimeters and a hand lens.

A narrow band along the distal side of the equatorial ridge of the megaspore is called a girdle. It is obscure when it is textured like the rest of the distal hemisphere. When it is textured differently it is distinguishable and may serve as a diagnostic character state.

Interspecific hybrids may occur where 2 or more species grow together. They can be distinguished by their irregularly shaped megaspores, which may be small to large flattened hemispheres, curved wedges, or dumbbell-shaped bodies.

The velum, a thin flap of tissue completely or partially covering the adaxial wall of the sporangium, is helpful in diagnosis. In some species the velum covers the entire sporangium wall; in others the velum covers less than 1/3.

Almost all North American *Isoëtes* species are aquatic and marsh plants, especially during their growth season. They have structural adaptations to aquatic life such as large air cavities in leaves and roots. Some tropical species have leaves permanently above water; some grow in seasonal pools that survive hot and dry seasons as leafless tubers. The Michigan species are either totally aquatic or semiemergent.

The 3 species found in Michigan may often be recognized by leaf length and texture (*I. engelmannii* has long and pliable leaves; *I. lacustris* has shorter but rigid leaves; and *I. echinospora* has shorter, pliable, grasslike leaves).

Key to Species of *Isoëtes* Found in Michigan

1. Megaspores spiny to spinulose, ornamented with thin, sharp spines; leaves 15–25 (–40) cm long; mostly flaccid; occasionally emergent .... *I. echinospora*

1. Megaspores cristate to reticulate with elongated ridges; leaves 10–30 or 60 (–90) cm long; pliable or stiff; occasionally emergent or totally submerged (2).

2(1). Leaves pliable, gradually tapering to tips; Up to 60 (–90) cm long; submerged or emergent; megaspores less than 560 μm diam., ornamented with reticulate ridges producing a regular honeycomb pattern, equatorial band obscure; velum covering less than 1/4 of sporangium ...... *I. engelmannii*

2. Leaves rigid, abruptly tapering to tips, 10–30 cm long; totally submerged in permanent lakes, ponds, or streams; megaspores 550–900 μm diam.;

ornamented with irregular, branching to anastomosing ridges, forming an irregular network; equatorial band consisting of a dense narrow band of tall, coarse spines; velum covering 1/4 to 1/2 of sporangium . . . *I. lacustris*

Isoëtes echinospora Durieu

Isoëtes braunii Durieu; *I. echinospora* var. *braunii* (Durieu) Engelmann; *I. echinospora* var. *muricata* (Durieu) Engelmann; *I. muricata* Durieu

Spiny-Spored Quillwort

Etymology. Greek, *echinos*, spiny, + *spora*, spore, referring to the spiny surface of this species' megaspores.

Growth habit. Evergreen or more or less deciduous.

Rootstock. Subglobose; small 2-lobed.

Leaves. 15–25 (–40) cm long, circa 1 mm diam., narrow, gradually tapering toward tips, pliant, spirally arranged, sometimes recurving; bright green to yellow-green, pale toward base.

Velum. Covers 1/4 to 1/2 of sporangium.

Sporangium wall. ± brown-streaked.

Megaspores. 400–550 µm diam., ornamented with numerous thin, sharp spines, not fused into ridges, girdle obscure; white; mature in late summer.

Microspores. 20–30 µm long, smooth to spinulose; gray to light brown in mass; mature in late summer.

Habitat. Aquatic, up to more than 1 m deep, occasionally emergent in shallow, cool, slightly acidic lakes, ponds, and slow-moving streams.

Distribution. Circumboreal. Greenland. Canada: throughout. USA: Alaska, West Coast states, Idaho, isolated populations in Colorado and Utah, Minnesota to New England south to Pennsylvania and New Jersey.

Chromosome #: $2n = 22$, diploid

Isoëtes echinospora is a distinct species that has considerable variation, especially in size, color, and shape of leaves. In North America it has stomata and has been called *I. muricata* or *I. echinospora* var. *braunii* to distinguish it from the European plants of *I. echinospora*, which lack stomata.

I. echinospora, **Michigan's most common quillwort, may be recognized by its narrow, bright green to yellow-green, sometimes recurving leaves, and megaspores with a spiny surface. (The megaspores of** *I. lacustris* **are covered with curved ridges.) It may be found submerged or emergent, whereas** *I. lacustris* **is always submerged.**

Isoëtes engelmannii A. Braun
Isoëtes valida (Engelmann) Clute
Engelmann's Quillwort

Etymology. Name honors George Engelmann (1809–84), a physician and botanist of St. Louis, Missouri, who collected many *Isoëtes* specimens, including this one near St. Louis, and revised the genus. He was influential in convincing Henry Shaw to establish the Missouri Botanical Garden.

Growth habit. Aquatic, becoming terrestrial during dry years.

Rootstock. Subglobose, 2-lobed.

Leaves. Up to circa 60 (–90) cm long (up to 48 cm in some Michigan collections), spirally arranged, pliable, bright green, pale toward base.

Velum. Covering less than 1/4 of sporangia.

Sporangia. Usually unpigmented, occasionally ± brown-streaked.

Megaspores. 400–560 µm diam., ornamented with tall, evenly reticulate, anastomosing ridges producing a honeycomb pattern, girdle obscure, white.

Microspores. 20–30 µm long, smooth to papillose; gray.

Habitat. In Michigan found in the wettest areas of a sandy swale between dune ridges. Elsewhere it is aquatic, emergent in shallow, soft-water lakes, ponds, streams or along their edges, intermittent wetlands, and ditches. It may become terrestrial in dry years.

Distribution. Canada: extreme southern Ontario. USA: New Hampshire in the northeast, west to New York, south along the Appalachians to extreme northern Florida, west in mountains to the northeastern tip of Mississippi, northeastern tip of Arkansas, and southeastern Missouri, north to southern portions of Indiana, Illinois, and Michigan.

Chromosome #: $2n = 22$

Isoëtes engelmannii is the most widely distributed quillwort of eastern North America. It is a Michigan threatened species but is globally secure. Seldom looked for, its real distribution in Michigan is unknown.

I. engelmannii, **Michigan's largest quillwort, is readily differentiated from the other Michigan quillworts by its much longer, broad, pale green, grasslike leaves. Its megaspores should be examined for confirmation. It is often found in a seasonably moist habitat and not submerged. It has only recently been found in Michigan and may well be recognized in other areas.**

Isoëtes lacustris **Linnaeus**
Isoëtes hieroglyphica A. A. Eaton;
I. macrospora Durieu
Lake Quillwort

Etymology. Latin *lacus*, lake, + *-estris*, a suffix indicating place of growth: the aquatic habitat of this species.

Growth habit. Evergreen, aquatic.

Rootstock. Subglobose, 2-lobed.

Leaves. 10–30 cm long, spirally arranged, arching to erect, abruptly tapering to tips, stout, stiff, dark green (sometimes almost black) to olive green or reddish-green, pale brown toward base.

Velum. Covers 1/4 to 1/2 of sporangium.

Sporangium wall. ± brown-streaked.

Megaspores. 550–900 µm diam., cristate to subreticulate with irregular branching to anastomosing ridges forming an irregular network, girdle a dense narrow band of tall, coarse spines (rarely smooth); white.

Microspores. 33–45 µm, papillose, gray.

Spores. Mature in late summer.

Habitat. Aquatic, totally submerged (up to 6 meters) in cool, slightly acidic lakes and streams and nutrient-poor sandy or gravelly bottoms, usually without other associated plant growth.

Distribution. Europe. Greenland. Canada: Saskatchewan to Prince Edward Island. USA: Minnesota to New England and south to New York. Disjunct populations have been found in New York, Virginia, and Wisconsin.

Chromosome #: $2n = 110$, decaploid

Isoëtes lacustris is a totally submerged aquatic. Plants have been found at depths of more than 3 m.

W. Carl Taylor, in *Flora of North America North of Mexico*, vol. 2 (1993), treats *I. macrospora* as synonymous with *I. lacustris*, and it is treated here as *I. lacustris*. The species are geographically separated, but both are decaploids and cannot be distinguished by their leaf or spore morphology.

I. lacustris **may be recognized by its mostly stout and stiff leaves and megaspores with a reticulate ridged surface. It is almost always submerged, sometimes in water more than 1 m deep.**

ISOËTES HYBRIDS

Isoëtes ×*hickeyi* W. C. Taylor & N. T. Luebke (*I. echinospora* × *I. lacustris*) has been collected in Alger County. Its spores are shriveled and abortive.

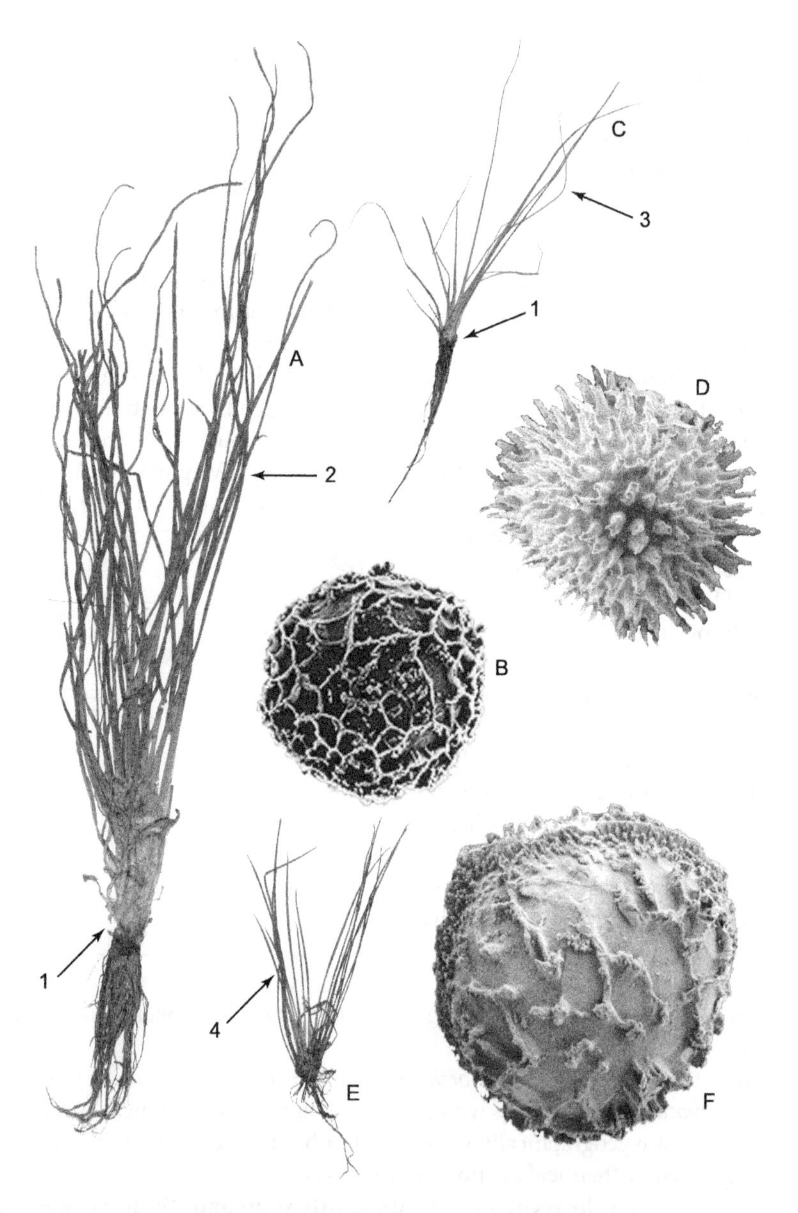

Isoëtes engelmannii: Engelmann's quillwort: A. Plant with leaves, corm, and roots. B. Megaspore showing ornamentation.

Isoétes echinospora: Spiny-spored quillwort: C. Plant with leaves, corm, and roots. D. Megaspore showing ornamentation.

Isoëtes lacustris: Lake quillwort: E. Plant with leaves, corm, and roots. F. Megaspore showing ornamentation.

Informative arrows: 1. Leaves arising from bulbous base in spiral fashion. 2. Leaves pliable. 3. Leaves mostly flaccid. 4. Leaves rigid.

LYCOPODIACEAE Clubmoss Family

Etymology. Greek *lykos*, wolf, + *podo*, foot, alluding to the way the branch tips of the type species of the genus *Lycopodium* resemble a wolf's paw.

Plants. Terrestrial.

Stems. Erect, trailing, unbranched to highly dichotomously branched.

Roots. Emerging near origin or growing from stem tips through the cortex and emergent some distance from origin.

Leaves. Small, simple, needle- or lancelike, 1-veined, spirally arranged or arranged in rows.

Strobili. If present, sessile or stalked.

Sporangia. Borne in axils of specialized leaves of terminal strobili or in axils of vegetative leaves.

Sporangia. Reniform or globose, 2-valved.

Gametophytes. Subterranean and nonphotosynthetic or surficial and photosynthetic.

Chromosome #: $x = 23, 34, 78, 134$

The treatment of Lycopodiaceae in this book is restricted to and appropriate for the Michigan genera. It will be inadequate for other areas of the world, especially the tropics. Worldwide there are 10 to 15 genera, and 350 to 400 species, which are most commonly found in mid- to high elevations in the tropics but also present in temperate and boreal areas.

Seven genera are treated in *Flora of North America North of Mexico*, vol. 2 (1993). They are represented in Michigan by 6 genera as treated here: *Dendrolycopodium, Diphasiastrum, Huperzia, Lycopodiella, Lycopodium,* and *Spinulum* (in *Flora of North America, Dendrolycopodium* and *Spinulum* were included in genus *Lycopodium*).

The monophyletic family Lycopodiaceae has been treated as a single genus or divided into subfamilies, genera, subgenera, or groups by various authors. The recognition of genera within this family is subject to much debate. In the past all species in the family Lycopodiaceae were placed in the single genus *Lycopodium*. In recent decades it has been shown that several genera must be recognized within the family.

The lineages of natural groups within Lycopodiaceae have been separated for tens to hundreds of millions of years. They have numerous significant differences in morphology, reproduction, gametophyte morphology, chromosome numbers, and developmental differences, as well as the inability to hybridize between genera.

Because of their decorative and evergreen characteristics some Lycopodiaceae species, *Dendrolycopodium obscurum* especially, are used by florists for winter seasonal decorations.

The six genera of clubmosses found in Michigan are distinguished by their general form, their leaves, and their strobili.

Dendrolycopodium: The tree clubmosses have an obvious upright stem, a treelike appearance, bristly leaves, and distinct sessile strobili (lacking stalks). Three species are found in Michigan.

Diphasiastrum: The ground cedars have flat branches and scalelike leaves. Three species are found in Michigan.

Huperzia: The firmosses have bristly leaves, lack strobili, bear sporangia in axils of otherwise typical vegetative leaves, and bear gemmae. Three species are found in Michigan.

Lycopodiella: Bog clubmosses have soft leaves, a light color, stobili that are only slightly wider than the erect stems, and grow in bogs. Three species are found in Michigan.

Lycopodium: These ground pines have bristly leaves bearing extended hairlike tips and distinct stobili on upright stems. Two species are found in Michigan.

Spinulum: These ground pines have bristly leaves and distinct sessile strobili without stalks. Two species are found in Michigan.

KEY TO THE GENERA OF *LYCOPODIACEAE* IN MICHIGAN

1. Strobili lacking, sporangia borne in the axils of regular leaves; flattened bilobed gemmae borne along stems; annual constrictions present . . . . *Huperzia*

1. Sporangia born in terminal differentiated strobili; gemmae lacking; annual constrictions present or absent (2).

2(1). Sporophyll leaves imbricate, soft, herbaceous; strobilus stalks always present, unbranched (rarely forked) with abundant herbaceous sterile leaves; erect stems unbranched (rarely branched); horizontal stems superficial; dying back in winter to tuberous tips. . . . . *Lycopodiella*

2. Sporophyll leaves yellowish, scalelike; strobilus stalks present or absent, branched or unbranched, leaves rigid; erect stems branched or unbranched; horizontal stems ± trailing on surface or deeply buried; perennial (3).

3(2). Stobili sessile (4).

3. Strobili stalked (5).

4(3). Erect leafy stems treelike with much treelike branching and definite main stem; horizontal shoots subterranean; strobili, when present, several per upright shoot; annual constrictions absent . . . . *Dendrolycopodium*

3. Erect leafy stems not treelike, horizontal on surface, simple or occasionally forked with ascending branching; strobilus 1 per upright shoot; annual constrictions present. . . . . *Spinulum*

5(3). Branches flat or squarish; leaves in 4 ranks, appressed, imbricated partially fused with the stem, ± scalelike, in opposite pairs; strobilus stalks branching dichotomously . . . . *Diphasiastrum*

5. Branches round, (flat undersurface of branches in *Dendrolycopodium obscurum*); leaves in 6 or more ranks, imbricated or not; strobilus stalks branching, falsely appearing to have a main axis from which the branches arise (6).

6(5). Leaves spreading, acuminate, with long, fine hairlike tips . . . . *Lycopodium*

6. Leaves acuminate to acute with stiff pointed tips. . . . . *Spinulum*

ILLUSTRATED GLOSSARY FOR LYCOPODIACEAE

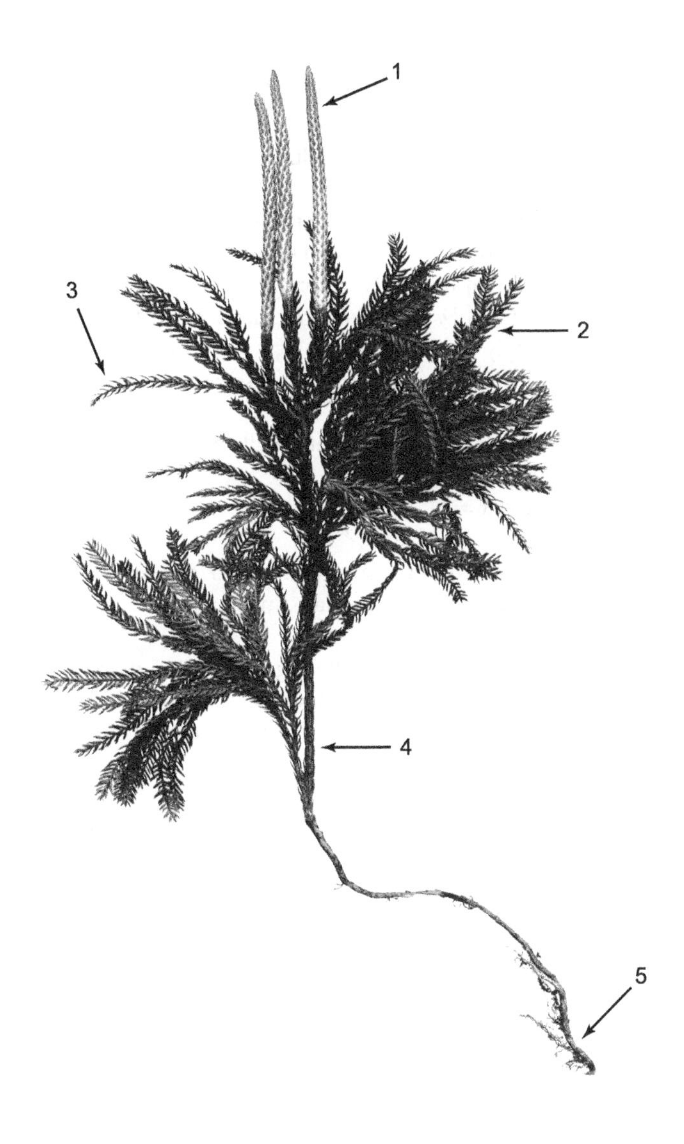

Lycophyte Illustrated Glossary: 1. Strobili. 2. Branches. 3. Leaves. 4. Upright stem. 5. Rhizome.

DENDROLYCOPODIUM A. Haines Lycopodiaceae

Ground Pines

Etymology. Greek, *dendro-*, tree, a treelike genus of Lycopodiaceae.

Plants. Evergreen.

Rhizomes. Subterranean, clothed with sparse scalelike leaves; branching annually to produce 1 upright shoot and 1 weak or dead secondary rhizome.

Upright shoots. 8–19 cm tall (exclusive of strobili); branching 3–4× with lateral branches, resembling small trees.

Branches. Round or flat on undersurface; each branch branching dichotomously 3 to 4× to produce 8–16 lateral branchlets.

Sterile leaves (trophophylls). Monomorphic or weakly dimorphic on main stems of *D. hickeyi* and *D. obscurum*; in 6 ranks; acute-tipped, lacking spinelike or hairlike tips; strongly decurrent.

Strobili. 1–7 (–10) per upright shoot; sessile; borne at tips of main stems or on dominant branches.

Sporangia. At adaxial base of sporophyll; with 2 equal valves.

Gemmae. Absent.

Chromosome #: $x = 34$

Three species are treated in *Flora of North America North of Mexico*, vol. 2 (1993) (there treated as *Lycopodium* sensu lato, in the *Lycopodium dendroideum* group). These 3 species are found in Michigan.

Aerial shoots of *Dendrolycopodium* grow for 4–5 years, with leaves showing annual winter constrictions. Most branching occurs in the second year, with occasional branching in the third. Strobili appear in the third or fourth growing season.

KEY TO THE SPECIES OF *DENDROLYCOPODIUM* FOUND IN MICHIGAN

1. Leaves on main stem below branches spreading at a 45° to 90° angle from the stem, at least on the upper portion of the axis, stiff, prickly; branches round in cross section. . . . . *D. dendroideum*

1. Leaves on main stem below branches all appressed to strongly ascending, spreading at less than a 30° angle from the stem, soft, not prickly; branches round in cross section or flat on underside (2).

2(1). Ultimate lateral branches flattened on underside; leaves unequal in size, lateral leaves longer than the somewhat appressed lower leaves . . . . *D. obscurum*

2. Lateral branches round on underside; leaves all equal in size, all more or less equally ascending in orientation . . . . *D. hickeyi*

Dendrolycopodium dendroideum (Michaux) A. Haines
Lycopodium dendroideum Michaux; *Lycopodium obscurum* Linnaeus var. *dendroideum* (Michaux) D. C. Eaton
Round-Branched Ground Pine, Prickly Tree Clubmoss

Etymology. Latin *dendro-*, tree, + *-oideum*, a suffix indicating resemblance, alluding to the treelike appearance of this species.

Growth habit. Evergreen; resembling small evergreen trees.

Upright shoots. Up to 26 cm tall; main stems with repeatedly forked, alternate lateral branches.

Lateral branches. 5–8 mm diam., **round**; winter bud constrictions inconspicuous.

Sterile leaves (trophophores). 2.4–5.5 × 0.5–1.2 mm; pale green; **leaves on main stem stiff, prickly,** margins entire, tips acuminate, **spreading at a 45° to 90° angle;** branch leaves spreading to ascending; 6 ranks—arranged at 1–3–5–7–9–11 o'clock on branches (2 dorsal, 2 lateral pairs, and 2 ventral ranks), appearing equal in size, linear.

Strobili. 1.2–5.5 cm long, 2–7 per upright shoot, sessile; single at shoot tips or on tips of ascending to erect dominant branches.

Fertile leaves (sporophylls). 3.3–3.9 × 3.2–3.6 mm, abruptly narrowing to tips, all the same size and shape.

Habitat. Second-growth shrubby fields, hardwood forests, mixed hardwood-conifer forests, and open fields.

Distribution. Asia. Russia. Japan. Canada: Yukon east to Newfoundland. USA: Alaska, Washington, Montana to the northeastern United States, south to Missouri, Tennessee, Alabama, and North Carolina.

Chromosome #: $n = 34$

The 3 species _Dendrolycopodium dendroideum, D. hickeyi,_ and _D. obscurum_ at first glance look very much alike, but attention to only 2 easily observed details separates them. These are the stiffness and degree of spreading of the leaves on the main stem and whether the lateral branches are round or flat on the undersurface in cross section.

Dendrolycopodium dendroideum **has stiff, prickly, needlelike, spreading (at a 45–90° angle) leaves on the upright stems, and its lateral branches are round when looked at from the tips.**

Dendrolycopodium obscurum **has leaves that are soft and appressed (at angles less than 30°) to the upright stems and branches that have a flat underside when looked at from the branch tips.**

Dendrolycopodium hickeyi **has main stems with the appressed leaves similar to those of** *D. obscurum* **and round branches similar to those of** *D. dendroideum.*

Dendrolycopodium hickeyi (W. H. Wagner, Beitel & R. C. Moran) A. Haines
Lycopodium hickeyi W. H. Wagner, Beitel & R. C. Moran; *Lycopodium obscurum* Linnaeus var. *isophyllum* Hickey
Hickey's Clubmoss, Hickey's Tree Clubmoss

Etymology. Name honors James Hickey, a professor of botany at Miami University of Ohio, past editor of the *American Fern Journal*, and author of many papers about ferns and fern-allies.

Plants. Evergreen, small treelike.

Upright shoots. Treelike, up to 16 cm tall.

Lateral branches. 5–8 mm diam., **round;** winter bud constrictions inconspicuous.

Sterile leaves. 3–5 × 0.5–0.8 mm, needlelike, linear, pale green, widest in middle margins entire; tips pointed; **leaves on main stem soft; appressed to ascending, spreading at angle less than 30°**; leaves on branches spreading to ascending; in 6 ranks equal in size—1 on top, 1 on underside, and 2 on each side (arranged at 12-2-4-6-8-10 o'clock on branches).

Strobili. 1.5–6.5 cm long, 1–7 per upright shoot, sessile, single at shoot tips or occasionally on tips of ascending to erect branches.

Fertile leaves (sporophylls). 3.2–4 × 2–2.7 mm, gradually to abruptly narrowing to tips; lowermost leaf shorter than the rest.

Habitat. Hardwood forests, mixed hardwood–conifer forests, open fields, and shrubby areas, often on dry sandy soils.

Distribution. Canada: Saskatchewan, Ontario to Nova Scotia, and Prince Edward Island. USA: Washington, Minnesota, Wisconsin, Indiana, Kentucky, and Tennessee east to the Atlantic coast. In Michigan *Dendrolycopodium hickeyi* has been found scattered throughout the Upper Peninsula.

Chromosome #: $n = 34$

See the discussion under *Dendrolycopodium dendroideum*.

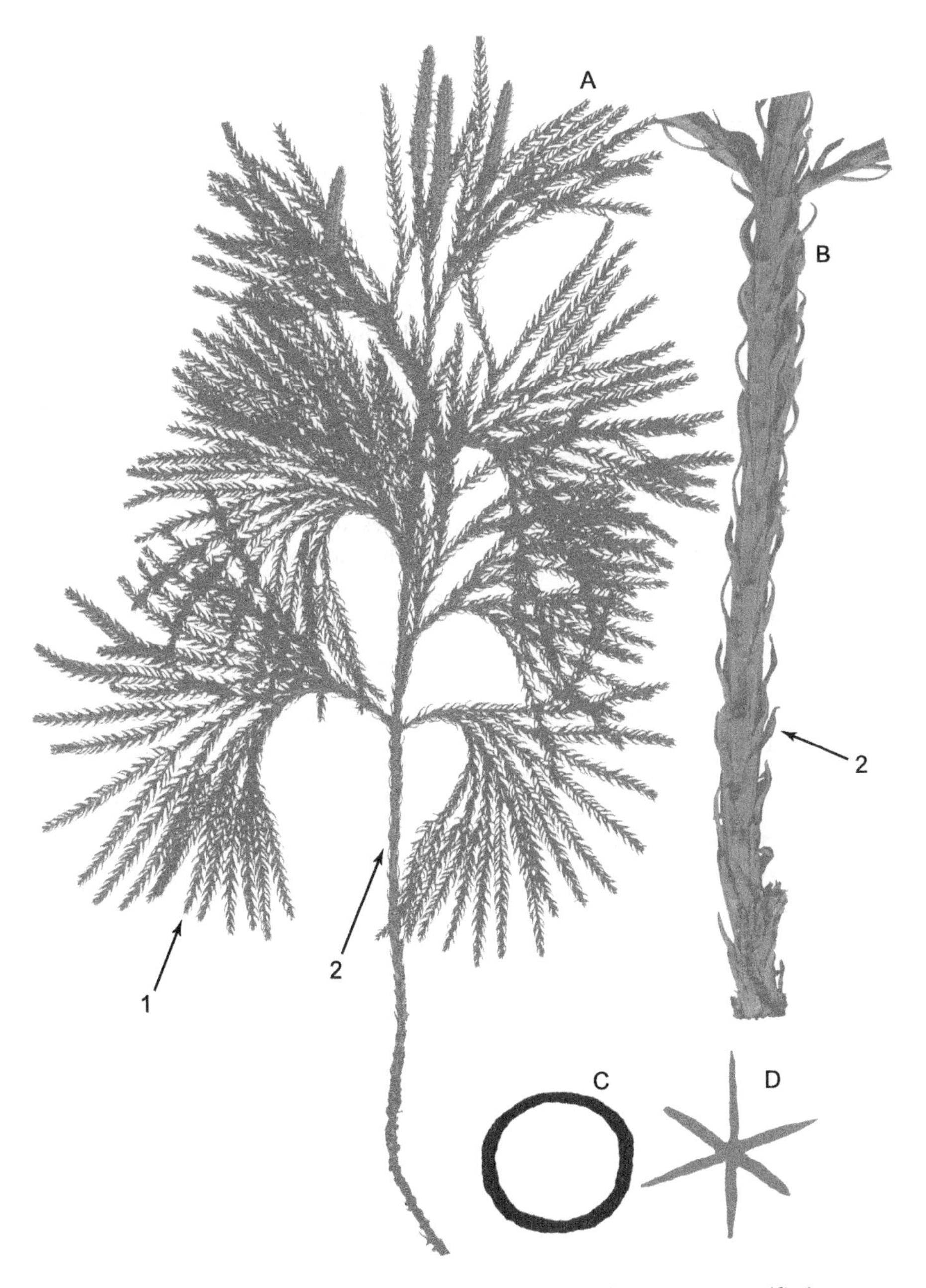

Dendrolycopodium hickeyi: Hickey's tree clubmoss: A. Upright stem. B. Magnified portion of main stem. C. Symbol representing circular appearance of branches when looked at from the tips. D. Arrangement of the 6 leaves of the branches.

Informative arrows: 1. Branches round when viewed from tips (like *D. dendroideum*). 2. Leaves ascending, soft, appressed (like *D. obscurum*).

***Dendrolycopodium obscurum* (Linnaeus) A. Haines**
Lycopodium obscurum Linnaeus
Flat-Branched Clubmoss, Eastern Tree Clubmoss

Etymology. Greek *obscurus*, hidden; the meaning is uncertain. Linnaeus indicated that the "fruits" were unknown to him.

Plants. Evergreen; **small treelike.**

Upright shoots. Up to 22 cm tall; main stem with repeatedly forked alternate lateral branches resembling a small tree.

Lateral branches. Flat on undersurface; annual bud constrictions conspicuous.

Sterile leaves (trophophylls). 1.3–5.5 × 0.5–1.2 mm or smaller only on lower surface; **leaves on main stem appressed to ascending, spreading at angle less than 30°; soft;** margins entire; leaves on branches arranged in 6 ranks at 12–2–4–6–8–10 o'clock on stems, 1 on top, 2 on each side, and **1 much shorter tightly appressed leaf on underside giving a flat undersurface,** ascending.

Strobili. 1.2–6 cm tall; 1–6 per upright shoot; sessile; single at shoot tips or on tips of ascending to erect branches.

Fertile leaves (sporophylls). 3.5–4.4 × 0.6–0.7 mm; all same size and shape.

Habitat. Mixed second-growth shrubby areas, hardwood forests, mixed hardwood–conifer woodlands, and open fields.

Distribution. Southeastern Canada. USA: northeastern and eastern Midwest to as far south as Alabama.

Chromosome #: $n = 34$

See the discussion under *Dendrolycopodium dendroideum.*

HYBRIDS

No confirmed hybrids between *Dendrolycopoium* species have been found.

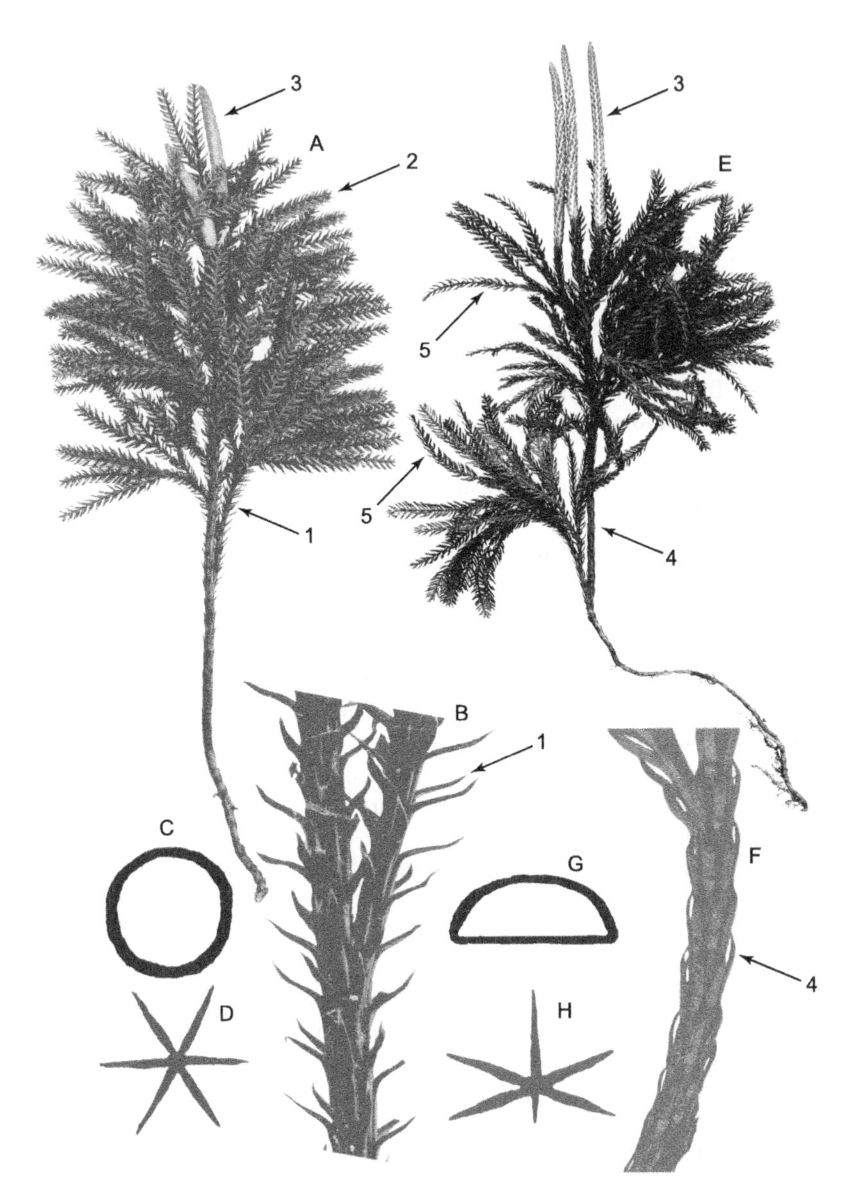

Dendrolycopodium dendroideum: Prickly tree clubmoss: A. Upright stem. B. Magnified portion of main stem. C. Symbol representing circular appearance of branches when looked at from the tips. D. Diagram showing arrangement of the 6 leaves of the branches.

Dendrolycopodium obscurum: Flat-branched tree clubmoss: E. Upright stem. F. Magnified portion of main stem. G. Symbol representing flat undersurface of branches when viewed from the tips. H. Arrangement and size of the 6 leaves of the branches.

Informative arrows: 1. Stiff, protruding, prickly, horizontally arranged leaves. 2. Branches round when examined from the tips. 3. Sessile strobili. 4. Leaves ascending, appressed, soft. 5. Branches flat on the undersurface.

DIPHASIASTRUM Holub Lycopodiaceae

Ground-Cedars

Etymology. Greek *Diphasium*, an older name for *Lycopodium*, + Latin *-astrum*, a suffix suggesting partial resemblance, thus resembling a *Diphasium* (*Lycopodium*). The vernacular name is easily understood because of the resemblance of these plants to small cedar trees.

Plants. Terrestrial; evergreen; miniature cedar-tree-like.

Horizontal shoots. Trailing on ground under leaf litter or subterranean; producing erect shoots

Erect shoots. 2–6 mm diam., usually with 2–5 lateral branches; leaves scale-like.

Branches. Flat or square in cross section, leaves small, scalelike, appressed.

Leaves. In 4 rows 90° apart, leaves on sides of branches larger and more spreading than those on top and undersurface; or leaves in all 4 rows nearly same size.

Strobili. 1.3–7.5 cm long, multiple, strobilus stalks dichotomously branched, in upright candelabralike clusters, cylindrical; blunt, pointed, or with sterile elongated apical tips, sporophylls (fertile leaves) straw-colored.

Gemmae. Absent.

Gametophytes. Subterranean; mycorrhizal; nonphotosynthetic; carrot-shaped with sexual organs on the upper crownlike part.

Habitat. Mostly in dry upland areas.

Distribution. South America. New Guinea. Marquesas Islands. Circumboreal, north temperate, and subarctic.

Chromosome #: $x = 23$

A genus of 15–20 species worldwide. Five species are treated in *Flora of North America North of Mexico*, vol. 2 (1993). Three species and 3 interspecific hybrids are found in Michigan.

A collection of *Diphasiastrum alpinum* found in Keweenaw County in 1895 may be inaccurately labeled; in any event, it has not been collected in Michigan since. A hybrid of *D. sitchense* × *D. tristachyum* (*D.* ×*sabinifolium*) was long known from a site in Chippewa County, now overgrown and from which it is now absent. *Diphasiastrum sitchense* is well know from the north shore of Lake Superior in Ontario, but has not yet been found in Michigan.

KEY TO THE SPECIES OF *DIPHASIASTRUM* IN MICHIGAN

1. Branches nearly square in cross section, cordlike; underside leaves approximately equal in size to lateral and upper-side leaves; rhizomes deeply subterranean; erect stems usually bluish-glaucous, sometimes green . . . . *D. tristachyum*

1. Branches flat in cross section; underside leaves much smaller than lateral and upper leaves; rhizomes on soil surface or under sparse litter; erect stems yellow to green (2).
2(1). Branches irregular, spreading in more than 1 plane, alternate; annual bud constrictions conspicuous; strobilus stalks regularly forked; strobili mostly 1.5–2.5 cm, lacking sterile pointed tips. . . . . *D. complanatum*
2. Branches very regularly fan-shaped and flat (in 1 plane); annual bud constrictions lacking; strobilus stalks mostly branching abruptly at base to produce false whorl of strobili; strobili mostly 2–3.5 cm long, often with sterile tips . . . . *D. digitatum*

Diphasiastrum complanatum (Linnaeus) Holub

Lycopodium complanatum Linnaeus

Northern Running-Pine, Northern Ground-Cedar

Etymology. Latin, *complanatus*, flattened, alluding to the flattened branches of the erect stems.

Plants. Evergreen, miniature tree-like.

Horizontal stems. 1.1–2.2 mm diam., **trailing on ground or under thin layer of litter,** producing erect shoots; branching repeatedly and sometimes extending to form circular clones several m wide.

Upright stems. 8–40 (mostly less than 20) cm tall, forking alternately up to 5×, branching 1–3 times the first year; leaves on upright shoot 1.2–3.2 × 0.5–1.1 mm, tips acuminate, appressed with the leaf base prolonged down the stem as a winged expansion.

Lateral branches. 1.8–4.4 mm wide; conspicuously flattened (resembling cedar branches); bearing 4 leaves in pairs—1 above, 1 below, and 1 on each side of branches; **branches** forking alternately up to 5×, **irregularly divided with branchlets not all in 1 plane,** indeterminate; **annual bud constrictions abrupt and conspicuous** at the juncture between consecutive years' growth; green and often shiny above, pale green below.

Sterile leaves (trophophylls). 4 rows; fused to branches for more than 1/2 their length; upper surface leaves flat, appressed, somewhat overlapping with free portion 0.7–2 × 0.5–1.2 mm; lateral leaves appressed, somewhat overlapping, keeled, 2.6–7.3 × 0.8–2.1 mm, lower surface leaves small, flat, 0.7–1.5 × 0.4–0.9 mm.

Strobilus stalks. 8.3–8.5 cm long × 0.5–1 mm diam.; 1 or 2 arising from the upper lateral branches, each bearing 1–4 strobili; green, becoming straw-colored before spores released in fall.

Strobili. 0.83–2.5 (–3.4) × 2–3 mm; in upright candelabralike clusters of 1–4, on distinct dichotomous branches; cylindrical; tips rounded, **lacking narrow sterile tips.**

Fertile leaves (sporophylls). 2–3 × 2–2.4 mm; deltate to cordate with abruptly tapering tips.

Habitat. Primarily in dry hardwood and coniferous forests but also found in dry open fields.

Distribution. Circumboreal. Greenland. Across Canada. USA: Alaska, northern tier of United States excluding North Dakota.

Chromosome #: $2n = 46$

The lateral branches extend each spring; an annual winter constriction separates the growth of the new growth from the previous year's.

Diphasiastrum complanatum **is distinguished from** *Diphasiastrum digitatum* **by annual constrictions on its lateral branches (annual constrictions, once recognized, are easily seen), strobili with round tips that lack elongated sterile tips, and strobilus stalks that turn straw-colored in the fall before the spores are shed.**

In the field the alternate branches of *D. complanatum* **have a somewhat 3-dimensional, irregular, tangled appearance, whereas the branches of** *D. digitatum* **are in 1 plane, forming fans with regular distal branchlets.** *D. digitatum* **strobilus stalks remain green until the spores are shed. Once these characters are recognized the species are easily distinguished without stooping to examine the plants.**

The horizontal shoots of both *D. complanatum* **and** *D. digitatum* **are superficial or under the leaf litter, while those of** *D. tristachyum* **are deeply buried. The upright shoots of** *D. complanatum* **and** *D. digitatum* **are medium green, while those of** *D. tristachyum* **are often blue-green. The branches of** *D. tristachyum* **are squarish in cross section while the others are flat.**

Diphasiastrum digitatum (Dillenius) Holub

Lycopodium digitatum Dillen; *Diphasium flabelliforme* (Fernald) Blanchard; *Lycopodium complanatum* Linnaeus var. *flabelliforme* Fernald

Southern Running Pine, Ground-Cedar, Fan Clubmoss

Etymology. Latin *digitus*, finger, + *-atum*, a suffix indicating likeness, probably alluding to fingerlike branching of the branches.

Plants. Evergreen, miniature and treelike.

Horizontal stems. 1.3–2.7 mm diam.; **trailing on ground or under thin layer of litter,** branching repeat-

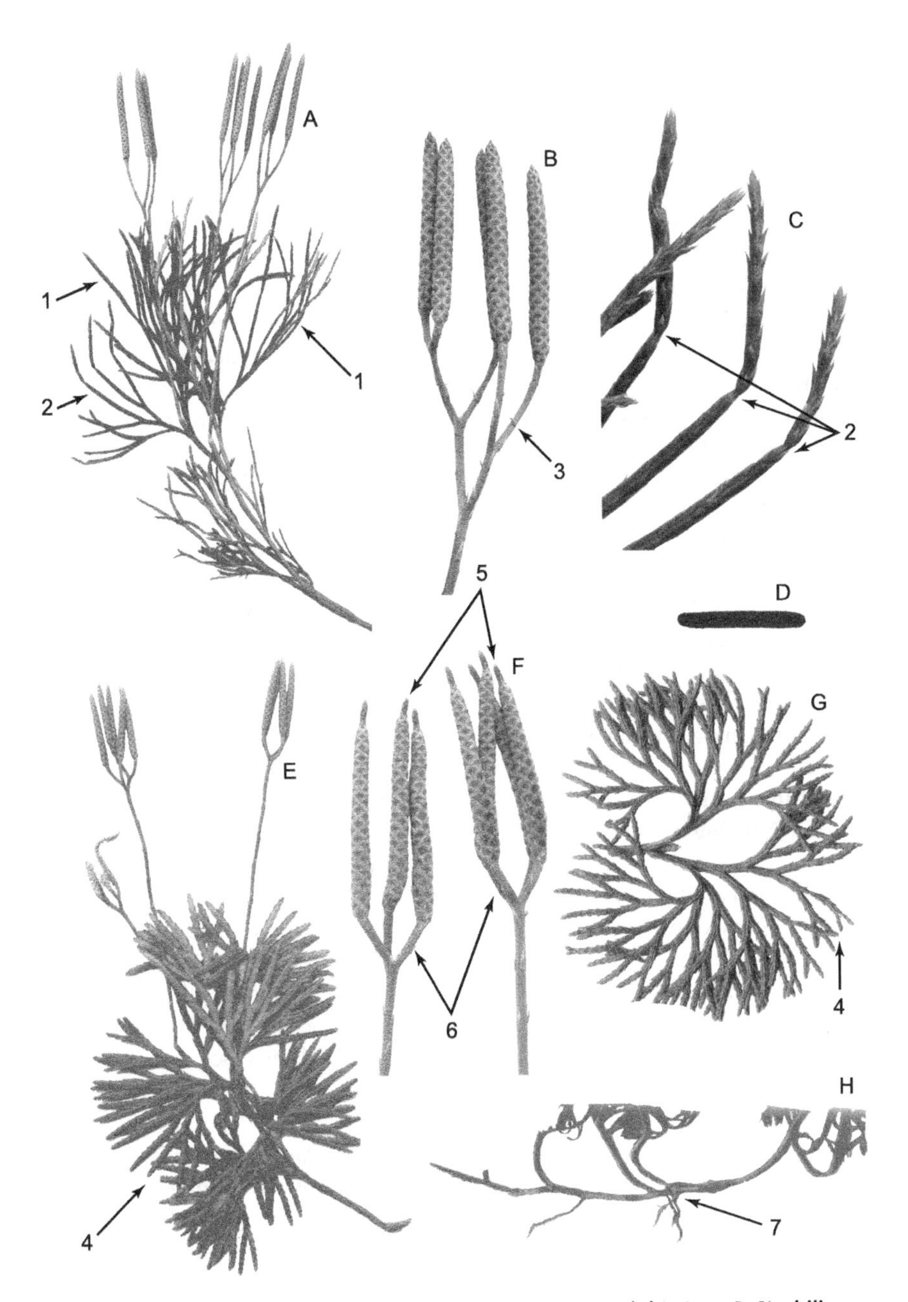

Diphasiastrum complanatum: Northern Ground-cedar: A. Upright stem. B. Strobili. C. Branch tips.

Diphasiastrum digitatum: Ground-cedar: E. Upright stem. F. Strobili. G. View of branches. H. Rhizome.

Informative arrows: 1 Look for scraggly, irregularly arranged branches. 2. Arrows showing annual constrictions. 3. Stobilus stalks with regular dichotomous branching. 4. Branches in 1 plane, annual constrictions lacking. 5. Sterile tips of strobili (in ca. 50 percent of plants). 6. Branching close to stalk tips appearing whorled (a variant, some plants have clear dichotomous branching). 7. Stems attached to rhizomes creeping on substrate surface (both species illustrated here).

edly and sometimes extending to form clones several m wide; producing numerous erect shoots.

Upright shoots. 12–35 cm tall, **treelike**, leaves on upright shoot 1.8–4.5 × 0.6–1.2 mm, appressed with decurrent base, linear with acuminate tips, often shed.

Lateral branches. 2.8–3.9 mm wide; **flat and wide** in cross section; **annual constrictions lacking**, determinate, **regularly branched,** branching 4–8× only in first year; **fan-shaped, flat, all branches in 1 horizontal plane,** undersurface dull, pale; upper surface green, shiny.

Sterile leaves. In 4 rows; upper-side leaves appressed, linear-lanceolate; lateral leaves appressed to spreading; underside leaves conspicuously smaller, spreading, narrowly deltate apex pointed.

Strobilus stalks. 4.4–12 cm × 0.6–1 mm; 1 or 2 arising from the upper lateral branches; mostly branching close to stalk tips, producing a pseudowhorl, or branching dichotomously, or combinations of these on same plant; remaining green until after spores shed.

Strobili. 1.4–4.0 × 0.2–0.3 cm exclusive of **narrow sterile tips that are up to 11 mm long (present on about 1/2 of specimens);** 4-stalked, strobili blunt if sterile tips absent; occasionally strobili are forked.

Sporophylls. 1.7–2.6 × 1.8–2.8 mm; deltate with abruptly tapering tips.

Habitat. Dry to moist sandy soils, hardwood and coniferous forests, young second-growth forests, and open fields.

Distribution. Eastern Canada. USA: Minnesota, Iowa, Illinois, south to Tennessee, Alabama, Georgia, and the Carolinas.

Chromosome #: $2n = 46$

Diphasiastrum digitatum is the most abundant species of *Diphasiastrum* in North America. It was long confused with *D. complanatum* and was treated as a subspecies of it.

The annual growth of upright shoots is produced by the lengthening of the erect stems and the production of new branches on the extension. The previous year's branches do not extend in the second season and thus lack annual winter constrictions.

See also the discussion under *D. complanatum.*

Diphasiastrum digitatum **is distinguished from** *D. complanatum* **by opposite lateral branches lacking annual constrictions, strobili with elongated sterile tips in about 1/2 the plants, and peduncles that remain green in the fall. In the field the lateral branches of** *D. digitatum* **appear flat and fan-shaped with regular distal margins and are all in 1 plane parallel to the ground, while the alternate branches of** *D. complanatum* **have a 3-dimensional, irregular appearance. Once these characters are recognized, the species are easily distinguished without stooping to examine the plants.**

The horizontal shoots of both *D. complanatum* **and** *D. digitatum* **are superficial and easily seen, while those of** *D. tristachyum* **are deeply buried.**

Diphasiastrum tristachyum (Pursh) Holub

Lycopodium tristachyum Pursh; *Diphasium tristachyum* (Pursh) Rothmann

Blue Ground-cedar, Deeproot Clubmoss, Northern Ground Pine

Etymology. Greek *tris-*, three, + *stachys*, spike, referring to plants bearing 3 strobili as originally described (but not the most common pattern for this species).

Plants. Evergreen, evergreen, miniature treelike.

Horizontal stems. 1.5–3.2 mm diam., **subterranean (5–12 cm deep);** long-creeping, branching well below ground, producing erect shoots.

Upright shoots. 15–30 cm tall, erect, clustered; treelike; branching near ground; leaves on upright shoot appressed with decurrent bases, linear with acuminate tips, 1.9–3.4 × 0.6–1 mm.

Lateral branches. 1–2.2 mm wide, **ascending, 4-sided, cross section square with rounded angles,** branching 4–7 times in first year, **annual winter constrictions conspicuous and abrupt. In young growth and sun-exposed areas, lateral branches with a more upright orientation and bluish color.**

Leaves. In 4 rows; leaf bases prolonged down the stems as a winged expansion on branches for more than 1/2 their length, tips acute, leaves of the underside of the branches nearly the same size as the lateral and upper leaves.

Strobilus stalks. 4–15 cm long × 0.4–1 mm wide, 2 or 3 arising from the upper lateral branches, remaining green until spores shed.

Strobili. 10–24 × 2–3 mm; borne in candelabralike clusters of 3–4; distinctly dichotomously stalked; cylindrical; tips rounded, sterile tip lacking.

Fertile leaves. 2.2–3.5 × 1.6–3 mm, deltate, tips gradually tapering.

Habitat. Dry open areas, sandy soils, and hardwood and coniferous forests.

Distribution. Europe. Asia in West China. Canada: eastern provinces west to Manitoba. USA: northeastern United States south to Georgia and Alabama, north to Michigan, Minnesota, Missouri, and Wisconsin.

Chromosome #: $2n = 46$

Diphasiastrum tristachyum **is distinguished from** *D. complanatum* **and** *D. digitatum* **by its stems with nearly square cross sections, its deeply subterranean rhizomes, and its tightly ascending branches with a bluish color in exposed sites.** *Diphasiastrum tristachyum* **has annual constrictions while** *D. digitatum* **lacks them.**

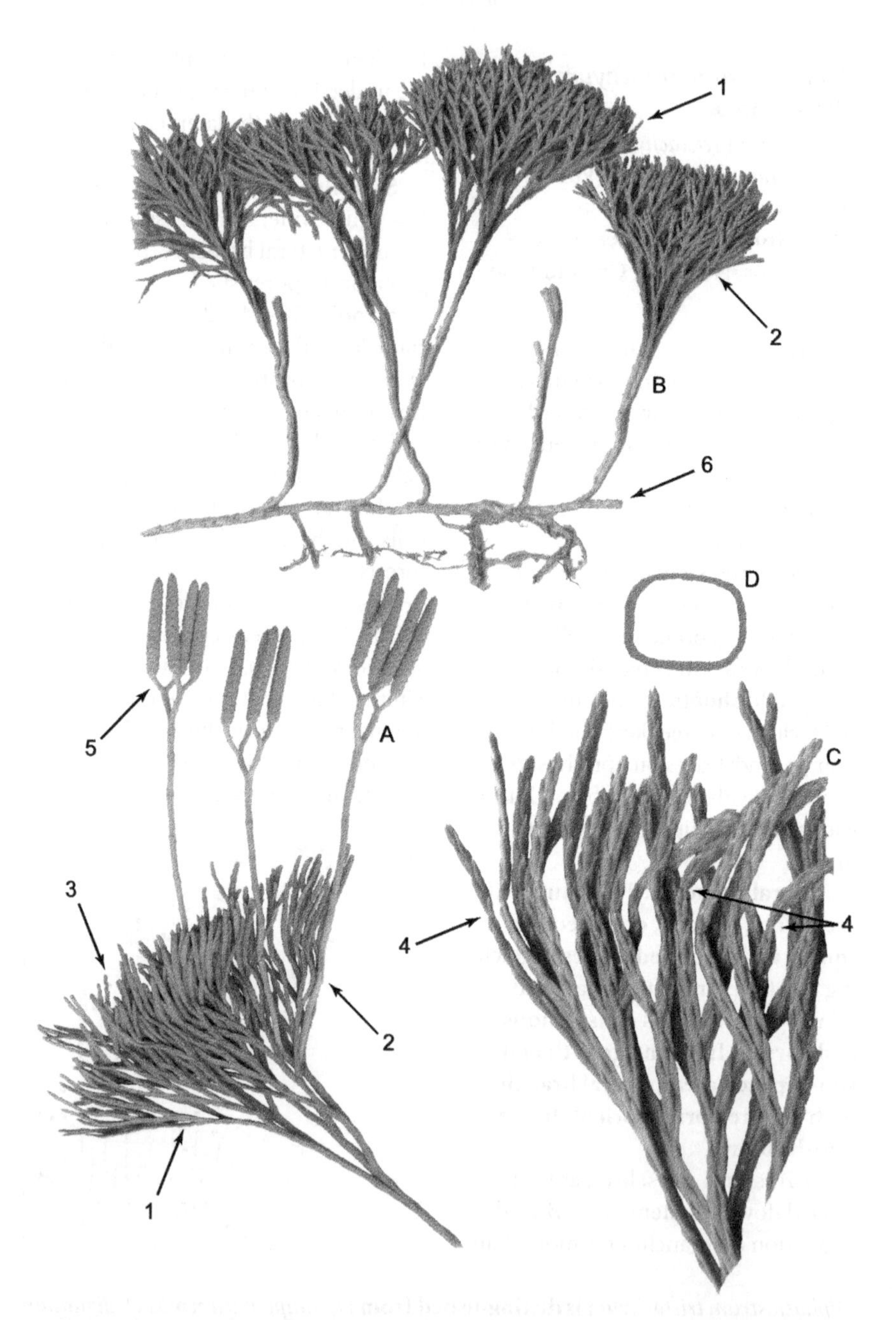

***Diphasiastrum tristachyum*: Blue ground-cedar: A. Upright stem. B. Stems growing from underground rhizome. C. Enlargement of branches. D. Symbol indicating somewhat square cross section of branches.**

Informative arrows: 1. Look for bluish color when sun exposed. 2. Ascending branches. 3. Branches with flat-topped growth form. 4. Annual constrictions. 5. Strobilus stalks with regular dichotomous forking. 6. Rhizomes deeply subterranean.

A tug on an upright shoot easily shows that the horizontal stem is deeply buried in the soil—the only Michigan *Diphasiastrum* species with this characteristic, a character shared by its hybrids.

HYBRIDS

Three *Diphasiastrum* species and 3 hybrids are now found in Michigan. Introgession between these hybrids and between them and the parent species may be found.

It is not unusual to find the species of *Diphasiastrum* and their hybrids growing together along with other species in the various genera of the family Lycopodiaceae.

SPORULATING TIMES

Sporulating times could influence the frequency with which certain hybrids are found. Early and late spores would produce gametophytes from which the sperm would seldom meet. Hybrids with the early and late sporulating species, *D. complanatum* × *D. digitatum,* should be rare, and are.

The following sporulating times are observations by Robert Preston.

Earliest: *Diphasiastrum complanatum*
Middle: *D. tristachyum*
Latest: *D. digitatum*

	Diphasiastrum digitatum	*Diphasiastrum digitatum* × *D. tristachyum*	*Diphasiastrum tristachyum*
Rootstock	Superficial	Mostly subterranean	Subterranean
Branches	Regularly fan-shaped, spreading horizontally	Spreading irregularly, some horizontally, others ascending	Fastigiate, ascending
Growth of branches	Without annual constrictions	Variable, mostly with annual constrictions	Annual constrictions prominent
Width of branches	2–3 mm	1–2 mm	1–1.5 mm
Color of leaves	Lustrous, bright green	Green, more or less lustrous	Blue-green, glaucous
Lateral leaves	Appressed or somewhat spreading	Spreading only in juveniles	Mostly appressed
Number of strobili	2 to many, averaging about 4	Usually none; 3–4 when present	1–6

Diphasiastrum digitatum × tristachyum
Diphasiastrum ×habereri (House) Holub
Lycopodium habereri House; *Lycopodium tristachyum* var. *habereri* (House) Victorin
Haberer's Clubmoss

Etymology. Hybrid name honors Joseph Valentine Haberer (1855–1925), a practicing physician, resident of Utica, and an enthusiastic student of the flora of Herkimer, Oneida, and Madison counties, New York. The hybrid description was based on his collection. He possessed one of the most complete herbariums in America, later donated to the state of New York.

Distribution. Probably present where the distribution of the 2 parent species overlap. It has been identified in Alcona, Barry, Charlevoix, Cheboygan, Emmet, Mackinac, Menominee, Midland, Oakland, and Ottawa Counties.

Diphasiastrum ×habereri **may be identified by horizontal stems that are often shallowly buried in the soil (vs. superficial in** *D. digitatum*) **and by its lateral branches with annual winter bud constrictions (mostly near branch tips).**

Diphasiastrum complanatum × digitatum
Diphasiastrum ×verecundum A. V. Gilman
Bashful Clubmoss

Etymology. Latin, *verecundus,* bashful, shy. The author of the name states that it alludes "to the bashful nature of the taxon which is rare, shows no outstanding feature, and is consequently difficult to come to know."

Habitat. Dry shallow organic soils overlying sandy or loamy soils.

Distribution. New England. In Michigan it has been found in Marquette and Mackinac counties.

A rare hybrid found where the parental species distributions overlap. It is difficult to recognize, particularly because its spores are usually well formed. Identification is often difficult on herbarium sheets.

Diphasiastrum ×verecundum **resembles both** *D. complanatum* **and** *D. digitatum*. **It differs from the former by its longer, less flexuous more robust peduncles and often more than 4 strobili on each peduncle. Its peduncles are thinner than those of** *D. digitatum.*

	D. complanatum	*D. ×verecundum*	*D. digitatum*
Branches	Irregular, ascending to spreading, straggly	Somewhat irregular, straggling. Fanlike branching lacking or weak	Regularly fan-shaped, spreading horizontally
Annual constrictions	Present, prominent	Present but not consistent	Lacking
Color of leaves	Dull green	Bright emerald green	Lustrous, bright green
Length of peduncles	8.3–32 cm	Intermediate	4.4–12 cm
Number of strobili	Usually 1–2 (–3)	Intermediate	(2–) 4–5 (–7) to many, averaging about 4
Strobilus	Becoming yellow to straw-colored before spores are shed in fall	Stout, green until spores are shed	Remains green until spores are shed
Strobilus length	0.83–2.5 (–3.4) cm	Intermediate	(1.4–) 2–4.0 cm

Diphasiastrum complanatum × tristachyum

Diphasiastrum ×zeilleri Rouy; *Lycopodium complanatum* race *zeilleri* Rouy; *Diphasium zeilleri* (Rouy) Damboldt; *Diphasiastrum complanatum* subsp. *zeilleri* (Rouy) Kukkonen

Zeiller's Clubmoss

Etymology. Name honors Charles René Zeiller (1847–1915), a French paleobotanist and professor of plant paleontology at the National Superior School of Mines, Paris.

Distribution. In Michigan it has been found scattered where the distributions of its parental species overlap (mostly above the 45th Parallel or north of an east-west line running through Benzie County).

Diphasiastrum ×zeilleri **differs from** *D. complanatum* **by the branches being ascending, erect-fastigiate; thin; secondary branches very close and somewhat compressed; almost equally 4-sided; leaves all the same or lateral leaves scarcely larger than those on upper and lower surfaces; and stobili numerous.**

	Diphasiastrum complanatum	*Diphasiastrum complanatum × D. tristachyum*	*Diphasiastrum tristachyum*
Rootstock	Superficial or under leaf litter	Shallowly buried	Subterranean
Branches	Irregular, ascending to spreading, flat in cross section	Irregular, ascending	Erect, ascending, clustered, parallel, and cross section square
Width of branchlets	1.5–2.5 mm	Intermediate	1–1.5 mm
Color of leaves	Green, dull	Green, weakly glaucous	Blue-green, glaucous
Lateral leaves	Appressed or somewhat spreading	Intermediate	Mostly appressed

HUPERZIA Bernhardi Lycopodiaceae

Gemma Firmosses

See Lycopodiaceae in appendix for a description of the family and a key to the Michigan genera.

Etymology. Name honors Johann Peter Huperz (1771–1816), a German physician, fern horticulturist, and author of a book on fern propagation.

Growth habit. Evergreen, spikelike; terrestrial, **strobili absent.**

Horizontal stems. Absent.

Stems. 2–16 mm diam., clustered, erect to decumbent, cross sections including leaves round, equally **dichotomously branched, lacking true horizontal shoots, all branches similar**; annual constrictions present or not; roots form near tips of stems and grow downward through the entire stem parenchyma to emerge at soil level.

Leaves. Many indistinct ranks, spirally arranged, spreading to appressed; triangular, lanceolate to oblanceolate, leaves same size or if varying in size smaller at winter annual constrictions, lowermost leaves mostly larger than terminal leaves; margins entire to irregularly dentate, sometimes with marginal papillae.

Sporangia. Reniform; bivalved; borne at adaxial bases of unmodified or only slightly modified sporophylls; **found in distinct zones on stems.**

Gemmae. 2.5–6 × 3–6 mm, with 4 leaves flattened into 1 plane, 2 large lateral leaves, 1 abaxial and 1 adaxial leaf, deltoid; **borne on modified leaves (gemmiphores) found among unmodified leaves.**

Sporangia. Reniform; borne individually at upper surface of base of leaves; fertile leaves in distinct regions or scattered along stem; **strobili absent.**

Gametophytes. Subterranean; mycorrhizal, nonphotosynthetic; "hot dog in bun" shape.

Habitat. Sandy soils in damp mesic forests, on rock faces, and on moist, often mossy ground.

Distribution. *Huperzia* is the largest genus in the Lycopodiaceae family, with more than 300, mostly tropical, species. It is found worldwide in tropical and high-elevation tropical forests. Most of the tropical species are lax and pendulous epiphytes in mountain forests. Another large group is terrestrial in mostly open pioneer habitats in north temperate to boreal areas. In these northern areas possibly 15–20 species are found.

Seven species are treated in *Flora of North America North of Mexico*, vol. 2 (1993). They are represented in Michigan by 3 species, and 2 hybrids.

Chromosome #: $x = 67, 68$

Huperzia lucidula is fairly common and widespread throughout the state. *H. appressa* (Upper Peninsula only) and *H. selago* (Upper Peninsula and upper Lower Peninsula) are rare.

The genus *Huperzia* is quite distinct from the other genera in the Lycopodiaceae as treated here. It has been placed in its own family, Huperziaceae, by some authors. This is based on several morphologic features. The stems bifurcate dichotomously (equally). It lacks true horizontal stems, and the roots are concentrated at the base of stem clusters. The roots begin as vascular bundles originating at the growing tips of the stems and elongate through its cortex until they emerge at ground level. The roots of other genera of Lycopodiaceae arise directly from creeping rhizomes.

The sporophylls (fertile leaves) of *Huperzia*, found in annual zones along the stem, are very similar to the trophophores (sterile leaves), and remain unchanged after spores are shed. In other Lycopodiaceae genera the sporophylls, which are usually different from the trophophylls, are aggregated in strobili, and shrivel after spore release.

Species of *Huperzia* reproduce both sexually by spores, and asexually by gemmae and clone formation. They have no strobili. Asexual gemmae that are produced late in the season are a major mode for its reproduction. They are not present in the other Lycopodiaceae genera. The gemmae are borne on a gemmiphore, which acts as a taut spring forcefully expelling the gemma when a raindrop, or a light touch, forcibly discharges it. (Try it, it's fun!) They have an aerodynamic shape and when thrown into the wind land and form a new plant at a distance. The colonies produced by the gemmae are clones of the original plants.

KEY TO THE SPECIES OF *HUPERZIA* IN MICHIGAN

1. Leaves long, narrow, widest above the middle, dark green with 1–8 spreading to reflexed, irregular, 7.5–11 mm long teeth; distinct annual constrictions present; gemmae in 2–3 pseudowhorls at end of annual growth; fairly common . . . . *H. lucidula*

1. Leaves narrowly triangular; green to yellow-green, with 0–3 mostly ascending, 3.5–7.5 mm long teeth; annual constrictions weak or absent; gemmae throughout mature stems or in single whorls at tips at end of growing season; rare (2).

2(1). Gemmae scattered throughout upper stem; annual constrictions lacking; leaves on lower stems longer and more spreading than those toward tips; annual constrictions absent; plants of talus slopes and cliffs . . . . *H. appressa*

2. Gemmae in single whorl at end of annual growth; leaves all about same size; weak annual constrictions on the stems; plants of wet soils . . . . *H. selago*

Huperzia appressa (Desvaux) Á. Löve & D. Löve

Lycopodium selago Linnaeus f. *appressum* Desvaux; *Huperzia appalachiana* Beitel & Mickel; *Urostachys selago* (Linnaeus) Herter var. *appressus* (Desvaux) Nessel

Huperzia appalachiana, published by Beitel and Mickel in 1992, is an oft-used synonym, but the name *Huperzia appressa* was created by Desvaux (as *Lycopodium selago* Linnaeus f. *appressum* Desvaux in 1827 (described from South Greenland and Newfoundland) and raised to specific level by Löve and Löve in 1961. Since these are all names for the same plant, the name *Huperzia appressa* has priority.

Mountain Firmoss, Appalachian Fir Clubmoss

Etymology. Latin, *appressa,* appressed, probably alluding to the ascending leaves of this species. The vernacular name derives from a favored habitat.

Growth habit. Evergreen, small clubmoss, determinate (plant dies after 12–15 years).

Stems. 5–10 cm long, 6–10 cm tall × 3–7 mm wide, including leaves, erect, clustered; decumbent portion up to 1 cm long; **annual constrictions absent; strobili absent.**

Leaves. Leaves of mature upper parts much smaller than leaves of juvenile lower parts, leaves in upper portion narrowly triangular, ca. 1/2 size of leaves on lower stem, 2–3.5 mm long; **lower leaves narrowly triangular,** 4–6 mm long, with smooth edges or occasional papillae, widest at base, spreading more than those on upper stem; all leaves ascending, **dull, green to yellow-green;** stomates present on both surfaces, numerous (35–60 per half leaf) on upper surface.

Gemmae. 3–4 × 2.5–3.5 mm; **scattered along mature upper stems;** lateral leaves 0.5–1 mm wide, narrowly acute.

Sporangia. Scattered in distinct regions on upper stem.

Habitat. Damp igneous rocks, rock crevices in basalt outcrops, cliffs, and talus slopes.

Distribution. A boreal, alpine and low arctic species found in several noncontiguous areas in eastern North America and northern Europe. Greenland. Iceland. Canada: along Atlantic coast in Newfoundland and Quebec. USA: New England, southern Appalachians, and the shores of Lake Superior in Michigan and Minnesota.

In Michigan known only from exposed basaltic outcrops and talus slopes on the shoreline of Isle Royale in Lake Superior.

Chromosome #: $2n = 264$ or 272 (6×?)

Huperzia appressa is considered imperiled in Michigan and secure or apparently secure globally.

Huperzia appressa **may be distinguished from** *H. lucidula* **by its narrowly triangular versus obovate leaves, shorter leaves (3 mm vs. 10 mm), shorter stems (10 cm vs. 25 cm), smaller leaves on the distal portion of the stems, and lack of annual constrictions (***H. lucidula* **has prominent ones). Its leaves have a dull greenish-yellow color in contrast to the dark green, shiny leaves of** *H. lucidula.*

It is distinguished from *H. selago* **by its single whorl of gemmae at the tips at end of annual growth (vs. scattered throughout mature upper stems on** *H. selago***) and by its lack of annual constrictions (***H. selago* **has weak ones).** *Huperzia appressa* **has larger leaves at the base of the stem and smaller ones toward the tips (vs. leaves uniform in size on** *H. selago***).** *Huperzia appressa* **has determinate stems (the entire stem dies and turns yellow) while** *H. selago* **stems are indeterminate—growing indefinitely).** *Huperzia appressa* **is associated with rocky sites (***H. selago* **grows in wet soils).**

Huperzia lucidula (Michaux) Trevisan

Lycopodium lucidulum Michaux; *Huperzia selago* (Linnaeus) Bernhardi subsp. *lucidula* (Michaux) A. Löve & D. Löve; *Urostachys lucidulus* (Michaux) Herter

Shining Firmoss, Trailing Evergreen Clubmoss

Etymology. Latin *lucidus*, shining, + *-ulum*, a possessive suffix, "with shiny leaves," in reference to this species' glossy dark green leaves.

Plants. Evergreen, indeterminate growth.

Upright stem. Stems growing in loose tufts, 14–25+ cm, 10–20 mm diam. including leaves, erect; sparsely branched dichotomously; bases becoming decumbent with long, tailing older portions that become senescent and brown; **annual constrictions prominent; strobili absent.**

Leaves. 7–11 mm long, spreading to ascending; **oblanceolate to obovate, broadest above the middle, dark lustrous green**; spirally arranged but appearing whorled, leaves at annual constrictions much smaller, narrowly lanceolate; 1–8 irregular large teeth on each margin; stomates only on abaxial surface.

Sporangia. Kidney-shaped; in axils of leaves; **fertile leaves** similar to sterile leaves **borne in distinct zones on upper stem**; yellow.

Gemmae. Appearing at end of annual growth in 1–2 whorls near tip, making stems appear thicker at the tips, 4–6 × 3–6 mm long, broadly obtuse with abruptly pointed tips.

Habitat. Shaded rich forest hab-

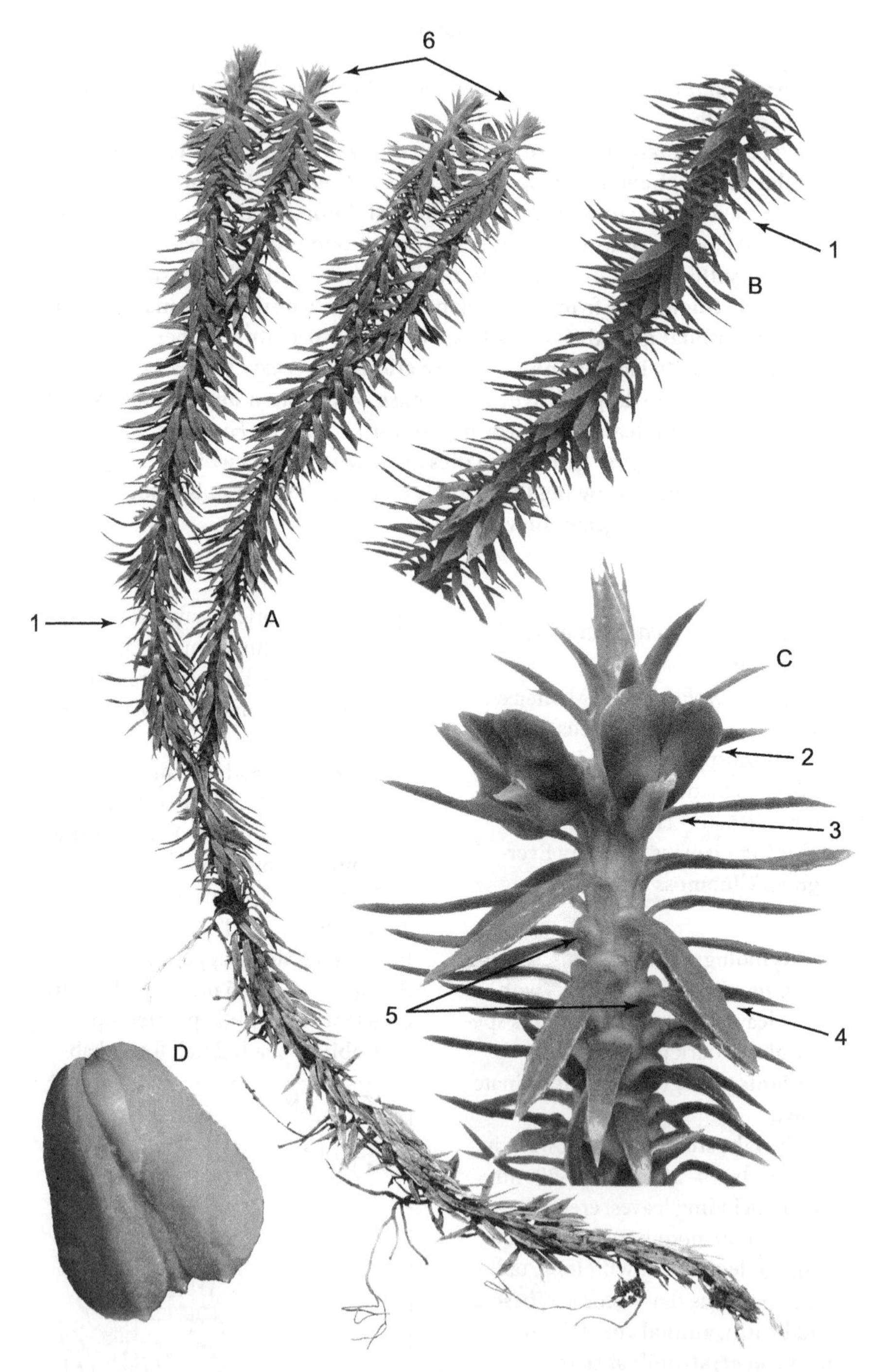

Huperzia lucidula: **Shining firmoss: A. Stem. B. Portion of stem. C. Tip of stem with gemmae. D. Gemma undersurface.**

Informative arrows: 1. Arrows showing annual constrictions. 2. Gemmae. 3. Gemmaphore. 4. Stiff, dark, lustrous leaves widest above the middle. 5. Sporangia. 6. Look for absence of strobili.

itats of diverse types, from boreal to hardwood, to mixed hardwood-conifer woods, as well as bog and stream edges

Distribution: Southeastern Canada, northeastern and midwestern United States extending south to northern Missouri and the Carolinas.

Chromosome #: $2n = 134$

Huperzia lucidula **may be distinguished from** *H. appressa* **by the width of its leaves, which are wider toward their tips (vs. triangular in** *H. appressa*)**, by longer leaves (7–10 mm vs. 2–3.5 mm in** *H. appressa*)**, by longer stems (25 cm vs. 10 cm in** *H. appressa*)**, and stems with annual constrictions (lacking in** *H. appressa*)**. Its leaves have a lustrous dark green color (vs. the dull greenish-yellow leaves of** *H. appressa*)**, and are toothed (vs. almost smooth in** *H. appressa*)**.**

H. selago **is distinguished by its nearly smooth margined leaves, which are largest and triangular on the lower stems. Leaves are green to yellow-green.**

It may resemble *Spinulum annotinum* **but is distinguished from it by its lack of a strobilus, by leaves that are less stiff and without a prickly tip, and by its gemmae (lacking in** *S. annotinum*)**.**

Huperzia porophila (Lloyd & Underwood) Beitel, is known from Wisconsin and the north shore of Lake Superior in Ontario, and could be found in Michigan. It grows on sandstone or basaltic cliffs and resembles a compact *H. lucidula*, but with leaves widest at or below the middle and stems without conspicuous annual constrictions.

Huperzia selago (Linnaeus) Bernhardi

Lycopodium selago Linnaeus; *Urostachys selago* (Linnaeus) Herter

Northern Firmoss, Fir Clubmoss

Etymology. *Selago*, an ancient, pre-Linnaean name for plants that look like savin (*Juniperus sabina*). Before Linnaeus's time the name *selago* was used for other clubmoss species. The vernacular name fir clubmoss alludes to the resemblance of its needle-like leaves to those of the firs (*Abies* spp.).

Plants. Evergreen, indeterminate.

Erect shoots. 8–13 cm tall, 7–14 mm diam. including leaves; cluster-forming, creeping, low-growing, becoming decumbent, growing in close dense clusters of nearly equal length with flat tops, branches few, **annual constrictions indistinct or absent; strobili absent.**

Leaves. Dense, small, spreading-ascending (in shade) to appressed-ascending (in sun); **leaves of mature upper portion same size or slightly smaller than leaves of juvenile lower portion,** largest leaves triangular, widest at base, 3.5–7.5 mm, smallest leaves

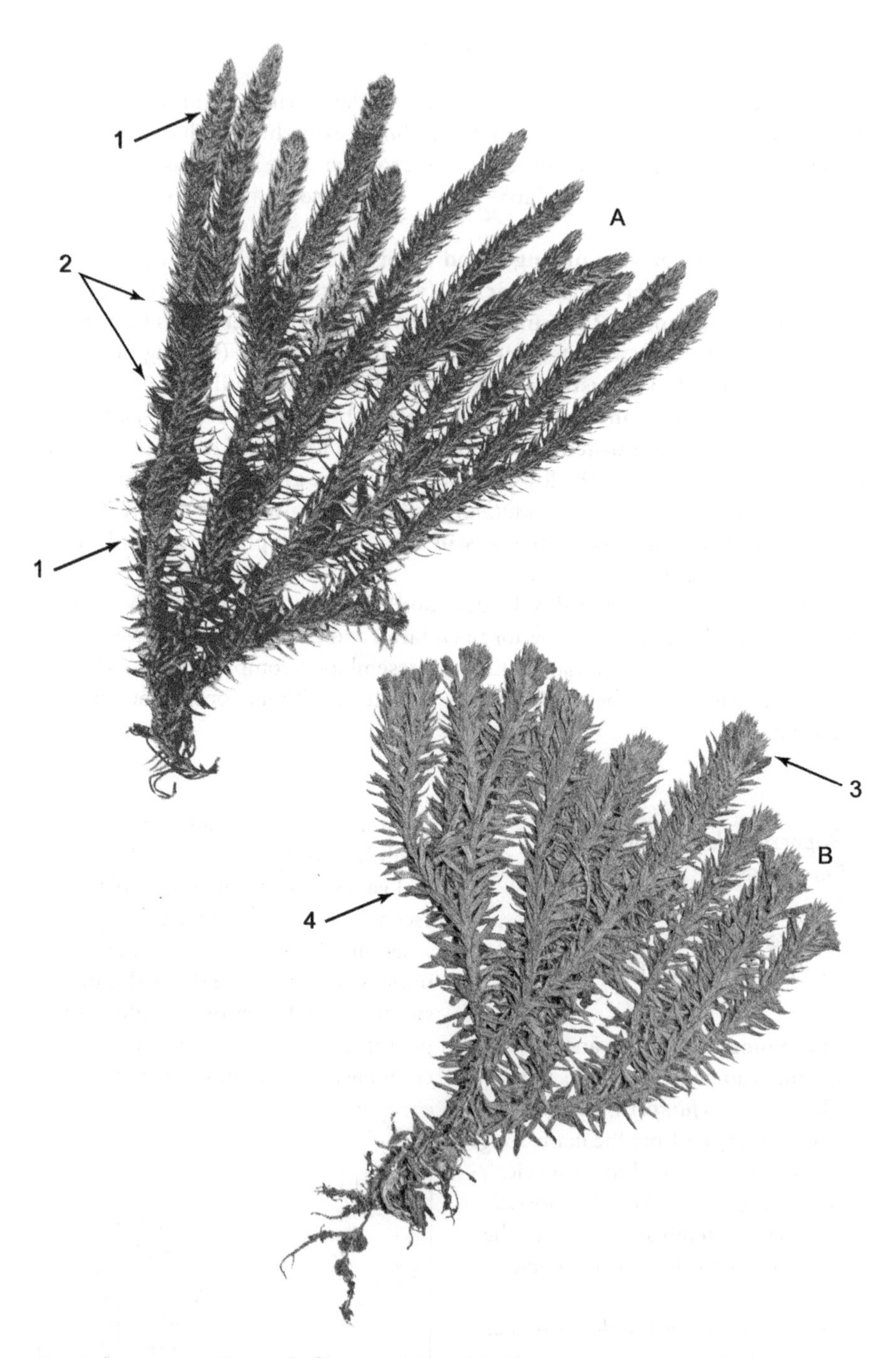

Huperzia appressa: Mountain firmoss: A. Stem cluster.

Huperzia selago: Northern firmoss: B. Stem cluster.

Informative arrows: 1. Leaves on lower stems longer and more spreading than leaves on upper stem. 2. Look for gemmae scattered throughout upper stem. 3. Look for gemmae in single whorl at end of annual growth. 4. Leaves all about same size.

lanceolate, 3.5–5 mm; margins almost smooth with uncommon papillae; stomata on both surfaces; lustrous, green (shade) to yellow-green (sun).

Sporangia. Reniform, in leaf axils in discrete zones on upper stems.

Gemmae. 4–5 × 3–4.5 mm; **restricted to a single whorl at the end of the annual growth,** dull, yellowish-green lateral leaves 1.5–2 mm wide, tips broadly acute to narrowly obtuse.

Habitat. Cool damp soils, sandy borrow pits, ditches, light-open forests, low areas near lakeshores, coniferous swamps, seasonally inundated wetlands, and roadside disturbed areas.

Distribution. Circumboreal. Europe. Asia. Greenland. North America: Labrador to Alaska, south to Connecticut and Massachusetts, and west to Michigan and Minnesota, with isolated disjuncts in Ohio.

Chromosome #: $2n = 268$

Huperzia selago **resembles** *H. appressa* **but may be distinguished from it by the presence of weak annual constrictions on the stems (***H. appressa* **lacks annual constrictions) and gemmae restricted to a single whorl at the end of each year's growth (***H. appressa* **has gemmae scattered along the mature stem).**

The leaves of *H. selago* **are more uniform in size whereas** *H. appressa* **has larger leaves at the base of the stem and smaller ones toward the tips. Its stems continue to grow at the tips indefinitely (indeterminate).** *H. appressa* **has determinate stems (the entire stem dies and turns yellow). It is associated with wet soils throughout much of its range (***H. appressa* **is associated with rocky sites).**

HUPERZIA HYBRIDS IN MICHIGAN

Huperzia ×buttersii (Abbe) Kartesz & Gandhi (*Huperzia lucidula × H. selago*)
Butters' Clubmoss

Etymology. Name honors Frederick King Butters (1878–1945), a Minnesota botanist and avid mountaineer, who worked out the characteristics of this hybrid before he died. He was a professor of botany at the University of Minnesota and coauthored botanical books.

Stems. 15–35 cm long, annual constrictions indistinct.

Leaves. Linear-tapering to slightly linear-lanceolate, nearly entire with occasional minute teeth, initially appressed, later spreading, and eventually becoming reflexed; stomata on both surfaces.

Gemmae. Borne in 1 whorl at tips of annual growth, slightly obtuse (intermediate between those of parent species).

Spores. Abortive.

Habitat. Coniferous and wet forests.

Distribution in Michigan. Rare in Marquette, Schoolcraft, Luce and Chippewa counties.

Huperzia ×buttersii **resembles a slender** *Huperzia lucidula.* **It has abortive spores. Its trophophylls have more or less parallel margins, obscure papillaelike teeth, and few stomates.**

	H. lucidulum	*H.* ×*buttersii*	*H. selago*
Shoot	Annual constrictions prominent	Annual constrictions indistinct	Annual constrictions lacking
Leaf shape	Oblanceolate to obovate	Intermediate	Linear-tapering
Leaf margins	Denticulate	Intermediate	Entire
Leaf stomata	Undersurface only	Intermediate both surfaces	Both surfaces, more on undersurface
Gemmae	Rounded to obtuse, slightly apiculate	Intermediate	Acute or subacute

Huperzia ×josephbeitelii A. Haines (_Huperzia appressa_ × _H. selago_)
Non *Huperzia beitelii* Øllgaard–nom. illeg.
Beitel's Clubmoss

Etymology. Name honors Joseph M. Beitel (1952–91), known to all as Joe, a brilliant scientist who died too young. He was first to discuss this hybrid. He was a well-rounded botanist, a competent scientist, an enthusiastic and effective teacher, and a good friend. He was a horticultural taxonomist at the New York Botanical Garden and an authority on ferns and lycopodiums,

He studied with W. H. "Herb" Wagner Jr. at the University of Michigan and conducted intensive studies of the genus *Huperzia*—the gemma firmosses. He published 25 scientific papers, coauthored the "Pteridophyte Flora of Oaxaca, Mexico," and assisted Herb Wagner in the Lycopodiaceae treatment in *Flora of North America North of Mexico*, vol. 2 (1993). He was an expert on the flora of Long Island and Michigan, where he led hundreds of field trips. He carried out fieldwork in Canada, China, Mexico, the British Virgin Islands, Costa Rica, Venezuela, and many parts of the United States, including Hawaii and Michigan.

Stems. 7–23 cm tall, upper portions 7–10 mm wide, including leaves; annual constrictions absent, growth indeterminate.

Leaves. In basal portion 6–7.5 cm long, entire, spreading, lanceolate, tips acute to acuminate; leaves in apical portion 3.5–4.5 mm long, narrow-triangular, spreading-ascending to ascending.

Gemmae. Found throughout upper stems; lateral leaves 1.1–1.5 (–1.7) mm wide, acute to broad-acute at tips.

Spores. Abortive.

Distribution. Canada: from Ontario, to the Atlantic Ocean. USA: Maine, Michigan, and New Hampshire. In Michigan found on the Isle Royale archipelago, including Passage Island.

Huperzia ×josephbeitelii **is difficult to identify. It resembles a lax, robust shade form of** *H. appressa*. **It has abortive spores. It is more robust than** *H. appressa,* **has wider stems (7–10 mm wide with trophophlls vs. 3–7 mm), and has larger gemmae with larger lateral leaves (1.1–1.5 mm wide vs. 0.5–1.0 mm wide). The gemmae are found throughout the upper stem and on leaves from the upper and lower portions of the stems that are dissimilar in size and orientation.**

	H. appressa	*Huperzia ×josephbeitellii*	*H. selago*
Upper stem	3–7 mm diam.	7–10 mm diam.	7–14 mm diam.
Leaf size	Upper and lower stem with different size leaves	Upper and lower stem with different size leaves	Upper and lower stem with ± same size leaves
Gemma width	3–4 × 2.5–3.5 mm; lateral leaves 0.5–1 mm wide, tips narrowly acute	3.4–4 × 2.5–3.5 mm, lateral leaves 1.1–1.5 (-1.7) mm wide, tips acute to broadly acute	4–5 × 3–4.5 mm, lateral leaves 1.5–2 mm wide, tips broadly acute to narrowly obtuse
Gemma distribution	Throughout upper stems	Throughout upper stems	Restricted to a single whorl at end of annual growth

LYCOPODIELLA Holub Lycopodiaceae

Bog Clubmosses

See Lycopodiaceae in appendix for a description of the family and a key to the Michigan genera.

Etymology. Greek *lycos*, wolf, + *podos*, foot, wolf's foot (referring to the genus *Lycopodium*), + *-ella*, a diminutive suffix—the "little wolf's foot." Probably referring to the smaller size of *Lycopodiella* species compared to *Lycopodium* species.

Plants. Deciduous, small; erect shoots die back each year to thickened stem tips.

Horizontal Stems. Long-creeping, prostrate or arching on wet ground, roots originating on undersides of stems near points where upright shoots branch.

Upright stems. Unbranched, terminating in a single unstalked strobilus, round, leaves crowded, linear to linear-lanceolate, spreading at bases, curved inward at tips, not in regular rows, margins toothed or untoothed.

Strobili. Formed on upper part of upright stems, stalkless, solitary, fully differentiated from upright stems; tips blunt to more or less acute, **fertile leaves (sporophylls)** less than 1 cm long, generally equal to or longer than leaves of strobilus stalks.

Sporangia. Borne in axils of sporophylls, nearly globose; thick-walled, anisovalvate.

Gemmae. Lacking.

Habitat. Moist to wet calcareous habitats, swamps, meadows, pastures, open woods, rarely on rock, full sun to semishade.

A pantropical, temperate to subtropical genus of 10–15 species. Six species are treated in *Flora of North America North of Mexico*, vol. 2 (1993). It is represented in Michigan by 3 species.

Chromosome #: $x = 78$

Two tetraploid species found in Michigan, now recognized as *Lycopodiella subappressa*, and *L. margueritae*, were previously treated as *L. appressa*.

Lycopodiella inundata is diploid. *Lycopodiella marguetitae* and *L. subappressa*, recently described species, are probably fertile allotetraploid species. *Lycopodiella margueritae* may be a hybrid between *L. inundata* and *L. alopecuroides*, and *L. subappressa* may be a hybrid between *L. inundata* and *L. appressa*. (*Lycopodiella alopecuroides* and *L. appressa* are species of more the southern and eastern United States and are not found in Michigan.) The full ranges of the recently described species still have to be established.

Hybrids are likely to be found at sites in southern Michigan where the Michigan species grow together. Hybrids between the diploid *L. inundata* and Michigan's tetraploid species will produce abortive spores. Finding abortive spores from a plant with intermediate characteristics would suggest a sterile triploid hybrid. Hybrids between tetraploid species produce a high percentage of normal spores and are harder to detect.

KEY TO THE SPECIES OF *LYCOPODIELLA* AS FOUND IN MICHIGAN

Species of this genus are often difficult to identify, especially the juvenile plants. They frequently grow together, and when they hybridize intermediate forms are produced.

1. Horizontal stems (excluding leaves) less than 1 mm diam.; erect stems usually 3.5–9 cm long; horizontal stem leaves mostly less than 6 mm long, margins entire; strobili mostly 33 to 50 percent of the total length of the entire upright stem, sporophylls spreading . . . . *L. inundata*

1. Horizontal stems (excluding leaves) 1–3 mm diam., erect stems 4–13 (–17) cm, often more than 10 cm long; horizontal stem leaves 4–13 mm long, margins toothed or bristled; strobili 1/10 to 2/3 of the total length of the entire upright stem, sporophylls spreading or appressed (2).

2. Horizontal stems 1.8–2.2 (–3) mm diam. without leaves, horizontal stem leaves spreading to perpendicular, 6–13 mm long, strobili 5–8 cm long, comprising 33 to 66 percent of the total upright stem length. . . . . *L. margueritae*

2(1). Horizontal stems 1.0–1.7 (–2) mm diam. without leaves, horizontal stem leaves appressed, 4–6 mm long; strobili 2–4 cm long, comprising 13 to 37 percent of the total upright stem length. . . . . *L. subappressa*

Lycopodiella inundata (Linnaeus) Holub
Lycopodium inundatum Linnaeus
Northern Bog Clubmoss

Etymology. Latin *inundates*, flooded, or flooded for part of the year, alluding to the habitat of this species. The common name alludes to its preferred boggy habitat.

Plants. Perennial, aerial stems die back in the fall, horizontal stems overwinter, producing new stems in the spring.

Horizontal stems. 3–12 cm × 3–4 mm (0.5–1 mm without leaves); fully prostrate; leaves monomorphic, spreading, upcurved, 5–6 × 0.5–0.7 mm, margins entire.

Upright stems. 3.5–9 cm long (including strobilus), 3–4 mm wide including leaves (0.5–1 mm diam. excluding leaves), 1 (rarely 2) per plant. unbranched, light yellowish-green; leaves narrow, long-triangular, 5–6 × 0.5–0.7 mm, margins entire, teeth rare, spreading to upcurved, spirally arranged.

Strobili. Formed on upper part of upright stems as "bushy tails," stalkless, 11–20 × 2.5–5.5 (—13) mm; wider than upright stem by (2–) 3–6 mm; more than 33 to 50 percent length of total upright stem length; leaves (sporophylls), 4–5.5 × 0.5–0.9 mm, crowded, wider at base, marginal teeth absent or rare, spreading to ascending, green.

Habitat. Moist to wet acidic soils, marshes, acidic peat bogs, sandy lake and pond shores, borrow pits, and wet sandy ditches.

Distribution. Interrupted circumboreal. Eurasia. Canada: British Columbia, Alberta, and Saskatchewan, in the east Newfoundland to Ontario. USA: in the east from the Canadian border south to Virginia and west to Wisconsin and Minnesota; in the west Washington south to California, east to Montana, and north to Alaska.

Chromosome #: $2n = 156$, diploid

Lycopodiella inundata is by far the most common Michigan *Lycopodiella.* It is a diploid and a hypothesized parent species of both Michigan tetraploid species, *L. margueritae* and *L. subappressa.*

Lycopodiella inundata survives the winter as a small bud on the tip of the creeping horizontal stem. When growth begins in the spring the bud first produces an upright stem. The first branching then produces a creeping vegetative stem. The creeping stems sometimes branch again, forming new creeping branches. These shoots also terminate the growth season with winter buds that become detached from the main shoot when the plant dies back. Over time large colonies may form.

It often occurs with one or both of the other Michigan species in the southern half of the Lower Peninsula and forms hybrids with them that are difficult to identify. In a few spots in southern Michigan, rare, very tall plants that are known to be diploid occur. These may represent a new taxon.

In winter it dies back to a small bud at the tip of the creeping shoot. When growth begins the following spring the bud first produces a horizontal shoot (sometimes an erect shoot). The first branching divides into an erect stobilus-bearing shoot and a creeping shoot continues. Some plants do not produce a strobilus-bearing shoot every year.

Lycopodiella inundata **is widespread in its favored habitats. It can be distinguished by its shorter uptight stems (1.7–8.7 cm); shorter strobili (1–2.5 cm); rhizomes having a smaller diam. (<1.0 mm without leaves); horizontal stem leaves widely spreading, and strobili that comprise only 1/3 to 1/2 of the upright stem.**

Lycopodiella margueritae J. G. Bruce, W. H. Wagner & Beitel

Northern Prostrate Clubmoss, Marguerite's Clubmoss

Etymology. Name honors Marguerita Bruce, the wife of J. G. Bruce, a coauthor of the species description.

Plants. Aerial stems die back in the fall, horizontal stems overwinter, producing new stems in the spring.

Horizontal stems. 10–18 cm long, 1.8–2.2 (–3) mm diam. (without leaves); creeping, prostrate, usually spreading to form extensive clones; leaves 6–13 × 0.8–1.2 mm, 3–4 marginal teeth on each side, nearly perpendicular to axis.

Upright stems. 13–17 × 1–1.6 cm, stems (excluding leaves) 1.8–2.2 mm diam., 1 (–2) per plant; leaves 6–13 × 0.4–0.8 mm, marginal teeth 0–2 on each side, mainly on proximal 1/2, initially nearly perpendicular to stem, then incurved at tips.

Strobili. Stalkless, 5–8 × 0.4–0.9 cm (2–) 3–6 mm; wider than strobilus stalk **(1 1/2 to 2 times as wide as stalk); mostly 33 to 66 percent as long as total upright stem length,** appearing in late summer and persisting through November; leaves (sporophylls) 4–6 × 0.4–0.5 mm, incurved, marginal teeth absent.

Habitat. Moist to acidic peaty sands, seasonally flooded wetlands, shallow depressions in glacial-lake-plain landscapes, ditches, sandy borrow pits, and lakeshore ponds.

Distribution. USA: Connecticut, Indiana, Ohio, Virginia, Pennsylvania, Michigan, and Wisconsin. In Michigan it has been found in the lower half of the Lower Peninsula especially along the west coast. It is uncommon and recently described. Its full Michigan range has not yet been identified.

Chromosome #: $2n = 312$, tetraploid

*Lycopodiella marguerita*e may be derived from a hybrid between *L. inundata* and *L. appressa* ($2n = 156$, not found in Michigan). It forms apparently fertile hybrids with *L. subappressa*, also a tetraploid species. Its hybrids with the diploid *L. inundata* are sterile.

Lycopodiella margueritae **resembles a very robust form of** *L. inundata* **with thicker stems and toothed leaves (***L. margueritae* **upright shoot 13–17 cm tall vs.** *L. inundata* **3.5–9 cm). It can be distinguished from** *L. subappressa* **by its rather thick erect stems and longer strobili, mostly 1/3–1/2 of the total length of upright shoots (vs. 1/13–1/3 for** *L. subappressa***); mostly longer fertile stems and more spreading leaves on its peduncles and rhizomes (leaves of the upright stems and strobili of** *L. subappressa* **are appressed). It also tends to have longer leaves on its rhizomes, and the diameter of its rhizomes, without leaves, is larger.**

Lycopodiella subappressa J. G. Bruce, W. H. Wagner & Beitel

Northern Appressed Clubmoss, Northern Bog Clubmoss

Etymology. Latin prefix *sub-*, below, under, + *appressa*, referring to *L. appressa*. The author stated that the name refers to its resemblance to *L. appressa* and its smaller stature. The vernacular name "northern" appressed clubmoss, reflects the fact that the species *L. appressa* has a generally more southern distribution.

Plants. Clone forming; horizontal stems overwinter and bear new shoots in the spring. Plants deciduous, overwintering with a thickened tip of stem. Horizontal shoots rooting (2–) 3.5–7.5 cm distal to the proximal-most upright shoot, 7–17 cm long, 1–1.5 (–1.7) mm thick exclusive of the trophophylls, producing 1 or rarely 2 upright shoots. Upright shoots 8–16.5 cm tall, 1–1.8 mm thick, with appressed ascending trophophylls. Trophophylls of horizontal shoots 4–6.8 (–7.8) × 0.7–1 mm, with (0–) 2–5 (–6) teeth per margin. Trophophylls of upright shoots varying from entire to toothed, usually those near the base of the shoot with abundant teeth, gradually becoming few toothed or even entire upward on shoot. Strobili 14–49 × 6–9 (–10) mm, representing 13 to 37 percent of the total upright shoot height. Sporophylls (4.6–) 4.9–6.7 mm long, loosely appressed to ascending, with 0–2 (–3) teeth per margin. Spores 53.4–53.6 μm (based on 2 measurements by Anton Reznicek, University of Michigan).

Horizontal stems. 7–17 cm × 1–1.5 (1.7) mm., 0.3–0.8 mm diam. without leaves, creeping, unbranched, flat on the ground; trailing stems usual-

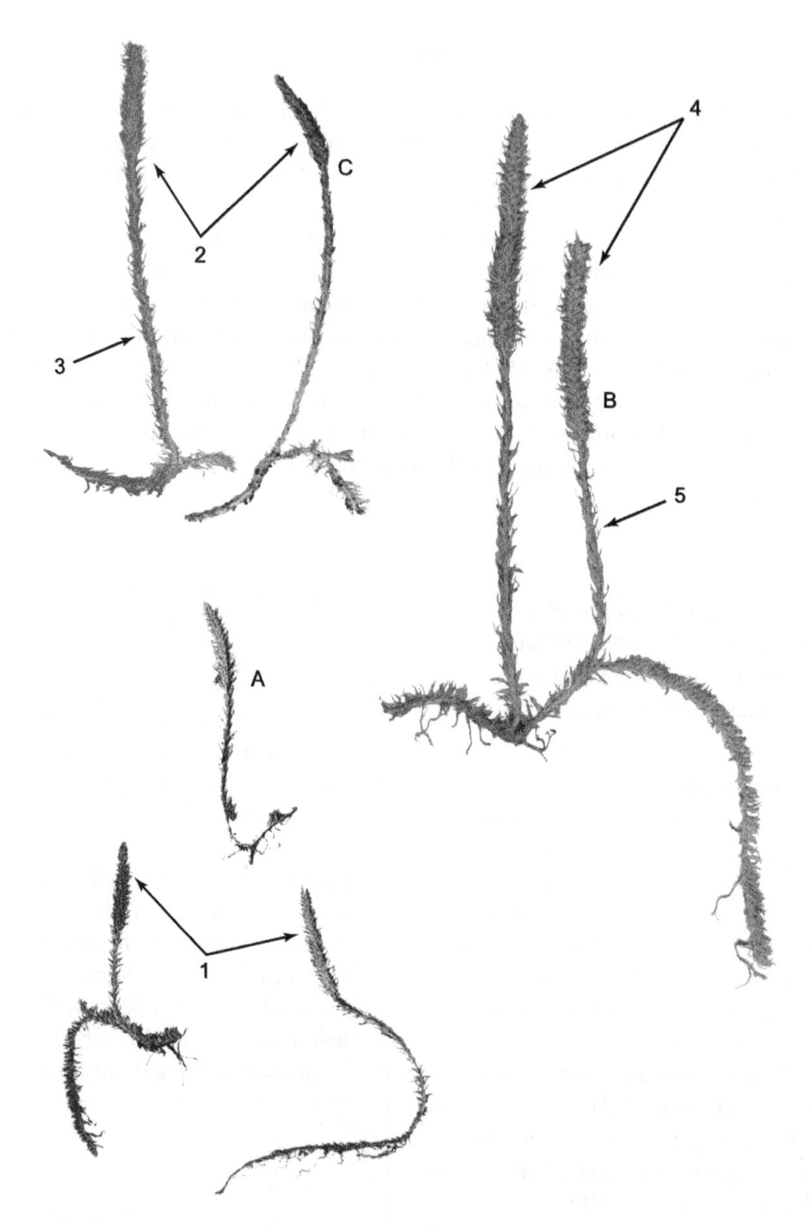

Lycopodiella inundata: Northern bog clubmoss: A.

Lycopodiella margueritae: Marguerite's clubmoss: B.

Lycopodiella subappressa: Northern appressed clubmoss: C.

Informative arrows: 1. Strobili 1/3 to 1/2 length of upright stem. 2. Strobili 1/8 to 1/3 length of, and only slightly wider than, upright stem. 3. Leaves of upright stem strongly appressed. 4. Strobili 1/3 to 2/3 length of upright stems, 1 1/2 to 2 times as wide as stalk. 5. Leaves initially nearly perpendicular to stem, then incurved at tips. Scale bar for all: The 3 plants in the lower left-hand corner are 5 cm long, and the rest are in proportion to that.

ly spread to form extensive clones; slender roots emerging on underside of stem; leaves linear, 4–6.8 (–7.8) × 0.7–1 mm, ascending, teeth absent.

Upright stems. 8–16.5 cm long, 1–1.8 mm diam. without leaves, 1 (rarely 2) per season, unbranched, green; leaves 4–6 × 0.7–1 mm, strongly appressed to ascending, close, narrow marginal teeth (0–) 2–5 (–6) teeth each side.

Strobili. 1.4–4.9 cm × 6–9 (–10) mm, **only slightly thicker (+ 0–2 mm) than the upright stem; comprises 13 to 37 percent of upright stem length;** appearing in late summer and persisting through November; **leaves** 3–4 × 0.2–0.5 mm, loosely appressed to ascending, margins vary with abundant teeth near the base of the shoot gradually becoming few toothed or entire upward on shoot.

Habitat. Moist sandy and pond shores, acidic peaty sands, seasonally flooded wetlands, sunny shallow sandy depressions in glacial lake-plain landscapes, ditches, and borrow pits.

Distribution. Indiana, Michigan, and Ohio. Because it was recently described, the full geographic distribution of *L. subappressa* is not known.

Chromosome #: $2n = 312$, tetraploid

In Michigan *Lycopodiella subappressa* is a species of special concern and uncertain status; it is not legally protected. It is considered imperiled in the state as well as globally.

Lycopodiella subappressa is believed to be a hybrid derived from *L. inundata* and *L. appressa* ($2n = 156$, not found in Michigan).

Specimens of this species were probably previously identified as *L. indundata.*

Species of *Lycopodiella* hybridize freely, and when the parent species have the same ploidy level the hybrids are fertile to some unknown degree. *Lycopodiella subappressa* can hybridize with *L. margueritae* and *L. inundata.*

Lycopodiella subappressa **is characterized by the combination of tall fertile shoots with short strobili and appressed sterile and fertile leaves that lack teeth.**

Lycopodiella subappressa **is similar to** *L. inundata.* **It differs in that its fertile and sterile leaves are appressed to strongly appressed. The fertile erect stems are taller (8–16 cm tall) than those of** *L. inundata* **(3.5–6 cm tall). The strobili are 20 to 33 percent of the total length of the erect stem versus 33 to 50 percent for** *L. inundata.*

Lycopodiella subappressa **can be distinguished from** *L. margueritae* **by its thinner and shorter strobili, mostly 1/5–1/3 of the total length of upright shoots (vs. 1/3 to 1/2 for** *L. margueritae*)**; shorter upright stems (8–16.5 cm long vs. 13–17 cm long for** *L. margueritae*)**; and leaves more ascending on its strobilus stalks and rhizomes. Its strobili are only slightly wider than the upright stem (vs. 1 1/2to 2× wider for** *L. margueritae*)**.**

	L. subappressa	*L. margueritae*	*L. inundata*
Horizontal stems			
Length × width without leaves	7–17 cm × 1–1.5 (1.7) mm	10–18 cm × 1.8–2.2 (–3) mm	3–12 cm × 0.5–1 mm
Leaf length	4–6.8 × 0.7–1 mm	6–13 × 0.8–1.2 mm	5–6 × 0.5–0.7 mm
Leaf teeth	(0)	3–4	absent to rare
Leaf orientation	strongly ascending	spreading (perpendicular)	spreading to upcurved
Upright fertile stem			
Height × width	3.5–9 cm × 1.8 mm	13–17 cm × 1–1.6 mm	3.5–9 cm × 3–4 mm
Number per plant	1	1 (–2)	1 (–2)
Upright fertile stem below strobilus			
Width without leaves	1–1.8 mm	0.3–0.7 mm	3–4 mm
Leaf length × width	4–6 (–7.8) × 0.7–1 mm	5–6 × 0.4–0.5 mm	3.5–6 × 3–4 mm
Leaf teeth	variable, abundant at base spare to absent above	absent	rare
Leaf orientation	strongly appressed	spreading (incurved tip)	spreading
Strobilus			
Length × width	1.4–4.9 cm × 0.6–0.9 mm	5–8 cm × 4–9 mm	1.1–2.0 cm × 2.5–5.5 mm
Length relative to upright stem	13–37%	33–66%	33–50%
Thickness relative to upright stem	minimally thicker	1 1/2 to 2× as thick 2–4 mm wider	1/3–1/2× as thick
Strobilus teeth	0–2	0–2	rare
Leaf orientation	loosely appressed to ascending	spreading, becoming incurved to appressed	spreading-ascending
Chromosome #	$2n = 312$, tetraploid	$2n = 312$, tetraploid	$2n = 156$, diploid
Range	MI, OH, IN	MI, OH, IN, PA, VA, CT	Circumboreal

LYCOPODIELLA HYBRIDS IN MICHIGAN

All species of Michigan *Lycopodiella* grow in similar habitats. Where they grow together they are likely to produce sterile triploid hybrids that would be recognized by intermediate morphology and abortive spores. Hybrids between the two tetraploid species could be found and might be fertile.

Three hybrids have been recognized.

L. inundata × *L. margueritae*
L. inundata × *L. subappressa*
L. margueritae × *L. subappressa*

LYCOPODIUM Linnaeus Lycopodiaceae

Common Clubmosses

Etymology. Greek, *lycos*, wolf, + *-poda*, foot or claw, alluding to the resemblance of the pawlike creeping stem of *L. clavatum* (the type species of the genus) to a wolf's foot. The common name derives from the clublike strobili and the mosslike quality of the stems.

Plants. Monomorphic; evergreen.

Horizontal stems. Unequally forked, resulting in both creeping horizontal rhizomes and upright branched stems, the upright stems appearing alternately right and left, **both covered with spirally arranged, linear leaves;** roots emerging sporadically from undersides of horizontal stems, not associated with upright branches.

Leaves. On upright and horizontal stems, monomorphic, narrow, linear, spreading at bases, then curving inward, with long hairlike tips, imbricate or not.

Strobili. Multiple, **arising from long, stalks;** unequally branched main stem or single; **fertile leaves (sporophylls)** aggregated into strobili, **shorter and narrower than sterile leaves, ovate-lanceolate, margins serrate to dentate,** gradually reduced to hairlike tips.

Sporangia. Reniform, borne in axils of fertile leaves.

Gemmae and gemmiphores. Absent.

Gametophytes. Subterranean; mycorrhizal; "walnut-shaped."

Chromosome #: $2n = 68$

Two species are treated in *Flora of North America North of Mexico*, vol. 2 (1993) within *Lycopodium sensu latu* in a *Lycopodium clavatum* group. Both species are found in Michigan.

Lycopodium is distinguished from *Huperzia* and *Lycopodiella* by its unequal forking and long trailing stems, which root on the undersurface and produce erect shoots alternately left to right on the main stem. The sporangia are born in cones with shield-shaped sporophylls very different from the leaves on the stems.

In the past, treatments of these 2 species relied on a single character, a single versus multiple stobili, to differentiate them. There are other characters, and the 2 species have been found growing together in southern Michigan while maintaining their distinctions.

Many treatments include the species of such genera as *Dendrolycopodium*, *Diphasiastrum*, and *Spinulum* in *Lycopodium*, making it difficult to establish a worldwide estimation of the number of species in the genus.

The spores of *Lycopodium* are full of oil droplets and when finely dispersed in air are very flammable. *Lycopodium* powder is a fine yellow powder mostly derived from the spores of *Lycopodium clavatum*. When a lighted match is dropped into a pile of this powder it does not burn. However, when the powder is dispersed into a fine mist near a flame, it ignites into a spectacular fireball. *Lycopodium* powder was used for flash photography before the advent of flashbulbs. It was used by magicians to create brief, explosive balls of fire in their theatrical pyrotechnics, a use now largely replaced with chemical flash powders. This phenomenon can be produced by collecting spores in the fall and, carefully, pouring or blowing them over a lighted candle or tossing them over a campfire. The substance was once called dragon's breath or, in Europe, witch's flower. (YouTube has several good demonstrations of this spectacle.)

In the past many medicinal uses were ascribed to the *Lycopodium* spores. They were used to treat bladder infections and inflammations, as a diuretic, for the "flushing of stones and gravel from the kidneys," and to treat gallbladder problems, gout, and rheumatoid arthritis—and that's just the beginning. None of the claimed benefits survives scientific scrutiny.

In Sweden it was believed that this plant would protect people and livestock from trolls if it was placed over stable doors or worn as a waistband on midsummer night.

Lycopodium powder was used to dust materials that might stick together. Rubber gloves were dusted when they were first used in surgery. The practice was discontinued when some patients developed foreign-body granulomas in the surgical wounds as a reaction to the spores. The powder was also used to dust early gelatin capsules so they would not stick together; in pancake makeup; and to dust babies' bottoms, nonlubricated condoms (granulomas have been reported with this use), and anal suppositories.

KEY TO THE SPECIES OF *LYCOPODIUM* FOUND IN MICHIGAN

1. Strobili (1) 3–6, each stalked and well separated along main stalk, not appearing to be fused at base. . . . . *L. clavatum*
1. Strobilus mostly single at tip of stalk or occasionally 2 strobili at tip of stalk appear to be fused at bases. . . . .*L. lagopus*

Lycopodium clavatum Linnaeus
Common Clubmoss, Running Pine, Running Clubmoss

Etymology. Latin *clavus*, club or cudgel, alluding to the shape of the fertile spikes with their stobili. The common name derives from the Latin name and alludes to the clublike strobili.

Growth habit. Evergreen, monomorphic.

Horizontal stems. Similar to upright shoots, low-creeping **on soil surface,** branching, overlapping, annual constrictions present, often extending a meter or more in a year, covering large areas.

Upright shoots. Alternate, to 25 cm long, 6–12 mm diam. including leaves; round, **annual constrictions obvious and abrupt.**

Lateral branches. 3–6 spreading to ascending branches; leaves 3–6 × 0.4–0.8 mm, long-triangular, in 10–20 ranks, **tapering to long, soft, thin, colorless hair tips 2.5–4 mm long;** spreading, becoming more appressed in distal 1/3 of branches; margins entire; green.

Strobilus stalks. 8–25 × 0.5–0.8 cm; naked; **3–6 alternate branches bearing strobili (rarely unbranched),** with few scattered appressed leaves.

Strobili. 15–25 × 3–6 mm; 2–5 on alternate stalk branches (occasionally single-stalked), cylindrical, plants in shade have fewer, smaller strobili; sporophylls 1.5–2.5 × 1.5–2.2 mm, straw-colored, abruptly tapered to colorless hair tips (these tips deciduous).

Habitat. Open fields and thickets, moist, shaded, well-drained woods, open coniferous forests, mixed woods, and swamp edges.

Distribution. Circumboreal. Europe. Africa. Asia. West Indies. Central and South America (in the tropics found on high mountains). Pacific Islands (not found in Australia). Canada: Alaska to Newfoundland. USA: Canadian border south to California, Idaho, Montana, Minnesota, and North Carolina. Generally more southern in its distribution than *L. lagopus.*

Chromosome #: $2n = 68$

Lycopodium clavatum **may be distinguished from** *L. lagopus* **by its branched (rarely unbranched) strobilus stalks, each branching with multiple strobili (rarely single). If 2 strobili are present on** *L. lagopus* **they are attached at their bases. (They are on separate strobilus stalk branches on** *L. clavatum.***)** *Lycopodium clavatum* **has more stem branches, 3–6 versus 2–3, and longer leaves, 4–6 mm versus 3–5 mm, than** *L. lagopus.*

Lycopodium lagopus (Laestadius) G. Zinserling

Lycopodium clavatum Linnaeus var. *lagopus* Laestadius; *L. clavatum* var. *monostachyon* Hooker & Greville.

One-Cone Clubmoss

Etymology. Greek *lagopus,* hare-footed.

Growth habit. Monomorphic; evergreen.

Horizontal stems. Horizontal stems trailing on or near surface;

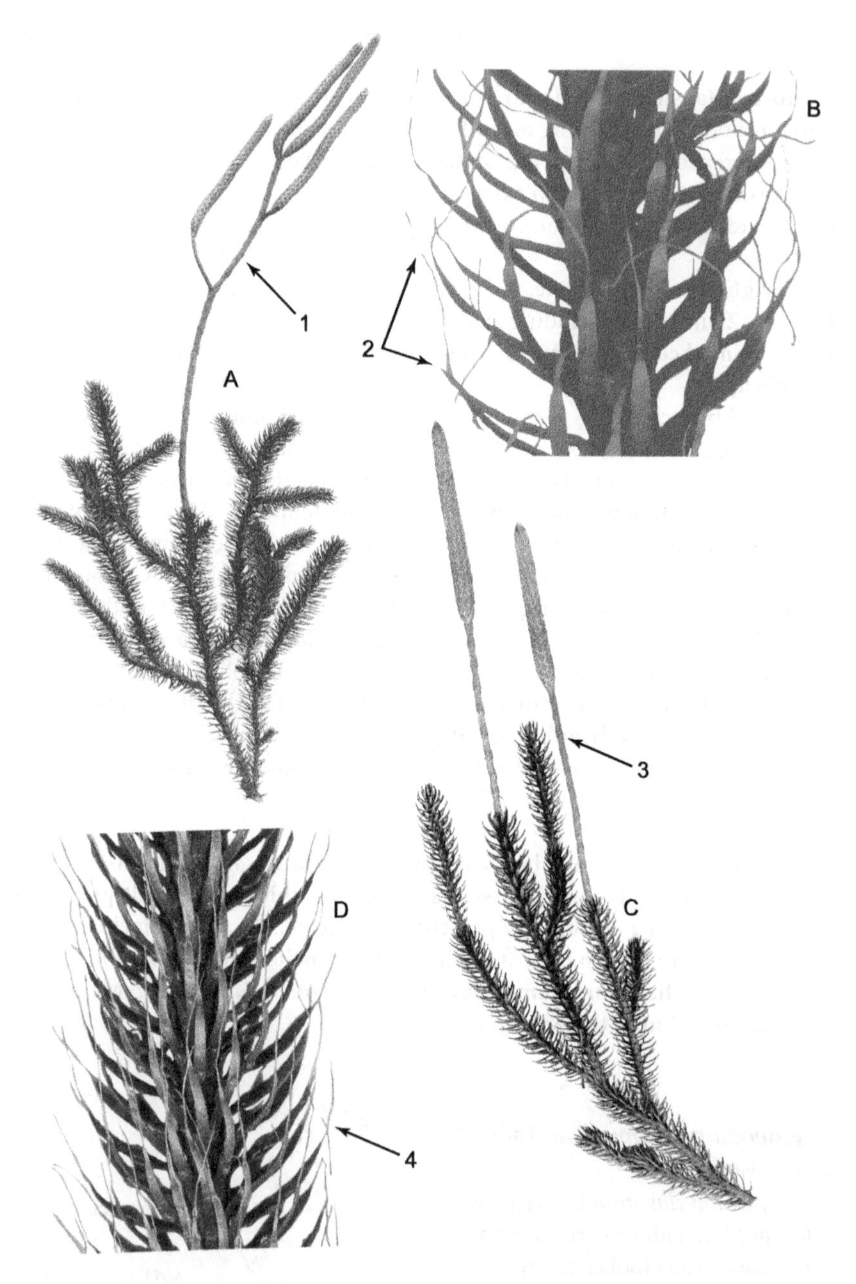

Lycopodium clavatum: Common clubmoss: A. Stem with branching strobili. B. Portion of stem magnified.

Lycopodium lagopus: One-cone clubmoss: C. Stem with solitary stobilus. D. Portion of stem magnified.

Informative arrows: 1. Strobili stalk-branching. 2. Somewhat spreading sterile leaves with hairlike tips. 3. Solitary strobilus on unbranched stalk. 4. Somewhat appressed sterile leaves with hairlike tips.

branching, branches overlapping; rooting at intervals; annual constrictions present.

Upright shoots. 10–25 × 0.5–0.8 cm; (1–) 2–3 branches mostly in lower 1/2; clustered; erect, abrupt and conspicuous annual constrictions.

Lateral branches. 1–3 (–4); ascending to erect; similar to upright shoots, annual constrictions present; leaves 3–5 × 0.4–0.7 mm; ascending to appressed; margins entire, **tapering to long, soft, thin, colorless** hair tips 1–3 mm long; medium green.

Strobilus stalks. 3.5–15 cm long; **unbranched,** with few scattered appressed leaves.

Strobili. 20–55 × 3–5 mm; **normally solitary, rarely double, if double attached (sessile) at base;** sporophylls 1.5–3 mm long, long triangular, gradually tapering to hairlike tips; yellow-green.

Habitat. More or less exposed grassy fields and woodland openings.

Distribution. Eurasia. Greenland. Canada: British Columbia to Labrador and Prince Edward Island. USA: Alaska and the upper tier of states from New England to Minnesota.

Chromosome #: $2n = 68$

Lycopodium lagopus grows generally farther north than *L. clavatum.* When the species grow side by side they maintain their distinctive characters.

Lycopodium lagopus, **while very similar to** *L. clavatum,* **may be distinguished from it by its unbranched strobilus stalks with single strobili. Occasionally when 2 strobili are present they are attached to each other at their bases (on** *L. clavatum* **they are rarely single and are on separate, distinct, strobilus stem branches).** *L. lagopus* **has fewer branches, 2–3 versus 3–6, and shorter leaves, 3–5 mm versus 4–6 mm, than** *L. clavatum.*

SPINULUM A. Haines Lycopodiaceae

Bristly Clubmosses, Interrupted Clubmosses, Stiff Clubmosses

Etymology. Latin, *spinula,* diminutive for spine (a spinule), alluding to the sterile leaves, each of which is tipped with a minute, firm spine. The common name refers to the bristly spine at the leaf tips or the annual winter constriction on the upright stems.

Plants. Evergreen; monomorphic, with superficial horizontal shoots and branched upright shoots.

Horizontal shoots. Superficial, linear, mostly unbranched, producing circa 8 alternate upright shoots a year.

Upright shoots. Forked near the base into more or less equal branches, stiff, prickly to the touch, round in cross section; leaves (trophophylls) in 8–10 ranks, **each tipped with a minute, firm spine.**

Strobili. Sessile at the tips of upright shoots; sporophylls stramineous.
Gametophytes. Subterranean; walnut-shaped.
Chromosome #: $x = 34$

One species of *Spinulum* is treated in *Flora of North America North of Mexico*, vol. 2 (1993) (as *Lycopodium annotinum*). *Spinulum canadense* is treated there as an environmentally induced form of *S. annotinum*. *Spinulum canadense* is now clearly recognized as a distinct species. Both are found in Michigan.

KEY TO THE SPECIES OF *SPINULUM* FOUND IN MICHIGAN

1. Leaves spreading or somewhat reflexed; longest leaves 6–9.8 mm long, toothed near tips (at least obscurely) .... *S. annotinum*
1. Leaves (all but the lowest) ascending; longest leaves 4.3–5.5 (–6) mm long, margins entire .... *S. canadense*

Spinulum annotinum (Linnaeus) A. Haines

Lycopodium annotinum Linnaeus
Bristly Clubmoss, Stiff Clubmoss, Interrupted Clubmoss

Etymology. Latin *annotinus*, a year old ("in yearly growths" in botanical Latin), referring to the constrictions between the annual zones of growth on the aerial branches. The common names derive from the stiff and prickly upright stems with annual constrictions (interruptions).

Plants. Evergreen, monomorphic; clonal.

Horizontal stems. 1.5–2.2 mm wide without leaves, creeping on surface or hidden under litter, branching; often with annual constrictions present.

Upright shoots. 15–28 cm tall × 1.0–1.8 cm diam., clustered, mostly unequally branched at base, upswept to erect, **stiff, bristly**, round in cross section, **annual constrictions conspicuous, abrupt,** green to yellowish-green; leaves 3–8 × 0.6–1.2 mm (longest leaves 6–9.8 mm long), much smaller at annual constrictions, linear-lanceolate to oblanceolate, bases narrowed, spreading, **stiff,** marginal teeth obscure to shallowly toothed, mostly in distal 1/2, **tips with sharp-pointed, small stiff spine,** leaves at annual constrictions with stomata on both surfaces, other leaves lacking stomata on the adaxial surface, vein inconspicuous.

Strobili. 15–43 × 4–7 mm, **solitary, sessile on stem tips,** abruptly narrowed to pointed tips, leaves 3–4.4 × 2–2.6 mm; ovate-triangular, abruptly tapered to pointed tips, yellow-tan when mature; mature from late July to early October.

Habitat. Dry to cool, moist, shady mature coniferous, northern hardwood, and mixed hardwood forests, open grassy sites, and wooded swamps.

Distribution. Circumboreal. Greenland. Canada: Labrador to Alaska. USA: Canadian border south to the western United States (excluding California and Nevada), Arizona, and New Mexico, northern tier of states excluding North Dakota, New England south to Virginia, Kentucky, and Tennessee.

Chromosome #: $2n = 68$

Spinulum annotinum is widespread and in varied habitats it has shown diverse, probably environmentally induced, forms.

It is sometimes confused with *Huperzia lucidula*. *Spinulum annotinum* **has strobili and spines on the tips of its leaves, while** *H. lucidula* **lacks strobili and spiny leaf tips and has gemmae that are lacking in** *S. annotinum*.

Spinulum annotinum **may be separated from** *S. canadense* **by its longer, reflexed sterile leaves and mostly longer strobili.**

Spinulum canadense (Nessel) A. Haines

Lycopodium canadense Nessel, *Lycopodium annotinum* Linnaeus spp. *pungens* Hultén

Northern Bristly Clubmoss, Northern Stiff Clubmoss, Northern Interrupted Clubmoss

Etymology. Canada, + *-ense*, a Latin suffix indicating country or place of growth, alluding to a frequent home of the species. The vernacular name derives from its more northern distribution compared to *S. annotinum*.

Plants. Evergreen; monomorphic; clone forming.

Horizontal stems. 1.2–2.3 mm wide without leaves, creeping on surface or hidden under litter, branching, annual constrictions present and prominent.

Upright shoots. 7–17 (–27) cm tall × 0.6–1.0 (–1.1) cm wide with leaves, round in cross section, clustered mostly unequally 1 to 4× branched at base, annual constrictions present and prominent; leaves 3–5.9 × 0.8–1 mm (longest leaves 4.3–5.5 [–6] mm long), less than 6× as long as wide, narrow-lanceolate to narrow-ovate or triangular, stiff, tips with a short stiff spine, margins entire or very obscurely

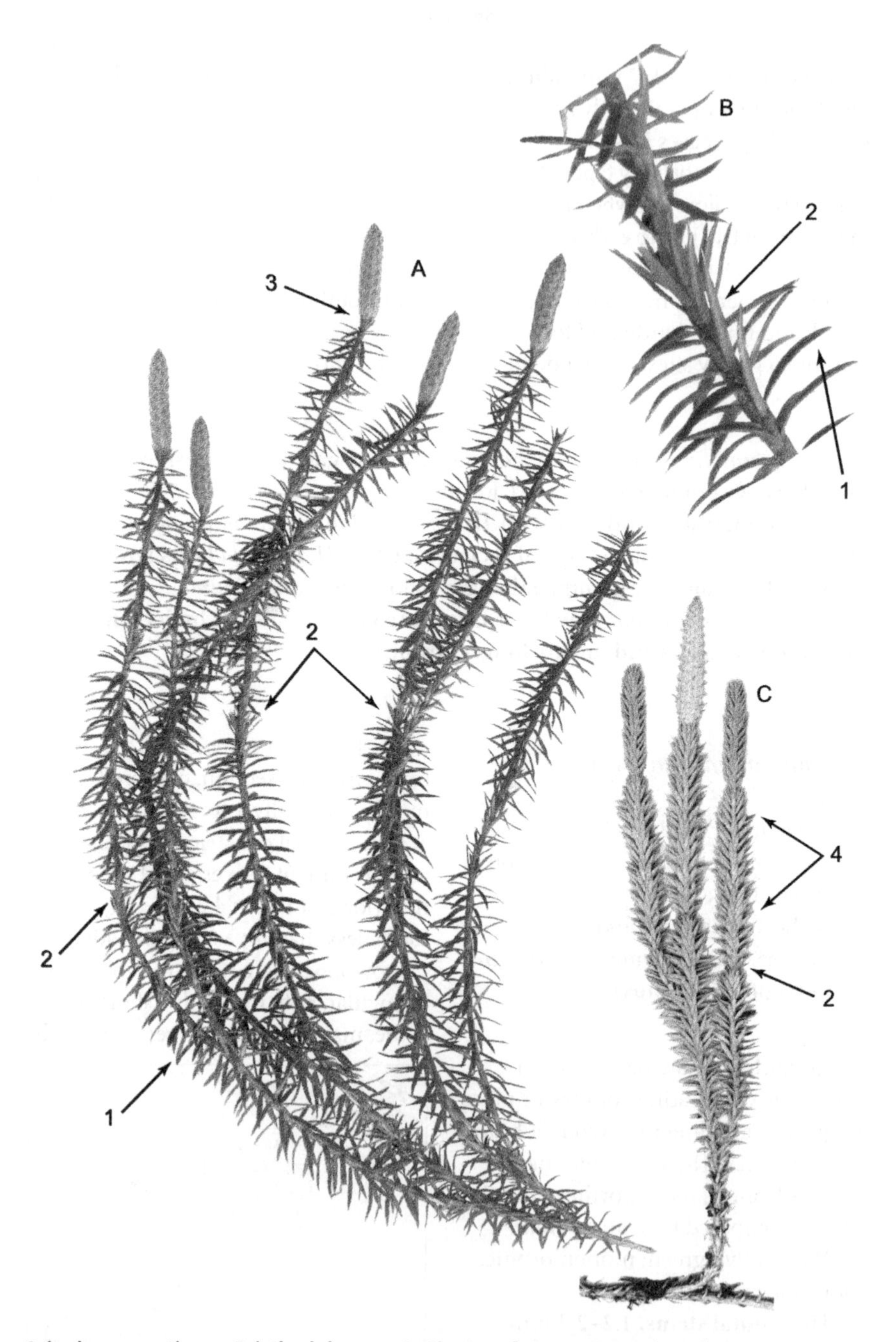

Spinulum annotinum: **Bristly clubmoss: A. Cluster of stems. B. Portion of stem magnified.**

Spinulum canadense: **Northern bristly clubmoss: C. Cluster of stems.**
Informative arrows: 1. Leaves spreading with stiff, pointed, spiny tips. 2. Annual winter constrictions 3. Strobili sessile. 4. Leaves appressed.

toothed, all but the lowest ascending, stomata present on both adaxial and abaxial surfaces of all leaves (reported to have mostly more than 25 stomata on the upper surfaces while *S. annotinum* usually has none).

Lateral branches. Branched 1 to 4×, round.

Strobili. 8–17 (–21) × 4.2–6 mm, **sessile**, sporophylls 2.9–3.7 × 2.2–2.5 mm, abruptly tapering to tips.

Habitat. Dry, often open, sandy or rocky woods and shores, rarely in cool peatlands. Boreal, alpine, and subalpine open habitats in northern New England.

Distribution. Canada: Yukon, Alberta to the Atlantic Ocean. USA: Alaska, Minnesota, Wisconsin, Michigan, West Virginia, Pennsylvania, New York, and New England.

Chromosome #: ?

Spinulum canadense is closely related to *S. annotinum*, which is usually found usually at higher elevations and latitudes.

The appressed leaves and narrower diam. of the erect stem, as well as shorter strobili, distinguish it from *S. annotinum.* **A more difficult character to utilize is the presence of mostly more than 25 stomata on the upper surfaces of the leaves while** *S. annotinum* **usually has none.**

HYBRIDS

Hybrids between *Spinulum annotinum* and *S. canadense* have been recognized in New England. They have not yet been recognized in Michigan.

SELAGINELLACEAE Spikemoss Family

SELAGINELLA P. Beauvois Selaginellaceae

See Selaginellaceae in appendix for a description of the family and a key to the subgenera.

Spikemosses

Etymology. *Selaginella* from the Latin *selago*, an ancient name for some species of *Lycopodium*, + *-ella*, a diminutive suffix—a small *Lycopodium*.

Plants. Heterosporous, herbaceous, annual or perennial, terrestrial or epipetric.

Stems. Prostrate, leafy, creeping, ascending or erect; slender, producing rhizophores (leafless descending shoots producing roots at bases) or roots intermittently, sometimes tufted, slightly to greatly branched, rhizophores absent or present with roots branching from tips, borne on upper surface of stems throughout or confined to stem bases; leaves on aerial stems (trophophylls) dimorphic or monomorphic, appressed or ascending, if monomorphic all alike, arranged spirally, small, simple, circular to linear-lanceolate, overlapping, with single vein; if dimorphic of 2 kinds and arranged in 4 ranks (2 lateral and spreading, 2 medial, linear-lanceolate to ovate); small ligules present at abaxial bases.

Roots. Branching several times dichotomously from rhizophore tips (if present).

Strobili. Composed of differentiated fertile leaves (sporophylls), quadrangular, flattened or cylindrical.

Sporangia. Short-stalked, solitary, of 2 kinds borne at adaxial bases of sporophylls, ovoid to reniform microsporangia containing thousands of male microspores, and 4-lobed megasporangia containing 4 female megaspores.

Spores. Megaspores (127–) 200–1,360 μm diam., ovoid or globose, variously ornamented; microspores 20–75 μm diam., variously ornamented; borne in separate sporangia but on same strobilus.

Habitat. Diverse, dry rocky ledges, limestone, granite, sandstone, wet places, mossy banks, bogs, lakeshores, sandy meadows, pastures, and open woods.

Chromosome # (in Michigan): $x = 9$

Selaginellaceae is a worldwide, primarily tropical and subtropical family with 1 genus containing over 700 species in 3 to 5 subgenera. Three subgenera and 38 species are treated in *Flora of North America North of Mexico*, vol. 2 (1993). It is represented in Michigan by 3 subgenera—subg. *Selaginella*, subg. *Tetragonostachys*, and subg. *Stachygynandrum*, each with 1 species.

Selaginellaceae has 2 distinct evolutionary lines that could possibly be treated as separate genera, one with species without rhizophores and with round strobili (subg. *Selaginella*) and the other with rhizophores and squarish strobili (all the other subgenera).

The Michigan *Selaginella* species are mostly small and inconspicuous and resemble mosses. They are difficult to find and usually noticed when one is looking for another plant.

Spores of the genus *Selaginella* are discharged by means of an ejection mechanism, the "compression and slingshot megaspore ejection."

Observation of spore ejection in 1 species (*Selaginella martensii*) showed the micro- and megasporangia opening slowly and separating into 2 ovoid valves, which subsequently dry and close suddenly in a quick movement that ejects the spores. The microspores reached up to 5–6 cm from the sporangia while megaspores were shot up to 65 cm.

KEY TO THE SPECIES OF *SELAGINELLA* FOUND IN MICHIGAN

1. Trophophylls dimorphic (with 2 forms), arranged in 4 ranks, 2 lateral ranks with larger leaves and 2 medial (above and below the stem); rhizophores present; stems spreading, forming thin, flattened mats on moist soils .... *S. eclipses*

1. Trophophylls monomorphic, spirally arranged, in distinct ranks or not; rhizophores present or absent; stems erect to ascending (2).
2. Strobili quadrangular; sporophylls usually appressed; trophophylls thick or fleshy (seldom thin), rigid, margins dentate, serrate with short bristle at tips; rhizophores present; found in dry sandy or rocky habitats . . . . *S. rupestris*
2. Strobili cylindrical; sporophylls spreading; trophophylls thin, soft, lax, margins short-spiny with scattered bristlelike teeth; rhizophores absent; found in wet habitats . . . . *S. selaginoides*

Selaginella eclipes **W. R. Buck (*Selaginella* subg. *Stachygynandrum* [Beauvois] Baker)**
Buck's Meadow Spikemoss, Hidden Spikemoss

Etymology. The epithet author stated that *eclipes* is of Greek derivation, meaning overlooked. He may have chosen this name because the species was "overlooked" and included in the treatments of *Selaginella apoda* or because it is an inconspicuous plant, easily overlooked.

Plants. Terrestrial.

Stems. Usually 2.5–15 cm long, up to 3.5 mm wide, short-creeping, usually highly branched, prostrate, forming loose to dense mats, branches 1-forked to 2-forked, not jointed, forming small mosslike mats; leaves In 4 ranks, 2 lateral and 2 median (1 upper and 1 lower), yellow-green, papery, *lateral leaves* 1–2 × 0.5–0.75 mm, nearly perpendicular to stem, ovate to ovate-elliptic, bases rounded, margins entire, dentate, serrate or ciliate, margins slightly transparent, tips acute; *upper and lower leaves* 1–1.8 × 0.4–0.8 mm, ovate to ovate-lanceolate or lanceolate, base rounded to oblique, margins serrate, tips abruptly tapered, long-acuminate to bristled, frequently transparent, midribs extend into tips, green.

Rhizophores. 06–0.1 mm diam., always present, translucent, long, thin, borne throughout stem length.

Strobili. 1–4 mm long, 4-sided, at base of some lateral stems, lax sometimes flattened, solitary or paired, sporophylls slightly larger than trophophylls, not closely packed, megasporophylls (usually lower on strobili), large sporophylls produce both megaspores and microspores in their axils, small sporophylls produce only microspores.

Megaspores. Reticulate with pitted surface and broad, low ridges, 330 μm diam., microspores diam. circa 22 μm.

Habitat. Moist to wet calcareous habitats, swamps, meadows, marly lakeshores and riverbanks, pastures,

and open woods, rarely on rock; full sun to semishade; disappears when overgrown with larger plants.

Distribution. Canada: Ontario and Quebec. USA: Arkansas, Illinois, Indiana, Iowa, Michigan, Missouri, New York, Oklahoma, and Wisconsin.

Chromosome #: unknown

Selaginella eclipes is placed in the subgenus *Stachygynandrum* with more than 650 species of mostly tropical and subtropical regions. Eleven species are treated in *Flora of North America North of Mexico*, vol. 2 (1993). It is represented in Michigan by 1 species.

Selaginella eclipes, a member of the *S. apoda* complex, is here treated, as in *Flora of North America*, as a species distinct from *S. apoda*. There is a taxonomic problem. It is very similar to *S. apoda*, and some would treat it as a subspecies of *S. apoda,* differing in dorsal leaf morphology and the sculpturing of the megaspores. Fortunately, the 2 species rarely grow in the same geographic range.

Hybrids between *S. eclipse* and *S. apoda* may occur.

S. eclipes **is a small, insignificant-looking plant that resembles a yellow-green moss and is frequently overlooked. It has dimorphic leaves arranged in 4 ranks. The 2 lateral ranks are larger than the upper and lower ranks. Its strobilus is 4-sided, and rhizophores are present. It is a mat former, but the mats are not large.**

Selaginella rupestris (Linnaeus) Spring (*Selaginella* subg. *Tetragonostachys* Jermy)

Lycopodium rupestre Linnaeus; *Stachygynandrum rupestre* (Linnaeus) P. Beauvois

Rock Spikemoss, Sand Spikemoss

Etymology. Latin *rupestris*, of rocks, alluding to a frequent habitat of this plant.

Plants. Monomorphic, evergreen, perennial.

Stems. 1–4 cm long, irregularly 1-forked to 3-forked, round, tufted, short-creeping, ascending to upright, stiff, main stem indeterminate, forming 1.5–6 cm high, cushionlike, spreading dense mats, sometimes quite large; leaves 2.5–4 (–4.5) × 0.45–0.6 mm, gray-green, bright green when fresh and moist, all the same size and shape, linear to lanceolate. (Leaves close to the soil surface are morphologically different from those on the side away from the soil surface. Here the leaves on the side of the axis away from the surface are called upper-side leaves, and those on the side toward the surface are called under-side leaves.) Bases wedge-shaped and de-

current on under-side leaves to rounded and adnate on upper-side leaves, tips slightly keeled, tightly appressed, overlapping, in spiral pseudowhorls of 6 on main stems to 4 on lateral branches, thick or fleshy (seldom thin); pubescent or glabrous; margins with transparent long cilia, tips bristled, white, or transparent; abaxial ridges distinct; green (occasionally reddish).

Rhizophores. 0.25–0.45 mm diam., thin, borne on upper side of stem lengths throughout.

Strobili. 5–35 mm long, sharply 4-sided, solitary at end of branches, sporophylls 4× as wide as sterile leaves, deltate-ovate to ovate-lanceolate, strongly tapering to long-bristled tips, or not, megasporophylls and microsporophylls same size, in 4 ranks; abaxial ridges well defined, base glabrous, margins ciliate to slightly dentate, appressed, base usually with 2 diverging flaps or auricles, auricles covering sporangia below, tips slightly keeled; sporulating July–October.

Megaspores. Yellowish-white.

Habitat. In open sun on well-drained rock exposures, rock crevices, frequently in barren situations that are only periodically moist, dry ledges, sea cliffs, limestone, sandstone, granite, sandy or gravelly soil, grassy meadows, and jack pine areas of the northern Lower Peninsula.

Distribution. Greenland. Canada: Alberta east to New Brunswick. USA: Canadian border south to all eastern states, Great Lakes states, midwestern states to Nebraska, south to northern Mississippi.

Chromosome #: $2n = 18$

Selaginella rupestris is in the subgenus *Tetragonostachys,* which has about 50 species worldwide. Of these, 26 species and 1 hybrid are treated in *Flora of North America North of Mexico*, vol. 2 (1993). The subgenus is represented in Michigan by a single species, *S. rupestris.*

The characters of *S. rupestris* are quite variable. Variation is seen in the hairiness of the leaf margins (hairs sometimes absent), the shape of sporophylls, sporangial distribution patterns in the strobili, the number of megaspores per megasporangium, and strobili with only megasporangia or also including microsporangia.

It tends to vanish when it is overgrown by woody species through succession.

Selaginella rupestris **very much resembles a thick, low-growing, mat-forming moss. It is easier to find when fresh, moist, and bright green than when it is desiccated and gray-green. It has rhizophores, which are lacking on mosses.**

Selaginella selaginoides (Linnaeus) Beauvois (*Selaginella* subg. *Selaginella* Beauvois)

Lycopodium selaginoides Linnaeus; *Selaginella spinosa* P. Beauvois

Northern Spikemoss, Club Spikemoss

Etymology. *Selaginoides* from the Latin *-oides*, a suffix suggesting similarity, referring to some resemblance of the plants to *Lycopodium selago*. The vernacular names describe its appearance.

Plants. Sterile stems perennial, fertile stems annual.

Horizontal stems. Up to 15 cm

long, filiform, 2 mm diam., including scattered leaves, indeterminate, tips not upturned, closely spaced, dichotomously branching, forming loose to dense mats; leaves 2.5–4.0 × 0.7–1.0 mm (smaller on horizontal stems, pointed, triangular, about 1–2 mm long, 1/3 size of those on upright stems), loosely spirally arranged, lanceolate to ovate-lanceolate, thin, soft, margins with soft, short spines, ascending, abaxial groove absent, base decurrent, forming saclike structure with stem, tips acuminate to awl-shaped, stomata scattered throughout abaxial surface; green.

Erect fertile shoots. Annual, unbranched, ending in simple strobili, round, bright green, later yellowish-green, more robust than the sterile stems, 2–6 (–10) cm tall × 2–6 mm diam., including leaves.

Rhizophores. Absent, roots attached laterally at the base of the shoots.

Strobili. (1–) 2–3 (–5) cm, cylindrical, solitary; leaves 4.5–6 × 1.15–1.5 mm; lanceolate-triangular, abaxial ridges absent, loosely spirally arranged, shape larger and slightly different from vegetative leaves; megasporophylls and microsporophylls present.

Megaspores. With flat warts on 3 flat surfaces, yellowish-white.

Habitat. Wet places, along mossy stream banks, lakeshores, bogs, rock and wet talus slopes, in neutral to alkaline soil, often on mossy hummocks among *Thuja occidentalis*.

Distribution. Eurasia. Canary Islands. Greenland. Throughout Canada. USA: Alaska, Montana east in upper tier of states to Maine and New York, southerly populations in Wyoming, Colorado, and Nevada.

Chromosome #: $2n = 18$

This species is in the subgenus *Selaginella*, which contains 2 species. It is represented in Michigan by 1 circumboreal species.

The second species, *S. deflexa*, is endemic to Hawaii and is very similar to *S. selaginoides*. It is infrequent but locally common and is found on all the major islands growing in open mossy bogs between 1,050 and 1,500 m elevation.

Selaginella selaginoides **may resemble spikes of moss. It is very different from the other Michigan** *Selaginella* **species and may be recognized by it smaller size; spikelike growth form; smaller, needlike leaves with marginal spines; strobili that are round and larger than the sterile stems; and its preference for damp habitats. It is often hidden in dense vegetation.**

Selaginella selaginoides: Northern spikemoss: A.

Selaginella eclipes: Hidden spikemoss: B.

Selaginella rupestris: Rock spikemoss or sand spikemoss: C.
Informative arrows: 1. Rhizophores. 2. Dimorphic leaves in 4 ranks: small upper and lower, larger lateral. 3. Needle-like leaves with marginal spikes. 4. 4-sided strobilus.

DESCRIPTIONS OF FERN FAMILIES AND KEYS TO THE GENERA AS THEY ARE FOUND IN MICHIGAN

These family descriptions include descriptive material that applies only to the families as seen in Michigan. In many cases these descriptions would have to be much changed to describe the families as seen worldwide.

The generic descriptions are found with the description of the genera and species in the text.

The families that contain only a single genus, where the generic and family treatments are identical, are treated in the main text under the generic treatment.

ASPLENIACEAE Newman Spleenworts

A family with a single genus, *Asplenium*. The treatment of the family is identical to the treatment of that genus. It will be found under the generic treatment of *Asplenium* in the main text.

A large family recognized here as with a single genus (some authors split the family into two or more genera) containing about 700 species genera worldwide, mostly in the tropics.

ATHYRIACEAE Roth Lady Ferns

Etymology. Greek, *athyros*, doorless, possibly referring to species in which the indusia opens slowly and tardily.

Plants. Monomorphic; clustered; medium-sized terrestrial ferns.

Rhizomes. Short to long-creeping, decumbent to erect, scaly.

Fronds. 1-pinnate-pinnatisect to 2-pinnate-pinnatifid; erect.

Stipes. Scaly; cross sections at the bases reveal ends of 2 ribbon-shaped vascular bundles (appearing as linear and parallel lines that unite upward, forming a single U- or V-shaped bundle in upper stipes.

Blades. 1-pinnate-pinnatifid to 2-pinnate-pinnatifid.

Veins. Free.

Sori. Linear or J-shaped along veins; indusia linear sporangia stalks 2-seriate to 3-seriate.

Flora of North America North of Mexico, vol. 2 (1993), treats the 2 genera recognized here as belonging to the family Dryopteridaceae. These 2 genera, each with 1 species, are found in Michigan.

In the past these 2 genera were also included in the family Woodsiaceae, but they are found segregated in all analyses.

1. Blades 1-pinnate-pinnatifid; grooves of pinna costae shallow, not continuous with grooves of rachises; costae with multicellular hairs; sori linear . . . . *Deparia* (*D. acrostichoides*) (Silvery spleenwort)
1. Blades 1-pinnate-pinnatisect to 2-pinnate-pinnatifid; grooves of pinna costae deep, continuous with grooves of rachises; costae lack multicellular hairs; sori linear or J-shaped, hooked over vein tip . . . . *Athyrium* (*A. filix-femina*) (Lady fern)

In the recent past the genus *Diplazium* (now *Homalosorus*) was included in the Athyriaceae. *Homalosorus* is now placed in the recently described family Diplaziosidaceae.

AZOLLACEAE Wettstein   Azollas

A family with a single genus, *Azolla*. The treatment of the family is identical to the treatment of that genus. It will be found under the generic treatment of *Azolla* in the main text.

BLECHNACEAE Newman   Chain Ferns

Etymology. Greek, *blechnon*, an ancient name for ferns in general.
Plants. Deciduous; monomorphic.
Rhizomes. Creeping subterranean.
Fronds. 30–180 cm long.
Stipes. With vascular bundles arranged in a U-shape.
Blades. Mostly pinnatifid to 1-pinnate, rarely 2-pinnate; rachises grooved, grooves not continuous with pinna rachises.
Veins. Anastomosing forming areoles near costae, free beyond; having 1 or more rows of areoles near costae when fertile.
Sori. Short and intermittent to long and continuous; often in paired lines on outer arches of areoles near costae; indusia covering sori, thin to thick, sometimes glandular.
Chromosome number. $x = 28–36$.

A worldwide family of about 10 genera. Two are treated in *Flora of North America North of Mexico*, vol. 2 (1993). It is represented in Michigan by 1 genus with 1 species: *Woodwardia*.

CYSTOPTERIDACEAE (Payer) Shmakov   Bladder Ferns

Etymology. Greek *cysto*, bladder, + Greek *pteris*, fern, from *pteron*, wing, feather—the bladder fern.

Plants. Deciduous; monomorphic; small to medium-sized.
Rhizomes. Slender long-creeping or short-creeping, or erect or ascending.
Blade. 1-pinnate to 3-pinnate.
Veins. Free; ending at the frond margins.
Sori. Small, round; *indusium* hoodlike, small, thin, ovate or oval, attached at bases, scalelike and often covered with mature sporangia, evanescent; or exindusiate.
Habitat. Forests or crevices.
$x = 40, 42$

KEY TO THE GENERA OF CYSTOPTERIDACEAE IN MICHIGAN

1. Indusia absent; blades mostly ternate (divided into 3 more or less equal parts) . . . . *Gymnocarpium* (Oak ferns)
1. Indusia hoodlike, opening outward, attached at bases, free at tips and margins, fragile and quickly disintegrating; blades pinnately divided . . . . *Cystopteris* (Bladder or brittle ferns)

A family with 4 genera worldwide. Two are treated in *Flora of North America North of Mexico*, vol. 2 (1993). These 2 genera are represented in Michigan by *Cystopteris* (5 species) and *Gymnocarpium* (3 species).

DENNSTAEDTIACEAE Ching Hay Scented Ferns

Etymology. Name honors August Wilhelm Dennstaedt (1776–1826), a German botanist.
Plants. Terrestrial, monomorphic.
Rhizomes. Long-creeping; often branching from buds on base of stipes; bearing jointed hairs; scales absent.
Fronds. Medium-sized to large.
Stipes. Close or remote, often with buds at base of petioles, vascular bundles U- or Ω-shaped.
Blades. Often large; 2-pinnate-pinnatifid to 3-pinnate-pinnatifid; hairy or glabrous; lacking scales.
Veins. Free; pinnate or forking in ultimate segment, ending in bulbous expansions short of margins or joined at margins in fertile segments.
Sori. Round or elongate; terminal on vein tips or on marginal connecting veins; marginal or submarginal.
Indusia. Tubular to cylindrical, formed by fusion of inner and outer laminar flaps to form a circular, small, delicate whitish cup; or marginal and linear covered by false indusia produced by a revolute margin, sometimes with true inner indusia sometimes partially covered by recurved projection of segment lobes.
Chromosome numbers. $x = 26, 29, 30, 31, 33, 34, 38, 46, 47, 48$.

The family Dennstaedtiaceae is characterized by submarginal or marginal sori usually with 2 indusia—an inner true indusium and an outer false indusium formed by the revolute, segment margin—indument usually of hairs rather than scales, and long-creeping rhizomes with rhizome buds on the bases of the petioles.

A nearly worldwide, mostly tropical, family variously circumscribed with between 8 to 20 genera with possibly 400 species. Four genera and 6 species are treated in *Flora of North America North of Mexico*, vol. 2 (1993).

The family is represented in Michigan by 2 genera, *Pteridium* (very common) and *Dennstaedtia* (very rare in Michigan), each with 1 species.

KEY TO THE GENERA IN THE DENNSTAEDIACEAE IN MICHIGAN

1. Sori continuous along margins of segments; inner indusium, if present, hidden by reflexed margin of blade and maturing sporangia; blades usually broadly triangular . . . . *Pteridium* (*Pteridium aquilinum*) (Bracken or Brake)
1. Sori distinct, round, not continuous along margins of ultimate segments; small, less than .5 mm in diameter, in sinus margins at vein tips, inner indusium present, outer indusium fused with inner one to form a cup or tubular structure; blades lanceolate to ovate-lanceolate . . . . *Dennstaedtia* (*Dennstaedtia punctilobula*) (Hay-scented fern)

DIPLAZIOPSIDACEAE X. C. Zhang & Christenhusz Glade Ferns

Etymology. Greek, *diplos*, twofold, + *opsis*, appearance, in reference to the back-to-back sori along veins in some species in this family.

A family with 2 genera. One *Homalosorus*, is found in Michigan. The treatment of that part of the family containing this genus is identical to the treatment of the genus found under the generic treatment of *Homalosorus* in the main text.

Diplzaiopsis, the other genus in this family, is found in Southeast Asia, and its description is not included here. It is quite different from that of *Homalosorus*.

Diplaziopsidaceae is a family with 2 genera, *Diplaziopsis* and *Homalosorus*. *Homalosorus* is treated in *Flora of North America North of Mexico*, vol. 2 (1993) as a *Diplazium* (*Homalosorus* had been customarily treated as *Diplazium pycnocarpon* in the family Athyriaceae). *Homalosorus pycnocarpos* is represented in Michigan by this monospecific genus and its single species.

Homalosorus pycnocarpos has been placed in the genus *Diplaziopsis* by Price (*Diplaziopsis pycnocarpa* [Spreng.] Price). The evidence for this change is strong, and the decision to decide which name to use was difficult. Future studies of the family may well confirm Price's choice. However, *Homalosorus pycnocarpos* varies from the other *Diplaziopsis* species by having non-anastomosing veins, narrow-creeping rhizomes and tapering pinnatifid tips, and different chromosome numbers.

DRYOPTERIDACEAE Herter Woodferns

Etymology. Greek *dryos*, oak or tree, + *pteris*, fern; the ferns of this genus often grow in woodlands.

Plants. Monomorphic or dimorphic; evergreen or deciduous.

Rhizomes. Creeping, erect or decumbent; often bearing old stipes or fronds, scaly; scales not clathrate.

Fronds. Small to medium-sized; erect to spreading, often forming rosettes.

Stipes. Usually close; cross section showing few to many small, round vascular bundles arranged in a semicircle.

Blades. 1-pinnate to 3-pinnate-pinnatifid; sometimes glandular or scaly; rachises grooved or not.

Veins. Free.

Sori. Round.

Indusia. Reniform or peltate.

Sporangia. Stalks with 3 rows of cells.

Chromosome number. $x = 41$.

A family of about 29 genera with worldwide distribution. It is represented in Michigan by 2 genera, *Dryopteris* (with 10 species) and *Polystichum* (with 3 species).

KEY TO THE GENERA OF DRYOPTERIDACEAE IN MICHIGAN

1. Pinnae and pinnules auriculate; margins usually sharply toothed with long, stiff, bristlelike projections, tips acute: grooves of rachises not connected with grooves of pinna costae; indusia peltate; blades stiff, 1-pinnate to 2-pinnate; never glandular . . . . *Polystichum*
1. Pinnae and pinnules not auriculate; with rounded margins and tips, with or without short teeth; grooves of rachis connected with grooves of pinna costae; indusia round or reniform; blades chartaceous; 1-pinnate-pinnatifid to 3-pinnate-pinnatifid; glandular or not. *Dryopteris*

EQUISETACEAE Michaux ex DeCandolle Horsetails, Scouring Rushes

A family with a single genus, *Equisetum*.

The treatment of this family is identical to the treatment of the genus. It will be found under the generic treatment of *Equisetum* in the main text.

LYGODIACEAE C. Presl Climbing Ferns

A family with a single genus, *Lygodium*.

The treatment of this family is identical to the treatment of the genus. It will be found under the generic treatment of *Lygodium* in the main text.

MARSILEACEAE Mirbel Water Clovers

A family with 3 genera. A single genus with 1 species, *Marsilea villosa*, is found in Michigan.

The treatment of this family includes only the genus found in Michigan. It will be found under the generic treatment of *Marsilea* in the main text.

ONOCLEACEAE Pichi Sermolli Sensitive Ferns

Etymology. Greek *onos*, vessel, + *kleiein*, to close, referring to the hard bead-like leaves enclosing the sori.

Plants. Strongly dimorphic; sterile fronds deciduous, fertile fronds becoming stiff and persistent over winter.

Rhizomes. Long- to short-creeping to ascending; sometimes stoloniferous (*Matteuccia*).

Fronds. Rhizomes long- or short-creeping to ascending.

Stipes. With 2 vascular bundles uniting distally into a gutter shape; blades pinnatifid to 1-pinnate-pinnatifid.

Veins. Free or anastomosing; lacking included veinlets.

Sori. Borne within inrolled hardened pinna margins.

Chromosome numbers. $x = 37, 39, 40$.

A family of 4 genera. Two genera and 2 species are treated in *Flora of North America North of Mexico*, vol. 2 (1993) (there treated as belonging to the family Dryopteridaceae). It is represented in Michigan by 2 genera, *Matteuccia* and *Onocleae*, each with 1 species. (Some authors would place both *Matteuccia* and *Onocleae* in the genus *Onoclea*.)

KEY TO THE GENERA OF ONOCLEACEAE IN MICHIGAN

1. Sterile leaves 1-pinnate-pinnatifid, 60–145 cm long, obovate, broadest toward the tips (ostrich-feather-shaped); fronds borne in dense clusters; veins free; fertile fronds 1-pinnate. . . . . *Matteuccia* (Ostrich fern)
1. Sterile leaves pinnatifid or 1-pinnate-pinnatifid at basal pinnae, 25–90 cm long, triangular, broadest near base; fronds scattered along rhizomes; veins netted; fertile fronds 2-pinnate . . . . *Onoclea* (Sensitive fern)

OPHIOGLOSSACEAE C. Agardh Adder's-Tongues

Etymology. Greek, *ophis*, snake, + *glossa*, tongue. The fertile spikes of some genera in this family suggest a snake's tongue.

Plants. Evergreen or deciduous; plants mostly small and fleshy, terrestrial, sometimes epiphytic.

Fronds. Usually 1 per rhizome, with common stalk divided into a sterile, laminate, photosynthetic portion (trophophore) and a fertile, spore-bearing portion (sporophore).

Trophophore blades. Compound to simple.

Sporophores. Pinnately branched or simple.

Upright stems. Short and mostly subterranean.

Roots. Fleshy, hairless, mycorrhizal.

Common stalk. Unbranched, upright.

Blades. Simple or divided.

Veins. Free, pinnate or fanlike, or anastomosing.

Sporangia. Relatively large with thousands of spores, opening by transverse slits, borne in 2 rows on fertile spikes arising from stipes or blades.

Gametophytes. Not green, subterranean, mycorrhizal.

Chromosome numbers. $x = 44, 45, 92, 1{,}200+$.

A family of about 6 genera with about 80 species in both tropical and temperate zones. Thirty-eight species in 3 genera (*Ophioglossum*, *Botrychium*, and *Cheiroglossa*) are treated in *Flora of North America North of Mexico*, vol. 2 (1993). *Botrychium* as treated there includes the genera *Botrychium*, *Botrypus*, and *Sceptridium* as treated in this book. (*Cheiroglossa*, a more tropical genus, is found only in Florida.) Ophioglossaceae is represented in Michigan by 4 genera, *Ophioglossum* (2 species), *Botrychium* (10 species), *Botrypus* (1 species), and *Sceptridium* (3 species).

The family Ophioglossaceae is not closely related to the other ferns treated in this book. Gross and microscopic anatomy and chemical (DNA and enzyme) studies have clearly demonstrated this separation. The subterranean gametophytes lacking chlorophyll are also distinctive.

In the past *Botrychium*, *Botrypus*, and *Sceptridium* were placed in the single genus *Botrychium*, but morphology and DNA studies have clearly shown that they should be treated as separate genera (Hauk, Parks, and Chase 2003).

KEY TO THE GENERA OF OPHIOGLOSSACEAE IN MICHIGAN

1. Sterile blades (trophophores) simple, ovate to elliptic-lanceolate, unlobed; veins netted; fertile blades (sporophores) erect, unbranched, arising from common stalks at bases of blades, with 2 rows of sporangia immersed in tissue of fertile spikes . . . . *Ophioglossum* (Adder's-tongue ferns)

1. Sterile blades lobed, up to 3-pinnate (or simple, less than 7 cm long in some *Botrychium simplex*); veins free; fertile spikes branched or unbranched, arising from ground level at the middle to distal portion of common stalk or at base of blades; sporangia mostly not immersed in tissue of fertile spikes (2).

2(1). Sterile blades (trophophores) mostly less than 15 cm long and less than 2.5 cm wide; simple to 2-pinnate; linear to ovate-lanceolate; erect; succulent; sporophores arising from middle to distal portion of common stalk, well above ground level; deciduous; sporophores almost always present . . . . *Botrychium* (Moonworts)

2. Sterile blades (trophophores) 5–45 cm long; up to 3-pinnate; deltate; bent horizontal to ground; leathery or papery; sporophores arising at ground level or at base of trophophores; sporophores present or not (3).

3(2). Fertile blades (sporophores) arise from base of trophophore blade high on common stalk; trophophore blade thin, herbaceous; deciduous . . . . *Botyrpus* (*Botrychium virginianum*) (Rattlesnake fern)

3. Fertile blades (sporophores) arise at or near the ground from a very short common stalk; trophophore blades herbaceous to leathery; leaf sheaths closed; evergreen . . . . *Sceptridium* (Grapeferns)

OSMUNDACEAE Berchtold & J. Presl Royal Ferns

Generic derivation of the family name is not clear. See the discussion of this name derivation for the genus in the *Osmunda* treatment in the main text.

A family of 3 genera and 16 to 36 species worldwide. One genus with 3 species and 1 hybrid is treated in *Flora of North America North of Mexico*, vol. 2 (1993). It is represented in Michigan by 1 genus, *Osmunda*, with 3 species.

POLYPODIACEAE Berchtold & J. Presl Polypodies

Etymology. Greek *polys*, many, + *podion*, little feet, alluding to short stumps left on the rhizome when the fronds abscise. (Some say the name alludes to its highly branched rhizomes, which resemble many little feet.)

The treatment of this family is identical to the treatment of the genus. It will be found under the generic treatment of *Polypodium* in the main text.

A family of about 40 genera and possibly 500 species worldwide, mostly in the tropics and subtropics. Seven genera with 25 species are treated in *Flora of North America North of Mexico*, vol. 2 (1993). A single genus, *Polypodium*, with 1 species, is found in Michigan.

PTERIDACEAE Reichenbach Maidenhair Ferns

Etymology. Greek *pteris*, fern, from *pteron*, wing, feather, in reference to the frond shape. An ancient name for ferns in general. In the past the name Adiantaceae has been used for this family.

Plants. Small to medium-sized terrestrial or epipetric.

Rhizomes. Erect to long-creeping; scales sparse, lanceolate, ovate-lanceolate or narrow.

Fronds. Monomorphic or dimorphic; usually erect; sometimes drought tolerant, curling up when dry.

Stipes. Green, dark brown, or black, glossy; 1 to several round vascular bundles variously arranged.

Blades. 2-pinnate to 2-pinnate-pinnatifid or pedate; indument on petioles, rachises, costae, and blades commonly of hairs or glands.

Veins. Free or joining at margins; forked.

Sori. Short-linear along margins at vein tips or long-linear along marginal commissural veins.

Indusia. Absent; sporangia often borne on undersides of recurved, revolute, or reflexed, often lighter-colored, membranous or leathery leaf margins, forming false indusia.

Chromosome numbers. $x = 29, 30$.

A major character defining the family Pteridaceae is its submarginal sori, which lack indusia and are protected by a reflexed or revolute blade margin.

A family of about 50 genera with about 1,000 species worldwide. Thirteen genera and ninety species are treated in *Flora of North America, North of Mexico*, vol. 2 (1993). It is represented in Michigan by 3 genera, *Adiantum* (1 species), *Cryptogramma* (2 species), and *Pellaea* (2 species).

KEY TO THE GENERA OF PTERIDACEAE AS FOUND IN MICHIGAN

1. Blades fan-shaped, semicircular; pinnules fan-shaped or with lower 1/2 (below vein) smaller or nearly absent; sori 3 mm or less long . . . .*Adiantum* (*Adiantum pedatum*) (Maidenhair fern)
1. Blades linear-triangular, deltate-oblong, or oblong-lanceolate; pinnules not fan-shaped, lower half of pinnae (below vein) equal to upper half; sori mostly more than 15 mm long (2).
2. Plants epipetric; fronds strongly dimorphic, fertile leaves obviously longer than the sterile; pinnules narrow, elongate, usually revolute; stipes green to straw-colored distally . . . . . *Cryptogramma* (Parsley ferns or Rock brakes)
2. Plants terrestrial; monomorphic, fertile leaves similar to sterile ones; pinnules narrowly elliptic to round; stipes dark reddish-brown or dark purple to nearly black . . . . *Pellaea* (Cliff brakes)

THELYPTERIDACEAE Ching ex Pichi-Sermolli   Marsh Ferns

Etymology. Greek *thelys*, female, + *pteris*, fern. The Greeks applied this name to a delicate fern, comparing it to *Dryopteris filix-mas*, the male fern.

Plants. Monomorphic; deciduous.

Rhizomes. Short to long-creeping to erect; bearing nonclathrate scales at tips.

Stipes. Cross sections at bases reveal ends of 2 ribbonlike vascular bundles (appearing as crescent-shaped more or less parallel lines) that unite to form a single U-shaped bundle in upper stipes; grooved.

Blades. Usually 1-pinnate-pinnatifid to 1-pinnate-pinnatisect.

Rachises. Grooved adaxially or not, grooves if present not continuous with grooves of costae; **transparent, needlelike, short, sharp-tipped hairs (acicular hairs) present.**

Pinna. Hairs on adaxial surfaces similar to those on rachises; sinuses between 2 pinna lobes sometimes with translucent, cartilaginous membrane at bases.

Veins. Free; pinnately branching in lobes but not forked beyond; lowest vein pair touching margin above sinus base.

Sori. Medial to submarginal, round to oblong; indusia reniform or absent.

Chromosome numbers. $x = 27–36$

One of the largest fern families with 5 to 30 genera, depending on how they are recognized taxonomically, with about 900 to 1,000 species worldwide. Three genera and 25 species are recognized in *Flora of North America North of Mexico*, vol. 2 (1993). It is represented in Michigan by 2 genera, *Phegopteris* (2 species) and *Thelypteris* (2 species).

Thelypteris differs from *Athyrium*, *Deparia*, *Homalosorus*, and *Dryopteris* by the presence of transparent needlelike hairs (vs. lack of such hairs), general absence of blade scales (vs. blade scales often present), stipes with 2 crescent-shaped bundles seen in cross section (vs. many round bundles arranged in an arc in *Dryopteris*) (*Athyrium*, *Deparia*, and *Homalosorus* are similar to *Thelypteris*), generally 1-pinnate-pinnatifid blades (vs. often more divided), veins usually not forking in the ultimate segments (vs. often forking), adaxial grooves discontinuous from rachis to costae or grooves lacking (vs. grooves often continuous), and chromosome base numbers from 27 to 36 (vs. generally 40 or 41).

KEY TO THE GENERA OF THELYPTERIDACEAE AS FOUND IN MICHIGAN

1. Indusia present; pinnae free, rachises not winged; costae grooved adaxially .... *Thelypteris* (Marsh ferns)
1. Indusia absent; pinnae mostly connected by wings along rachis, the wings often forming semicircular lobes between pinnae; costae not grooved adaxially .... *Phegopteris* (Beech ferns)

WOODSIACEAE Herter Cliff Ferns

Etymology. Name honors Joseph Woods (1776–1864), an English botanist and architect.

A family of 3 genera worldwide, only *Woodsia*, with 10 species, is treated in *Flora of North America North of Mexico*, vol. 2 (1993). It is represented in Michigan by this single genus, *Woodsia*, with 4 species.

The family treatment is identical to the generic treatment of *Woodsia* in the main text.

DESCRIPTIONS OF LYCOPHYTE FAMILIES AND KEYS TO THE GENERA AS THEY ARE FOUND IN MICHIGAN

These family descriptions include descriptive material that applies only to the families as seen in Michigan. In many cases these descriptions would have to be much changed to describe the families as seen worldwide.

The generic descriptions are found with the descriptions of the genera and species in the main text.

KEY TO THE LYCOPHYTE FAMILIES FOUND IN MICHIGAN

1. Leaves round, quill-like; spore-bearing structures embedded in expanded leaf bases at or below ground level; plants aquatic or semiaquatic; leaves arising from tightly compacted 2- to 3-lobed corms . . . . Isoetaceae

1. Leaves simple, lancelike or ovate; spores borne at adaxial base of smaller fertile leaves in strobili or at base of unmodified leaves; plants on dry or wet land, not aquatic; leaves arising from stems (2).

2(1). Sporangia borne in flattened or 4-sided strobili (except *Selaginella selaginoides* with cylindrical cones); spores (of 2 sizes) hidden beneath the expanded parts of green leaves (sporophylls) in fertile stems that are often the only erect part of the plant. . . . . Selaginellaceae

2. Sporangia borne singly in leaf axils; sterile and fertile leaves the same or different; aggregated in cylindrical strobili at branch tips stalked or not, present or absent: spores of 1 size . . . . Lycopodiaceae

ISOËTACEAE Reichenbach Quillworts

A family with a single genus, *Isoëtes*.

The treatment of this family is identical to the treatment of the genus. It will be found under the generic treatment of *Isoëtes* in the main text.

LYCOPODIACEAE Mirbel Club-Mosses

A family with 8 genera.

The treatment of this family and its genera will be found under the generic treatment of Lycopodiaceae in the main text.

SELAGINELLACEAE Willkomm Spike-Mosses

A family with a single genus, *Sellaginella*.

The treatment of this family is identical to the treatment of the genus. It will be found under the generic treatment of *Sellaginella* in the main text.

TABLES AND LISTS SUMMARIZING MICHIGAN FERN DATA

Numbers of Michigan Ferns and Lycophytes Treated in This Book

	Families	*Genera*	*Total Species*[a]	*Hybrids*	*Total Taxa*[b]	*Alien Species*
Ferns (excluding *Equisetum* and Ophioglossaceae)	13	24	55	4	59	1
Equisetum	1	1	9	4	13	0
Ophioglossaceae	1	4	18	0	17	0
Total ferns treated here (Polypodiosida)	15	30	82	8	89	1
Lycophytes	3	8	22	5	27	0
Total	20	38	104	13	116	1

Note: Uncommon or rare hybrids are not treated.
[a]All species, excluding hybrids and species with two subspecies.
[b]All species, subspecies, varieties, and hybrids.

Most Species-Rich Genera of Michigan Ferns

Genus	*Number of Species*
Botrychium	11
Dryopteris	10
Equisetum	9
Asplenium	7
Cystopteris	5
Woodsia	4
Sceptridium	4
Gymnocarpium	3
Osmunda	3
Polystichum	3

Rare or Endangered Michigan Ferns and Lycophytes

Asplenium rhizophyllum	Walking fern	T
Asplenium ruta-muraria	Wall-rue	E
Asplenium scolopendrium var. *americanum*	Hart's-tongue fern	E
Asplenium viride	Green spleenwort	T
Botrychium campestre	Prairie moonwort	T
Botrychium michiganense	Michigan moonwort	T
Botrychium mormo	Goblin moonwort	T
Botrychium pallidum	Pale moonwort	SC
Botrychium spathulatum	Spatulate moonwort	T
Cryptogramma acrostichoides	American rock-brake	T
Cystopteris laurentiana	Laurentian fragile fern	SC
Dennstaedtia punctilobula	Hay-scented fern	X
Diphasiastrum alpinum	Alpine clubmoss	X
Dryopteris celsa	Log fern	T
Dryopteris filix-mas subsp. brittonii	Male fern	SC
Dryopteris fragrans	Fragrant cliff woodfern	SC
Equisetum telmateia	Giant horsetail	X
Gymnocarpium jessoense	Northern oak fern	E
Gymnocarpium robertianum	Limestone oak fern	T
Huperzia appressa (as H. appalachiana)	Mountain fir-moss	SC
Huperzia selago	Fir clubmoss	SC
Isoetes engelmannii	Appalachian quillwort	E
Lycopodiella margueriteae	Northern prostrate clubmoss	T
Lycopodiella subappressa	Northern appressed clubmoss	SC
Lygodium palmatum	Climbing fern	E
Ophioglossum vulgatum	Southeastern adder's-tongue	T
Pellaea atropurpurea	Purple cliff-brake	T
Woodsia alpina	Northern woodsia	T
Woodsia obtusa	Blunt-lobed woodsia	T
Woodwardia areolata	Netted chain-fern	X

Note: X = extirpated, T = threatened, E = endangered, SC = of special concern.

Federal and State Statuses of Rare or Endangered Michigan Ferns and Lycophytes

Scientific Name	*Common Name*	*Family*	*US Status*	*State Status*	*Global Rank*	*State Rank*
Asplenium rhizophyllum	Walking fern	Aspleniaceae		T	G5	S2S3
Asplenium ruta-muraria	Wall-rue	Aspleniaceae		E	G5	S1
Asplenium scolopendrium var. *americanum*	Hart's-tongue fern	Aspleniaceae	LT	E	G4T3	S1
Asplenium viride	Green spleenwort	Aspleniaceae		SC	G4	S3
Botrychium campestre	Prairie moonwort or dunewort	Ophioglossaceae		T	G3G4	S2
Botrychium michiganense	Michigan moonwort	Ophioglossaceae		T	G4	S2
Botrychium mormo	Goblin moonwort	Ophioglossaceae		T	G3	S2
Botrychium pallidum	pale moonwort	Ophioglossaceae		SC	G3	S3
Botrychium spathulatum	Spatulate moon-wort	Ophioglossaceae		T	G3	S2
Cryptogramma acrostichoides	American rock-brake	Pteridaceae		T	G5	S2
Cystopteris laurentiana	Laurentian fragile fern	Dryopteridaceae		SC	G3	S1S2
Dennstaedtia punctilobula	Hay-scented fern	Dennstaedtiaceae		T	G5	S1
Diphasiastrum alpinum	Alpine clubmoss	Lycopodiaceae		X	G5	SX
Dryopteris celsa	Small log fern	Dryopteridaceae		T	G4	S2
Dryopteris filix-mas	Male fern	Dryopteridaceae		SC	G5	S3
Dryopteris fragrans	Fragrant cliff woodfern	Dryopteridaceae		SC	G5	S3
Equisetum telmateia	Giant horsetail	Equisetaceae		X	G5	SX
Gymnocarpium jessoense	Northern oak fern	Dryopteridaceae		E	G5	S1
Gymnocarpium robertia-num	Limestone oak fern	Dryopteridaceae		T	G5	S2
Huperzia appressa	Mountain fir-moss	Lycopodiaceae		SC	G4G5	S2
Huperzia selago	Fir clubmoss	Lycopodiaceae		SC	G5	S3
Isoetes engelmannii	Engelmann's quillwort	Isoetaceae		E	G4	S1
Lycopodiella margueritae	Northern prostrate clubmoss	Lycopodiaceae		T	G2	S2
Lycopodiella subappressa	Northern ap-pressed clubmoss	Lycopodiaceae		SC	G2	S2
Lygodium palmatum	Climbing fern	Schizaeaceae		E	G4	S1

Ophioglossum vulgatum	Southeastern adder's tongue	Ophioglossaceae	E	G5	S1
Pellaea atropurpurea	Purple cliff brake	Pteridaceae	T	G5	S2
Woodsia alpina	Northern woodsia	Dryopteridaceae	E	G4	S1
Woodsia obtusa	Blunt-lobed woodsia	Dryopteridaceae	T	G5	S1S2
Woodwardia areolata	Netted chain-fern	Blechnaceae	X	G5	SX

Federal (US) Status: LE = listed endangered; LT = listed threatened; LELT = partly listed endangered and partly listed threatened; PDL = proposed delist; E(S/A) = endangered based on similarities/appearance; PS = partial status (federally listed in only part of its range); C = species being considered for federal status.
State Status: E = endangered; T = threatened; X = extirpated; SC = of special concern.

Global Rank:
G1 = critically imperiled globally because of extreme rarity (5 or fewer occurrences range-wide or very few remaining individuals or acres) or because of some factor(s) making it especially vulnerable to extinction.
G2 = imperiled globally because of rarity (6 to 20 occurrences or few remaining individuals or acres) or because of some factor(s) making it very vulnerable to extinction throughout its range.
G3 = either very rare and local throughout its range or found locally (even abundantly at some of its locations) in a restricted range (e.g., a single western state or a physiographic region in the East) or because of other factor(s) making it vulnerable to extinction throughout its range; in terms of occurrences, in the range of 21 to 100.
G4 = apparently secure globally, though it may be quite rare in parts of its range, especially at the periphery.
G5 = demonstrably secure globally, though it may be quite rare in parts of its range, especially at the periphery.
GH = of historical occurrence throughout its range (i.e., formerly part of the established biota) with the expectation that it may be rediscovered (e.g., Bachman's warbler).

State Rank:
S1 = critically imperiled in the state because of extreme rarity (5 or fewer occurrences or very few remaining individuals or acres) or some factor(s) making it especially vulnerable to extirpation in the state.
S2 = imperiled in the state because of rarity (6 to 20 occurrences or few remaining individuals or acres) or some factor(s) making it very vulnerable to extirpation from the state.
S3 = rare or uncommon in the state (on the order of 21 to 100 occurrences).
SR = reported from the state, but without persuasive documentation that would provide a basis for either accepting or rejecting the report.
SX = apparently extirpated from the state.

MICHIGAN SPECIES WITH WIDELY DISJUNCT POPULATIONS

Dennstaedtia punctilobula
Dryopteris celsa
Lygodium palmatum

EVOLUTION OF TAXONOMIC CONCEPTS

Morphology and Cytology

Prior to the 1950s most taxonomic and phylogenetic decisions were based on often subjective interpretations of gross morphologic features such as veins, scales, hairs, sori, presence or absence of indusia, stipe vascular bundles, glands and spore ornamentation, and what minimal fossil evidence was available. All the objective characteristics important for evolutionary classification were still treated in a largely subjective manner until the 1950s when things began to change.

Irene Manton (1904–88), an English botanist and cytologist and a truly remarkable and productive woman, formalized the science of cytology and made it an accurate evolutionary tool. Her chromosome counts, studies of chromosome figures in meiosis, and chromosome pairing in hybrids revolutionized the study of ferns and lycophytes. The cytological studies of many others since, including the work of Michigan botanists Warren H. and Florence Wagner, represented the first systematic use of objective measurements to identify taxa by chromosome numbers, in contrast to previous studies, which relied primarily on observations made with the naked eye or under low magnification. A method for assigning the correct parentage of hybrids by observing chromosome pairing in meiosis and the accurate recognition of autopolyploid and allopolyploid taxa, as well as hybrids, became available.

The ability to accurately count chromosomes, the use of chemistry (flavonoids, scanning electron microscopy of spores, enzymes, DNA, etc.), and the recognition of new morphologic characters led to a more precise understanding of plant evolution and systematics. An ever-expanding collection of herbarium material, a thorough review of the literature, examination of type specimens, and publication of the findings have moved fern classification toward stability. Historically, authors frequently disagreed on the circumscription of taxa and proposed alternative hypotheses about the evolutionary relationships of related species. The availability of more data, and molecular phylogenetics in particular, has helped to resolve historical controversies and clarified many of these relationships. Although most disputes about evolutionary relationships can now be satisfactorily resolved, authors often disagree about whether to "lump" or "split" since choices still need to be made on whether to define any given taxon broadly or narrowly.

Robert Sokal (1926–2012), an Austrian-born entomologist who fled Hitler's Germany and spent time in Shanghai before moving to the University of Kansas, developed a system of numeric taxonomy as an objective means of arranging species by applying a number to every morphologic character (*with each character being given the same value*). By using numeric algorithms he could define a species, and by measuring similarities he attempted to order the families, genera, and species according to their evolutionary history. However, similarities do not necessarily reflect evolutionary relationships. The taxonomy developed by this system, called phenetics, in most cases resembled what the taxonomists using morphology alone had been saying all along.

In the 1950s and 1960s a method of phylogenetic systematics called cladistics (Greek *klados* = branch) was created by Willi Hennig (1913–76), a German entomologist (who while in the German army in 1945 was captured by the British). He developed another method of numeric taxonomy, *but instead of giving equal value to all characters he assigned value to shared evolutionary novelties*. This system was expanded by many botanists, including Warren H. Wagner at the University of Michigan.

In subsequent years, cladistics became the most accepted means of establishing evolutionary relationships. One of the objectives of this analysis is determining whether traits are shared from a single common ancestor, whether they converged (evolved in parallel), or whether a trait was secondarily lost. The most widely adopted method for developing phylogenetic hypotheses using morphological data has been *maximum parsimony*, which identifies the fewest number of steps (trait changes) needed to generate a tree. A major benefit of using molecular data for generating phylogenies is that not all morphological characters are equivalent and most involve multiple genes. Molecular data are more objective and provide many more characters for analysis since each variable nucleotide represents a single character. Although *Maximum Parsimony* can also be used for these analyses, most researchers favor more nuanced statistical approaches, such as *Maximum Likelihood* and *Bayesian* analysis, which allow for the incorporation of a priori knowledge about mutation rates, and data partitioning, since some genes evolve at different rates and certain mutations occur more frequently than others.

A very real problem now is that there are fewer taxonomists (those relying mostly on morphology and capable of identifying the plant itself to species) left and very few in training. The numeric and chemical systematists are very good at what they do, but some may lack the ability to clearly identify plants to species based on herbarium collections. This can lead to errors in their work if they perform their studies on the wrong plants. Competent taxonomists are fundamental to the success of systematists.

Phylogenomics

Over the years chemical taxonomy, involving cytochrome and hemoglobins in animals and the flavonoids and other chemicals in plants, was used to establish evolutionary relationships. Next in chemical taxonomy was the study of isozymes (slightly different variations of the same enzymes), thought to reflect the relationships among the plants tested. This was based on the discovery that natural populations of sexually reproducing organisms usually maintain significant genetic polymorphisms in enzymes, which will change as these populations evolve and differentiate into separate species. Therefore, similarities and differences in these variants of the enzymes should be particularly powerful tools for uncovering similarities and differences when used at low taxonomic levels (i.e., among closely related species). Most molecular work on ferns has been done on segments of chloroplast or nuclear DNA, employing Sanger sequencing to look at gene regions typically on the order of a thousand base pairs long. Current approaches involve entire genomes, expanding available data to include tens of millions of base pairs. By examining genes that mutate at different rates, we can understand deep evolutionary relationships as well as recent radiations. Transcriptomics, or the study of RNA, allows us to study gene expression, which allows a more nuanced understanding of how organisms interact with their environment through time. Technology has enabled a low-cost method of generating unprecedented quantities of data, and the primary challenges facing systematists are now computational. This exciting moment in time will surely yield many surprises involving the genome, which may yet change our contemporary understanding of evolution and again alter our concepts of taxonomic relationships.

A Short Discussion of Botanical Nomenclature

The term *basionym* is used in botany only in situations in which a previous name exists with a useful description, and the *International Code of Nomenclature for Algae, Fungi, and Plants* does not require a full description with the new name. Basionyms are regulated by articles 6.10, 7.3, 41, and others.

In the mid-nineteenth century among the fern genera and families, three were widely recognized. These included *Aspidium* with around 800 species, *Nephrodium* with more than 700 species, and *Lastrea*, which was sometimes recognized as a subgroup of *Nephrodium*, with over 500 species. Today there are no ferns bearing those generic or family names. What happened to all those ferns?

These three genera were poorly defined with very few and often inaccurately described characteristics separating them. *Aspidium*, according to Hooker in *Synopsis Filicum* (1868), had subglobose (round) sori with an involucre (indusia) similar in shape to the sorus. *Aspidium* was later split into two groups, *Aspidium* and *Polystichum*. The genus *Polystichum*, the shield ferns, was early separated largely because its species had perfectly round peltate indusia, although it was recognized that ferns in this group had other distinguishing characters. The remaining aspidiums had round indusia with a notch on one side, making them kidney shaped.

Nephrodium was identified by its reniform indusia attached at the sinus. Hooker used this as the sole identifying character for the genus and observed that its species vary widely in size, texture, cutting, and venation. He recognized that the shape of the indusium was the single character separating *Aspidium* and *Nephrodium*, that species assigned to either (many had been assigned to both) often had very similar ranges of variation in cutting and venation, and that it was by no means clear in which of the two genera several species should be placed.

Lastrea had been treated as a division of *Aspidium* by some authors and as a division of *Neprodium* by others. The main character separating it from those genera was the fact that the species in this group had free veins and veinlets.

As time passed and knowledge was gained, the importance of many other characters, such as scales, hairs, glands, frond cutting and shape, venation patterns, arrangement of vascular bundles in the stipe, and grooving of the rachis and costae, were recognized as important in defining genera. Newer, more clearly defined genera were re-created, and when this was done the species in *Nephrodium*, *Aspidium*, and *Lastrea* were gradually removed and placed more appropriate genera.

With additional help in the form of chromosome numbers and enzyme DNA studies, the process continues today. New families and new genera are appearing.

At one time the ferns in the genus *Thelypteris* were included in the genus *Dryopteris* because of the shape of the sorus and many other features that suggested a similarity. It was later recognized that many species treated as *Dryopteris* have two straplike vascular bundles in the stipe (vs. a C- or U-shaped collection of small round bundles seen in cross section). Other features found in *Thelypteris* and lacking in *Dryopteris* include very distinctive small, transparent needlelike hairs; veins not forking in the ultimate segments; a general lack of stipe scales; and other characters.

The chromosome numbers also differed, with 27 to 36 in *Thelypteris* and 40 or 41 in *Dryopteris*.

New families, such as Diplaziopsidaceae (2011), are still being recognized. Species are being placed in different genera from those recently used, and genera are being placed in different or new families.

GLOSSARY

Many terms used in this book can be found in standard dictionaries. Some of those listed below are less commonly used, and some may have special meanings when applied to ferns or lycophytes.

abaxial: The undersurface of a frond. *Compare* adaxial
acicular: (Hairs) sharp-tipped, translucent, needlelike.
acroscopic: Directed toward the tip of a frond, pinna, or any structure on which a feature is borne. *Compare* basiscopic
acuminate: Gradually tapering to a narrow, sharp tip.
acute: Sharp-pointed, with straight or somewhat convex distal margins that form an angle of between 40 and 90 degrees at the tip.
adaxial: The upper surface of a frond. *Compare* abaxial
adnate: Broadly attached; said of the attachment of bases of ultimate segments to the axis on which they are borne.
aerophores: Lines, swellings, or other specialized tissue, located along the stipes, rachises, costae, or pinnule bases; usually bearing abundant stomates and apparently associated with gas exchange.
ala, pl. alae: Greenish wings, usually composed of frond blade tissue, on the side of a stipe, rachis, or costa.
alate: Winged.
allopleisoapoparsimony: A word representing the confusion many readers of this book will experience when using this glossary.
allopolyploid: Polyploids with two or more complete sets of chromosomes derived from different species.
alternate: Alternating in staggered fashion, for example, pinnae, pinnules, segments, arising from rachis, costae, or midribs. *Compare* opposite
alternation of generations: The regular alternation of forms or mode of reproduction in the life cycle of an organism, such as the alternation between diploid and haploid phases.
anastomosing: Joining together and forming enclosed areoles; said of veins.
annulus, pl. annuli: a single row of thick-walled cells around the sporangia of many ferns that serves in spore dispersal.
antheridium: Male sex organ of spore-producing plants, producing spermatids.
apex, pl. apices: Tip of a structure, for example, blade tip, pinna, or ultimate segment.
apiculate: Abruptly terminating in a small, sharp tip.
apogamous: Producing sporophytes from gametophytes by asexual budding rather than sexual fertilization of an egg by a sperm.
appressed: Tightly pressed to or lying flat against a surface.
areole, pl. areoles: The spaces enclosed by veins in a vein network.
aristate: Bearing long, stiff, slender extensions of leaf tips.

ascending: Directed toward the tip of a structure, usually at an oblique angle, for example, pinnae.

attenuate: Narrowly tapering, usually to a long point.

auricle: An earlike lobe or projection at the base of a pinna, pinnule, or ultimate segment.

auriculate: Bearing auricles.

basionym: The oldest valid name of a taxon being used in a new combination because of a change in taxonomic status or rank.

basiscopic: Directed toward the base of the blade. *Compare* acroscopic

bicolorous: Having two colors, usually referring to scales; *Compare* concolorous

bifurcate: Forked, divided into equal or nearly equal branches.

bilateral: Symmetrical about a single axis.

canals: In *Equisetum* stems, longitudinal channels appearing as empty spaces on a cross section of a stipe; central, vallecular, and carinal.

cartilaginous: Thick and translucent, resembling cartilage.

caudate: Bearing a narrow, elongate, tail-like tip.

caudex, pl. caudices: A thick, erect, trunklike rhizome.

chartaceous: Having a texture similar to that of thick writing paper.

circinate vernation: A growth pattern in which the emerging bud is at the center of a coil, as in crosiers or fiddleheads of young ferns, which uncoil to become the fronds.

clathrate: Latticelike, having thick, lateral (adjacent) cell walls with clear centers (reminiscent of stained glass windows) and marginal cells with thin outer walls; said of scales; characteristic of *Asplenium* species.

clone: A group of genetically identical individuals originating by asexual means (vegetative reproduction) from a single individual; can be accomplished by long-creeping rhizomes, gametophyte gemmae, rachis budding, and so on.

close: Closely spaced, nearly touching; said of stipes, pinnae, or segments.

clustered: Said of fronds clustered at the tips of rhizomes.

cm: Centimeter(s).

common-garden study: The study of various forms or varieties of a taxon grown in a single environment under uniform conditions.

concolorous: Uniform in color. *Compare* bicolorous

confluent: Running together.

contiguous: Touching at a boundary.

cordate: Shaped like a Valentine's Day heart.

coriaceous: Thick and tough in texture, leathery in appearance.

costa, pl. costae: The major axis of a pinna, a prominent rib or vein.

costal: Near or adjacent to the costa, costule, or midrib. *See* Fronds Anatomy illustrated glossary.

costule: The major axis of a pinnule.

creeping: Lying on the ground and extending horizontally; said of rhizomes.

crenate: Having a round-toothed or scalloped edge; said of segment margins.

crenulate: Minutely crenate.

crisped: Irregularly curled; said of pinnae, pinnules, and ultimate segments.

crosier: The young, unexpanded tip of a fern frond, the fiddlehead.

cultivar: A cultivated variety; a plant originating in and persistent under cultivation.

cuneate: Narrowly triangular with the acute angle toward the base.

deciduous: Falling away, not persistent, mostly applying to leaf fall before winter.

decumbent: Lying or creeping on the ground with the apex erect or ascending; said of rhizomes.

decurrent: Extending downward to and adnate with the axis; said of ultimate segments of a blade.

deltate: Broadly triangular.

dentate: Toothed along the margins.

denticulate: Finely dentate.

determinate: Blade tips limited in growth to a predetermined size and shape.

diam.: Diameter.

dichotomous: More or less regularly forking with branches at each fork about equal in length and width.

dimidiate: Divided in two with one part, usually the lower, much smaller or nearly absent; seen in pinnae and ultimate segments in certain *Adiantum* and *Asplenium* species.

dimorphic: Having two forms or sizes; often said of fronds; fertile and sterile blades different in form or size. *Compare* monomorphic

discombobulated: The feeling an uninitiated fern amateur experience when encountering the technical terms in this text.

distal: Toward the tip. *Compare* proximal

distant: Similar parts that are well separated and not overlapping or touching; said of stipes, pinnae, or segments.

elater: Tiny, elongated, elastic structures that aid in the dispersal of spores by expanding or folding, caused by the absorption of moisture and drying; a band attached to the spores of *Equisetum* species.

elliptic: Having the outline of an elongate circle.

emarginate: Having margins notched by sinuses at the tips of usually round or truncate ultimate segments, lobes, or teeth.

entire: Having a continuous, even, or smooth margin, not cut or toothed, with or without hairs or cilia.

exindusiate: Lacking indusia.

falcate: Sickle-shaped, curved, and tapered toward a tip.

false indusium: A modified, reflexed blade, pinna, or ultimate segment margin that protects young sporangia, especially found in the Pteridaceae.

fertile: Producing spores.

flabellate: Fan-shaped, without a prominent central vein; said of ultimate segments or veins in an ultimate segment.

form: The lowest rank normally used by pteridologists for sporadic distinct variants that sometimes occur in populations; variations may be based on relatively minor genetic traits, but their effects can be conspicuous.

free: Simple or forked but without uniting branches; said of veins.

frond: A fern leaf, including blade and stipe.

gametophyte: An inconspicuous nonvascular plant bearing sexual organs (sperm and eggs) that with fertilization produce sporophytes; haploid.

gemma, pl. gemmae: A small, flattened, winged, readily detached, asexual propagule found on some sporophytes that, when mature, are ejected from gemmaphores when touched, blown by the wind, or shaken by raindrops and produce new sporophytes.

gemmiphore: A structure bearing the gemma and capable of projecting a gemma.

genus: A group of closely related species.

glabrous: Without scales, hairs, or glands.
gland: A club-shaped or globular structure containing or secreting often colored, resinous, or translucent substances; borne on blade surfaces, axes, or hair tips.
glaucous: Very thinly covered and not obscured by a light bluish-gray or bluish-white, waxlike coating; said of the undersurfaces of some blades.
globose: Nearly spherical.
grooved: Having one or more longitudinal indentations on an axis, especially on the adaxial surface; said of stipes, rachises, costae, or other axes.
hair: Epidermal outgrowth composed of one elongate cell or multiple cells in single file.
hairlike scale: A very narrow scale, 2 to 5 cells wide at the base, tapering to a long, uniseriate tip.
hastate: Having the shape of an arrowhead, with basal lobes pointed laterally.
herbaceous: Thin and soft in texture.
heterosporous: Bearing spores of two kinds, small microspores (male) and large megaspores (female).
hexaploid: Containing six sets of chromosomes, six times the haploid number.
homopolypoid: A polyploid with two or more complete sets of chromosomes arising from a species.
homosporous: Producing spores of a single kind that are not differentiated by sex.
hydathode: A swollen, dark, or pale vein ending, located near the margin of a blade, which may exude salts and water.
imbricate: Overlapping like shingles on a roof; said of leaves, scales, and so on.
imparipinnate: Having an uneven number of pinnae by virtue of having one terminal pinna the same shape as a lateral pinna.
incised: Cut more or less deeply; said of parts of a blade.
indeterminate: Not limited in longitudinal growth; continuously or intermittently elongating from the tip; said of blades.
indigenous: Native to a region but also naturally distributed elsewhere in the world.
indument: Hairs, scales, or glands.
indusiate: Bearing indusia.
indusium, pl. indusia: A thin epidermal outgrowth from a fern blade that covers the sorus.
infraspecific: Referring to taxonomic subdivisions within a species such as subspecies, varieties, and forms.
isozyme: Multiple molecular forms of enzymes with the same catalytic function.
lacerate: Torn or with deeply and irregularly cut or jagged margins; said of pinnae and ultimate segments or any flat surface.
lamina, pl. laminae: the blade (the expanded portion) of a fern frond.
lanceolate: Lance-shaped, broadest between base and middle and tapering toward the tip.
ligule: A small triangular flap of leaf tissue found on the adaxial surface of a leaf just above the sporangia; said of *Isoëtes* and *Selaginella* species.
linear: Narrow and long with parallel margins.
lobe (adj. lobed): One of the usually rounded segments of a blade, pinna, or ultimate segment with sinuses on either side.
long-creeping rhizome: One that grows horizontally and produce stipes arising distant from each other.
m: Meter(s).
macrophyll: A leaf with several large veins branching apart or running parallel and

connected by a network of smaller veins. (e.g., fronds of ferns and leaves of gymnosperms and angiosperms). *Compare* microphyll

marcescent: Remaining attached long past maturity but in a withered state.

marginal: Borne on the margin; at the edges of flat structures such as pinnae, pinnules, ultimate segments, or scales; said of sori and veins. *See* Sori and Indusia illustrated glossary.

medial: Positioned midway between the margin and the midrib, costa, midvein, or costule; said of soral position.

megaphyll: See macrophyll

megaspore: The large female spore of the heterosporous ferns and fern-allies *Azolla, Isoëtes, Marsilia,* and *Selaginella.*

megasporocarp: A sporocarp bearing megasporangia.

microphyll: A leaf with a single unbranched vein and no demonstrable gap around the leaf trace (small in lycopodiums, large in *Isoétes*).

microspore: The small male spore of the heterosporous ferns and fern-allies *Azolla, Isoëtes, Marsilia,* and *Selaginella.*

microsporocarp: Bearing microsporangia.

midrib: The central vein of a blade, pinna, pinnule, or ultimate segment, the midvein.

mm: Millimeter(s).

monolete: Having a single, linear, unbranched, thickened suture; said of spores.

monomorphic: Having one form; said of fertile and sterile blades when they are identical. *Compare* dimorphic

monophyletic: Said of taxa arising from diversification by a single ancestor (i.e., the several distinct species arising from a single ancestral species are monophyletic).

morphology: The form or structure of an organism or any of its parts.

nothospecies: A species produced by the hybridization of two or more species.

novanomenophobia: The hysterical feeling that many amateurs, and some professional botanists, experience when a necessary change in scientific names occurs.

oblanceolate: Lanceolate with the widest part toward the tip instead of the base.

oblique: Slanting at an angle from the structure to which something is attached, neither perpendicular nor straight.

oblong: Longer than wide with the long sides mostly parallel.

obtuse: Blunt, rounded, lacking sharpness.

opposite: Oppositely attached along the rachises or costae; said of pinnae, costules, and pinnules.

orthospecies: A species produced through the usual processes of selection and evolution.

ovate: Having an outline like that of a longitudinal cross section of a hen's egg with the basal end broader.

paraphysis, pl. paraphyses: A small, usually elongate, unicellular or multicellular, often glandular structure borne on a soral receptacle, sporangial capsule, or sporangial stalk.

pedicel: A stalk, usually short and narrow, supporting a gland or other larger structure.

peltate: Round or nearly so and attached by a central stalk, umbrellalike; often said of indusia and scales.

phyllopodium: In species with articulate stipes, that portion of the stipe proximal to the articulation that remains attached to the rhizome.

phylogeny: The evolutionary history of a taxonomic group.

pinna, pl. pinnae: The primary division of a compound blade that is sessile, narrowed or stalked at the base.

pinnate: With two rows of pinnae, pinnules, segments, or veins on either side of an axis. *Compare* pinnatifid
pinnate-pinnatifid: Pinnate with the pinnae pinnatifid (i.e., 1+ times divided).
pinnate-pinnatisect: Pinnate with the pinnae pinnatisect.
pinnatifid: Cut deeply (but not to the midrib) into lobes along an axis. *Compare* pinnate, pinnatisect
pinnatisect: Cut all the way to an axis with the segments not contracted at their bases. *Compare* pinnate, pinnatifid
pinnule: The secondary division of twice-divided blades; a division of a pinna that is sessile, narrowed, or stalked at the base.
polymorphic: Having several forms.
proliferous: Bearing plantlets or bulblets on the stipes, rachises, stolons, blades, or roots of ferns.
proximal, proximally: Toward the base. *Compare* distal
pteridologist: Someone who studies ferns and fern-allies.
pteridology: The study of ferns and lycophytes.
pubescent: Hairy.
rachis: The axis or midrib of a fern blade; the extension of the stipe through the blade.
recurved: Curved toward the base.
reduced: Becoming smaller.
reflexed: Bent abruptly downward.
remote: Distant, far apart, well separated.
reniform: Kidney-shaped; crescent-shaped with rounded margins and a central sinus.
reticulate: Forming a netlike or latticelike network; said of veins.
revolute: Lateral margins inrolled abaxially.
rhizome: A creeping, ascending, or erect stem of a fern from which roots and fronds arise.
rhizophore: An outgrowth from a stem that grows toward the substrate surface where it forms roots; found commonly in *Selaginella* species.
scale: A usually flat, two-dimensional, multicellular outgrowth that is a few to several cells wide, at least at the base; may be lanceolate to round and be attached at the base or from a lower surface.
sclerotic: Hard and stony.
segment: Part of a blade, pinna, or pinnule with deep sinuses on each side cut more than halfway from its tip to the axis to which it is attached.
septate: Divided by a partition or cross wall, having obvious transverse walls between cells; said of hairs.
serrate: Having usually sawlike, sharp, acroscopically directed teeth on the margins. *See* Pinna Forms and Margins illustrated glossary.
sessile: Attached directly at the base, lacking a free basal portion, lacking a stalk. *See* Pinna Forms and Margins illustrated glossary.
short-creeping rhizome: One that grows horizontally and produces fronds at short intervals.
shuttlecock: A conical object with a rubber nose and radiating feathers used in badminton.
simple: Blades without lobes or pinnae; not divided, compound, or branched; pinnae without lobes or segments.
sinuate: Having a wavy margin.
sinus: The indentation or recess between two lobes or segments.

sorophore: The gelatinous sporangium-bearing structure produced when the sporocarps of Marsileaceae open.
sorus: A cluster of sporangia.
span: That portion of a circle that is "spanned" by the outer circumference of the pinna.
spatulate: Spatula-shaped, narrow at the base and gradually widening distally to a rounded tip.
sporangium, pl. sporangia: The spore-bearing structure of ferns.
sporocarp: A short-stalked, globular structure containing sporangia embedded in a sorophore; found in *Azolla*, and *Marsilea* species.
sporophore: Division of the common stalk bearing numerous globose sporangia, the fertile blade.
sporophyll: An often smaller fertile, sporangia-bearing leaf.
sporophyte: The more conspicuous plant in the fern life cycle, the spore-producing generation, the plant most people recognize as the fern. *Compare* gametophyte
stalk: A short supporting axis, for example, the short, free, basal portion of some pinnae, pinnules, and segments; structure attaching a sporangium to its receptacle; the petiole of a frond.
stalked: Joined to an axis by means of a stalk. *Compare* sessile
sterile: Not producing spores.
stipe: The petiole of a fern frond, that is, the stalk of a frond that joins the rhizome to the blade.
stolon: An elongate, creeping stem, often subterranean, that intermittently forms new plants and roots along its length.
stoloniferous: Bearing stolons.
strobilus, pl. strobili: A spike or cone borne on a central axis at the tip of a branch consisting of closely spaced fertile leaves (sporophylls) with sporangia at their bases; seen in the lycophytes and Ophioglossaceae.
sub-: Almost, nearly, approaching.
submarginal: Slightly closer to the margin than the midrib. *See* Sori and Indusia illustrated glossary.
subopposite: Nearly but not quite opposite.
subspecies: A major division of a species; some pteridologists consider them a rank higher than a variety, while others view subspecies as equivalent to varieties.
superficial: On the surface.
sympatric: Occupying the same general locality; usually describing ranges of the populations of different species.
taxon, pl. taxa: any group of plants of the same rank, such as genus, species, subspecies, variety, or form.
ternate: ternately: Arranged in threes, having three equal or nearly equal parts often at nearly right angles to each other.
tetraploid: Having four sets of chromosomes, a chromosome number that is four times the basic or haploid number.
trilete: Having three radiating, thickened sutures; said of spores.
triploid: Having three sets of chromosomes, a chromosome number that is three times the basic or haploid number.
trophophore: One division of a common stalk bearing the expanded photosynthetic sterile blade. (Greek, *trophe*, food, nourishment, plus *phore*, carrier, alluding to the photosynthetic function of this blade.)
truncate: Appearing as if cut off perpendicular to an axis with the end square or even.

tuber: An enlarged, often globular storage body usually borne on a stolon or rhizome, often scaly.
type specimen: A specimen or illustration of a species or infraspecific taxon used by the author as the nomenclatural type. As long as the type specimen exists, the application of the name is fixed to that plant.
ultimate: Final, last, apical, terminal, used to describe segments that are the last and smallest divisions of a blade.
undulate: Wavy and not flat; said of margins.
uniseriate: Cells in a single line; said of multicellular hairs in which the cells are in a single line.
variety: A lesser subdivision of a species; some pteridologists view varieties as equivalent to subspecies, while others consider them a division of subspecies.
vascular bundles: Elongate strands of cells that conduct water distally (xylem) and sugars in solution proximally (phloem).
vein: A line of vascular tissue in the blade tissue of a frond or part of a frond.
vernation: The manner in which the tips of fronds uncoil or unfold in development.
wings: See ala

SELECTED BIBLIOGRAPHY

Literature citations in the main taxonomic section of this book are generally restricted to pertinent published taxonomic reviews that include Michigan families and genera.

Barnes, V. B., and W. H. Wagner. 2004. *Michigan Trees: A Guide to the Trees of the Great Lakes Region*. Ann Arbor: University of Michigan Press.

Billington, Cecil. 1952. *Ferns of Michigan*. Cranbrook Inst. Sci. Bull. 32.

Bridson, G. D. R., and E. R. Smith. 1991. *Botanico-Periodicum-Huntianum/Supplementum*. Hunt Institute for Botanical Documentation. Lawrence, KS: Allen Press.

Brown, R. W. 1985. *Composition of Scientific Names*. Washington, DC: Smithsonian Institution Press.

Christenhusz, M. J. M., X. Zhang, and H. Schneider. 2011. A linear sequence of extant families and genera of lycophytes and ferns. *Phytotaxa* 19:7–54.

Clausen, R. T. 1938. A monograph of the Ophioglossaceae. *Mem. Torrey Bot. Club* 19:5–177.

Clausen, R. T. 1943. Studies in the Ophioglossaceae: *Botrychium*, subgenus *Sceptridium*. *Amer. Fern J.* 33 (1): 11–27.

Clausen, R. T. 1944. Status of *Botrychium dissectum* var. *oneidense*. *Amer. Fern J.* 34(2): 55–60.

Cobb, Boughton, Elizabeth Farnsworth, and C. Lowe. *Ferns of Northeastern and Central North America*. 2005. 2nd ed. Boston: Houghton Mifflin.

Eaton, D. C. 1880. *The Ferns of North America: Colored Figures and Descriptions, with Synonymy and Geographical Distribution, of the Ferns (Including the Ophioglossaceae) of the United States of America and the British North American Possessions*. 2 vols. Salem, MA.

Evrard, C., and C. V. Hove. 2004. Taxonomy of the American *Azolla* species (Azollaeae) a critical review. *Systematics and Geography of Plants* 74:301–18.

Farrar, Donald. *Botrychium* web page. www.public.iastate.edu/~herbarium/botrychi um.html

Flora of North America North of Mexico. Vol. 2: *Pteridophytes and Gymnosperms*. 1993. Oxford: Oxford University Press.

Fraser-Jenkins, C. 2006. *Adv. Forestry Res. India* 29:139–60.

Gleason and Cronquist. 1991. *Manual of Vascular Plants of Northeastern United States and Adjacent Canada*, New York: New York Botanical Garden.

Hagenah, D. J. 1950. New records for the Bruce Peninsula, Ontario. *Rhodora* 52:41–45.

Hagenah, D. J. 1966. Notes on Michigan Pteridophytes. Pt. 2: Distribution of the Ophioglossaceae. *Amer. Fern J.* 56 (4): 150–62.

Hauk, W. D., C. R. Parks, and M. W. Chase. 2003. Phylogenetic studies of Ophioglossaceae: Evidence from *rbcL* and *trnL-F* plastid DNA sequences and morphology. *Molecular Phylogenetics and Evolution* 28:131–51.

Hauke, R. L. 1963. A taxonomic monograph of *Equisetum* subgenus *Hippochaete*. *Beib. Nova Hedwigia* 8:1–123.

Hauke, R. L. 1978. A taxonomic monograph of *Equisetum* subgenus *Equisetum*. *Nova Hedwigia* 30:385–455.

Hennig, Willi, 1966. *Phylogenetic Systematics*. Translated by D. Davis and R. Zangerl. Urbana: University of Illinois Press.

Hennipman, E. 1996. Scientific consensus classification of Pteridophyta. In *Pteridologia in Perspective*, edited by J. M. Camus, M. Gibby, and R. Johns, 191–212. Kew: Royal Botanical Gardens.

Hickey, R. J. 1977. The *Lycopodium obscurum* complex in North America. *Amer. Fern J.* 79:78–89.

Holttum, R. E. 1971. The family names of ferns. *Taxon* 20:527–32.

International Plant Names Index (IPNI). http://www.ipni.org

Lellinger, David B. 1985. *A Field Manual of the Ferns and Fern-Allies of the United States and Canada*. Washington, DC: Smithsonian Institution Press.

Linnaeus, C. 1753. *Species Plantarum* . . . 2 vols. Stockholm.

Manton, I. 1950. *Problems of Cytology and Evolution in the Pteridophyta*. London: Cambridge University Press.

Montgomery, J. D., and E. M. Paulton. 1981. *Dryopteris* in North America. *Fiddlehead Forum* 8 (4): 25–31.

Øllgaard, B. 1987. A revised classification of the Lycopodiaceae s. lat. *Opera Botanica* 92:153–78.

Pichi-Sermoli, R. K. F. 1977. *Fragmenta pteridologiae* VI. *Webbia* 31 (1): 237–59.

Pryer, K. M., H. Schneider, A. R. Smith, R. Cranfill, P. G. Wolf, J. S. Hunt, and S. D. Sipes. 2001. Horsetails and ferns are a monophyletic group and the closest living relatives to seed plants. *Nature* 409:618–22.

Pryer, K. M., E. Schuettpelz, P. G. Wolf, H. Schneider, A. R. Smith, and R. Cranfill. 2004. Phylogeny and evolution of ferns (monilophytes) with a focus on the early leptosporangiate divergences. *American Journal of Botany* 91 (10): 1582–98.

Pteridophyte Phylogeny Group. 2016. A community-derived classification for extant lycophytes. *J. Syst. Evol.* 54 (6): 563–603.

Rothfels, C. J., M. A. Sundue, L-Y Kuo, A. Larsson, M. Kato, E. Schuettpelz, and K. M. Pryer. 2012. A revised family-level classification for eupolypod II ferns (Polpodiidae: Polypodiales). *Taxon* 61 (3): 515–33.

Smith, A. R., K. M. Pryer, E. Schuettpelz, P. Korall, H. Schneider, and P. G. Wolf. 2006. A classification of extant ferns. *Taxon* 55 (3): 705–31.

Sokal, R. R., and P. H. E. Sneath. 1963. *Principles of Numerical Taxonomy*. San Francisco: Freeman.

Stearn, W. T. 1995. *Botanical Latin*. 4th ed. Portland, OR: Timber Press.

Stensvold M. C. 2007. A taxonomic and phylogeographic study of the *Botrychium lunaria* complex. Retrospective Theses and Dissertations. 15767. http://lib.dr.iastate.edu/rtd/15767

University of Michigan Herbarium. “Michigan Flora Online.” http://michiganflora.net

Voss, E. E. 1978. *Botanical Beachcombers and Explorers: Pioneers of the 19th Century in the Upper Great Lakes*. Contributions from the University of Michigan Herbarium, no. 13. Ann Arbor: University of Michigan Herbarium.

Wagner, W. H., Jr. 1959. American grapeferns resembling *Botrychium ternatum*: A preliminary report. *Amer. Fern J.* 49 (3): 97–103.

Wagner, W. H., Jr. 1960a. Evergreen grapeferns and the meaning of infraspecific categories as used in North American pteridophytes. *Amer. Fern J.* 50 (1): 32–45.
Wagner, W. H., Jr. 1960b. Periodicity and pigmentation in *Botrychium* subg. *Sceptridium* in the northeastern United States. *Bull. Torrey Bot. Club* 87 (5): 303–25.
Wagner, W. H., Jr. 1961. Roots and taxonomic differences between *Botrychium oneidense* and *B. dissectum*. *Rhodora* 63:164–75.

INDEX

Note to the Reader: Mentions of plant names in the keys and tables are not indexed as these appear within the main treatments identified by the bold page number ranges for genera. Infraspecific entities are also not indexed as they always appear within main treatments identified by the bolded page numbers. Plurals are not indexed separately. Families, genera, and species treated, and the pages on which the treatments occur, appear in boldface.